HANDBOOK OF ADHESION

POLYMER SCIENCE AND TECHNOLOGY SERIES

SERIES EDITORS: DR D M BREWIS AND PROFESSOR D BRIGGS

Published

I S MILES AND S ROSTAMI (eds), *Multicomponent Polymer Systems*

Forthcoming

D M BREWIS AND B C COPE (eds), *Handbook of Polymer Science*

F R JONES (ed.), *Handbook of Polymer–fibre Composites*

H R BRODY (ed.), *Synthetic Fibres*

E A COLBOURN (ed.), *Computer Simulation of Polymers*

R N ROTHON (ed.), *Particulate Filled Polymer Composites*

POLYMER SCIENCE AND TECHNOLOGY SERIES

SERIES EDITORS: DR D M BREWIS AND PROFESSOR D BRIGGS

HANDBOOK
OF ADHESION

EDITOR:

D E PACKHAM

SCHOOL OF MATERIALS SCIENCE
UNIVERSITY OF BATH

Longman
Scientific &
Technical

Longman Scientific & Technical
Longman Group UK Ltd
Longman House, Burnt Mill, Harlow
Essex CM20 2JE, England
and Associated Companies throughout the world

copublished in the United States with
John Wiley & Sons, Inc., 605 Third Avenue, New York, NY 10158

First published 1992

0 582 04423.5
British Library Cataloguing in Publica.:on Data
A British Library CIP record is available for this book

Library of Congress Cataloging-in-Publication Data
Handbook of adhesion / editor, D.E. Packham.
  p.   cm. — (Polymer science and technology series)
  Includes bibliographical references and index.
  ISBN 0-470-21870-3
  1. Adhesives. 2. Adhesion.   I. Packham, D. E. (David Ernest),
  1030– . II. Series.
  TP968.H34 1992
  668'.3—dc20   92-15139
  CIP

Typeset by 6JJ in 10/12$\frac{1}{2}$ pt Times

Produced by Longman Group (FE) Limited
Printed in Great Britain by The Bath Press, Avon

Contents

Foreword

Good adhesion is vital to a number of important technologies. These include adhesive bonding, lamination, metallization, printing, painting, coating and composite production. All these technologies are growing, in some cases rapidly, aided by the introduction of new materials and processes, and by a better understanding of the background science.

Adhesion is a multidisciplinary subject embracing surface science, chemistry, physics, materials science and mechanical engineering. A very large and diverse technical literature on adhesion exists. This includes a substantial number of books on various aspects of adhesion. However, the *Handbook of Adhesion* differs from other books in the wide range of technologies covered and in its format. About two hundred articles are arranged in alphabetical order; extensive cross-referencing enables the reader to study all the complementary subject material. Each article contains a select bibliography.

The *Handbook of Adhesion* is edited by Dr D E Packham who is a senior lecturer at Bath University. Dr Packham has worked for many years on various aspects of adhesion. He has drawn together a highly experienced team of scientists and engineers to cover the diverse range of topics.

The *Handbook of Adhesion* is part of the new series 'Polymer Science and Technology'.

D M BREWIS
Institute of Surface Science
and Technology
University of Technology
Loughborough
Leicestershire LE11 3TU

D BRIGGS
ICI Wilton Research Centre
PO Box 90
Wilton
Middlesbrough
Cleveland T56 8JE

Preface

There is a long and valuable tradition of members of academic staff of universities acting as authors and editors of reference works written for the benefit of the general public as a whole or of sections of it with specialized interests. This is one way in which a university repays its debt to the society which sustains it. Thus when asked by the Series Editors if I would act as editor for the *Handbook of Adhesion*, I agreed to do so. It was not without some misgivings that I gave my agreement. I was somewhat overawed at the prospect of commissioning some 200 articles from over 50 authors in industry and higher education, and then relating the articles to one another in a single book with consistent style. In practice the job was lightened by the enormous support given me by the publisher's staff, particularly Dr Michael Rodgers and Dr Paula Turner, and also by the series editors, Professor David Briggs and Dr Derek Brewis. I would also like to thank the individual authors themselves for their patience with my editorial idiosyncrasies. In an attempt to keep the style of the book it has sometimes been necessary to modify articles in a way that individual authors would not have chosen. I must emphasize that the responsibility for the final form of the text, including the errors, is mine.

This book, then, represents the fruits of fifty to sixty authors drawn from universities and industry. Such an authorship is unexceptional: there are hundreds of books resulting from collaboration like this in many areas of pure science, applied science and engineering.

In the past decade in the United Kingdom there has been an enormous emphasis on the importance of universities' collaborating with industry and serving its needs. This emphasis has been backed up by Government financial policies designed to make universities change their ways. Many of the practices of industry have been introduced into universities with the result that increasingly the performance of academics is being judged in terms of the *cash* they can bring in from research contracts and the *income* they can raise by selling of their services in the market place. It is ironical that the hard canons of market forces being imposed on universities will mean that collaborative ventures, such as this book, will become things of the past: they are bound to fail the market criterion of cost effectiveness.

The changes being urged on universities represent much more than a trimming of the sails to the winds of political change. They represent a determined effort radically to change their fundamental values, or in the more percussive language used recently by the leader of a powerful pressure group, 'to break the academic mould'. If this happens, books of this sort will be among the lesser casualties of the breaking of the academic mould.

D E PACKHAM
UNIVERSITY OF BATH
JANUARY, 1992

Introduction

Scope of the Handbook

The *Handbook of Adhesion* is intended as a book of reference in the field of adhesion. Adhesion is a phenomenon of interest in diverse scientific disciplines and of importance in a wide range of technologies. Therefore the scope of this *Handbook* includes the background science (physics, chemistry and materials science) and engineering and aspects of adhesion relevant to the use of adhesives, sealants and mastics, paints and coatings, printing and composite materials.

Intended readership

The book will be of value to professional people of many different backgrounds who need to have an understanding of various facets of adhesion. These will include those working in research, development or design, as well as others involved with marketing or technical service. This book is intended as a reference work for all those needing a quick, but authoritative, discussion of topics in the field of adhesion, broadly interpreted. It is intended for scientists and engineers qualified at national certificate or degree level. The aim has been to write it so that a detailed knowledge of individual science and engineering disciplines is not required.

Length of articles

The *Handbook* has been designed so that it is easy to retrieve the information required, whether this is confined to a single point or it is more extensive. Thus articles are arranged alphabetically and it has been editorial policy for each article to be, as far as possible, intelligible on its own, and to limit its length to around three pages which can be quickly assimilated. Many enquirers will want more extensive information than a single article can provide. For this reason there is copious cross-referencing to related articles elsewhere in the *Handbook*, and a comprehensive index. **Cross-references** are shown by giving titles to articles in **bold**.

Literature references

The literature references at the end of the articles are intended to give further information to the *general* enquirer, so, where possible, they list authoritative reviews, monographs or text books, rather than original research papers. Those who need access to original papers should easily find the reference through these secondary sources. The article on **Literature on adhesion** and the associated selected bibliography in **Appendix 3** give broad guidelines on the book and periodical literature in the field.

Broader study

As well as providing an answer to a specific query, the *Handbook* can be used for a discursive study of topics in adhesion, even as a starting point for an extended research project. Use might be made of the **Classified list of articles** (p. xxiii) where articles on related topics are grouped together and arranged, where appropriate, in a logical sequence for reading.

How to use the Handbook

Detailed instructions on use of the *Handbook* are given inside the front cover.

Remember **Cross-references** are shown by giving titles to articles in **bold**.

List of articles – alphabetical

Abrasion treatment   J F WATTS Removal of loose layers, roughening, improved adhesion

Accelerated ageing   A MADDISON Shear and wedge tests, humidity, corrosive environments

Acid–base interactions   K W ALLEN AND J R G EVANS Relation to work of adhesion, Drago equation

Acids   D E PACKHAM Concept of acid–Brönsted–Lewis, conjugate bases–electron donor/acceptor

Acoustic microscopy   M G SOMEKH Basis of technique; application to adhesion

Acrylic adhesives   F R MARTIN Basic material, setting mechanism, applications, advantages, disadvantages

Addition polymerization   J COMYN Initiation, propagation, termination; application to adhesives

Adhesion   D E PACKHAM Etymology, usage – qualitative, quantitative, practical, theoretical, bonds at interface

Adhesion – fundamental and practical   D E PACKHAM Relationship between joint strength and interfacial forces

Adhesion in medicine   M E R SHANAHAN Macroscopic level (e.g. prostheses) and cellular level

Adhesion under ultra-high vacuum   R G LINFORD Adhesion in ultra-high vacua oxide-free metals, cold welding, crystallographic effects

Adhesive classification   B C COPE Classified by setting mode and chemical nature

Adhesives for textile fibre bonding   A J G SAGAR Non-woven materials, flocking, tyres, belts and hose

List of articles — classified

Note: The classes are somewhat arbitrary. Some articles are listed in more than one class.

General

Adhesion D E PACKHAM Etymology, usage – qualitative, quantitative, practical, theoretical, bonds at interface

Adhesion – fundamental and practical D E PACKHAM Relationship between joint strength and interfacial forces

Literature on adhesion D E PACKHAM Guide to textbooks, reference books and journals

Appendix 3 A selected bibliography on adhesion

Health and safety D C WAIGHT Hazards associated with adhesives: sources of information

Statistics C CHATFIELD Data collection and analysis; quality control; reliability

Release D E PACKHAM Examples where low adhesion needed – internal and external release agents

Friction–adhesion aspects A D ROBERTS Influence of adhesion on friction; Schallamach waves

Autohesion J COMYN Contact theory, diffusion theory, development of bond strength

Internal stress K KENDALL Origin; effect on adhesion measurement; reduction of internal stresses

Background materials science

Secondary bonds, surface energy, interfacial tension

Dispersion forces   K W ALLEN Nature of dispersion forces, ubiquity, energy–distance relationships; Lennard-Jones potential

Polar forces   K W ALLEN Nature of Keesom and Debye forces, attraction constants; Lennard-Jones potential

Hydrogen bonding   D BRIGGS Nature and occurrence of hydrogen bonding, work of adhesion, examples where important

Surface energy   D E PACKHAM Thermodynamic definitions of surface tension and surface energy: connection with bond type

Wetting and spreading   M E R SHANAHAN Young's equation, work of adhesion and cohesion, spreading coefficient

Wetting kinetics   M E R SHANAHAN Spreading on a solid surface; capillary rise

Roughness of surfaces   D E PACKHAM Characterization of roughness, effect on adhesion

Wetting and work of adhesion   J F PADDAY Thermodynamic works of adhesion, wetting, spreading and cohesion; Young's equation

Contact angle   J F PADDAY Young's equation; nature of the contact angle; roughness

Contact angle measurement   J F PADDAY Techniques – where applicable; precautions needed to obtain reproducibility

Contact angles and interfacial tension   D E PACKHAM Young's equation; work of adhesion, interfacial tension and surface energy

Good–Girifalco interaction parameter   D E PACKHAM Definition of ϕ; evaluation for 'dispersion force' interface; interfacial tension; solid surface energies

Surface energy components   D E PACKHAM Dispersion and polar components; geometric mean relationships; solid surface energies

Acid–base interactions   K W ALLEN AND J R G EVANS Relation to work of adhesion, Drago equation

Surface characterization by contact angles – polymers   M E R SHANAHAN Critical surface tension, 'one-liquid' and 'two-liquid' methods

Surface characterization by contact angles – metals   M E R SHANAHAN Use of 'one-liquid' and 'two-liquid' methods

Polymer science

Addition polymerization   J COMYN Initiation, termination; application to adhesives

Condensation polymerization   J COMYN Application to adhesives

Compatibility   J COMYN Free energy of mixing; solubility parameter; diffusion; weak boundary layers

Glass transition temperature   D A TOD Property changes at T_g; measurement; effect of molecular structure and moisture

Viscoelasticity   D W AUBREY Transient and dynamic viscoelastic functions; Boltzmann superposition principle

Viscoelasticity – time–temperature superposition   D W AUBREY Shift factor, WLF equation

Thermal analysis   D E PACKHAM Differential scanning calorimetry – crystallinity, heat of reaction, glass transition temperatures

Mechanics

Fracture mechanics   A J KINLOCH Basis: energy balance and stress intensity factor approaches

Fracture mechanics test specimens   A J KINLOCH Test methods for flexible and rigid joints

Finite element analysis   A D CROCOMBE Principles and application to adhesive joints

Other

Humidity   J COMYN Relative humidity, laboratory control of humidity

Acids   D E PACKHAM Concept of acid – Brönsted–Lewis, conjugate bases – electron donor–acceptor

Acid–base interactions   K W ALLEN AND J R G EVANS Relations to work of adhesion, Drago equation

Statistics   C CHATFIELD Data collection and analysis; quality control; reliability

Theories of adhesion

Theories of adhesion   K W ALLEN General introduction to mechanical, adsorption, diffusion and electrostatic theories

Adhesive–substrate interface

Adhesive types

Classification

Broad categories

Structural adhesives   P. CULLEN Properties required, adhesive types; applications

Toughened adhesives   P CULLEN Mechanism of toughening; advantages and disadvantages

High-temperature adhesives   S J SHAW Comparison of different types; recent developments

High-temperature stability: principles   D E PACKHAM Molecular structure and thermal stability

Underwater adhesives   M R BOWDITCH Wetting problems; pretreatments; durability

Specific chemical types

Phenolic adhesives   S TREDWELL Resoles, novolaks; modifications; uses

Epoxide adhesives   J COMYN Basic material, setting mechanism, applications

Acrylic adhesives   F R MARTIN Basic material, setting mechanism, applications, advantages, disadvantages

Toughened acrylic adhesives   P CULLEN Setting mechanism; applications; advantages and disadvantages

Cyanoacrylate adhesives   J GUTHRIE Anionic polymerization, additives, applications, advantages and disadvantages

Alkyl-2-cyanoacrylates   J GUTHRIE Monomer synthesis; polymerization

Polyurethane adhesives   G PARKER Solvent-free and solvent-based systems, toxicology

Ethylene–vinyl acetate copolymers   D E PACKHAM Copolymer composition and properties; hot melt and emulsion adhesives

Polybenzimidazoles   S J SHAW Structure; use as high-temperature adhesive

Polyether ether ketone   D A TOD High-temperature stability; use as matrix for fibre composites

Polyimide adhesives   S J SHAW Condensation and thermoplastic polyimides, imide prepolymers; high-temperature stability

Polyphenylquinoxalines   S J SHAW Chemical constitution: use as high-temperature adhesive

Animal glues and technical gelatins   C A FINCH Origin and uses

Testing

Tests of adhesion   D A TOD Basic principles, effects of test parameters, peel, shear, tensile, wedge test

Testing of adhesives   D A TOD Survey of tests including mechanical, rheological and thermal

Fracture mechanics   A J KINLOCH Basis: energy balance and stress intensity factor approaches

Fracture mechanics test specimens   A J KINLOCH Test methods for flexible and rigid joints

Wedge test   B M PARKER Use of test for comparing joint durability

Blister test   A J KINLOCH Use, theory, variations

Peel tests   D E PACKHAM Peel force and peel energy; factors affecting peel energy; angle variation

Climbing drum peel test   K B ARMSTRONG Peeling of metal sheet, e.g. skin from honeycomb

Tensile tests   D E PACKHAM Description, uneven stress distribution

Shear tests   A D CROCOMBE Tests in tension and torsion

Napkin ring test   D E PACKHAM Calculation of shear stress: refinements

Scratch test   D E PACKHAM Assessment of thin film adhesion

Standards of adhesion and adhesives   G R DURTNAL Discussion of scope and background to national standards

Appendix 1   Standards concerned with adhesion and adhesives

Appendix 2   Standard test methods for adhesive joints

Non-destructive testing   G J CURTIS Acoustic wave techniques, resonance and pulse echo testers

Acoustic microscopy   M G SOMEKH Basis of technique; application to adhesion

Accelerated ageing   A MADDISON Shear and wedge tests, humidity, corrosive environments

Fibre—matrix adhesion – assessment techniques   F R JONES Direct and indirect test methods

Rubber to metal bonding – testing   J A LINDSAY Peel tests, tests in tension

Paint service properties and adhesion   J L PROSSER Internal stress; testing; weathering

Pretreatment of surfaces

Metals

Engineering surfaces of metals J F WATTS Practical metallic surfaces are oxidized, contaminated and rough: pretreatments; adhesion

Pretreatment of metals prior to bonding D M BREWIS Survey of mechanical and chemical treatments

Pretreatment of metals prior to painting J L PROSSER Need for pretreatment; survey of common methods

Degreasing J F WATTS Solvent, alkali and emulsion cleaners; efficiency

Abrasion treatment J F WATTS Removal of loose layers, roughening, improved adhesion

Anodizing A MADDISON As a pretreatment: different electrolytes and uses

FPL etch D E PACKHAM Sulphochromate treatment for Al; durability; topography

Microfibrous surfaces D E PACKHAM Examples of preparation Fe, Cu, Zn. Use as substrates in adhesion–energy dissipation

Conversion coating A MADDISON Phosphate, chromate and alkali oxide treatments

Pretreatment of aluminium D M BREWIS Chromic acid etch, chromic acid anodizing, phosphoric acid anodizing; relative bond durability

Pretreatment of copper D E PACKHAM Conventional and microfibrous surfaces

Pretreatment of steel J F WATTS Abrasion, pickling and conversion coatings

Pretreatment of titanium D M BREWIS Survey of important types of pretreatment and comparisons of bond durability

Rubber to metal bonding – pretreatments P M LEWIS Degreasing, mechanical cleaning, chemical cleaning

Polymers

Surface nature of polymers D BRIGGS Migration of additives and low-molecular-weight fraction to the surface

Pretreatment of polymers D M BREWIS Effects of solvent, mechanical, oxidative and plasma treatment

Processing and assembly

Welding and autohesion

Composite materials

Paint

Rubber

List of contributors

MR S G ABBOTT, SATRA Footwear Technology Centre, Rockingham Road, Kettering, NN16 9JH

PROFESSOR R D ADAMS, Department of Mechanical Engineering, The University, Bristol, BS8 1TR

MR K W ALLEN, Ranworth, Tydehams, Newbury, RG14 6JT

DR K B ARMSTRONG, 20 Homewaters Avenue, Sunbury on Thames, TW16 6NS

DR R ASHLEY, CMB Technology, Denchworth Road, Wantage, OX12 9BP

DR D W AUBREY, 2A Viga Road, Grange Park, London, N21 1HJ

DR G BATTERSBY, Coates Lorilleux Int., Cray Avenue, St Mary Cray, Orpington, BR5 3PP

MR J C BEECH, Building Research Establishment, Garston, Watford, WD2 7JR

MR M R BOWDITCH, Defence Research Agency, Holton Heath, Poole, Dorset, BH16 6JU

DR D M BREWIS, Institute of Surface Science and Technology, University of Technology, Loughborough, LE11 3TU

DR D BRIGGS, ICI, Wilton Research Centre, PO Box 90, Wilton, Middlesbrough, TS6 8JE

DR C CHATFIELD, School of Mathematical Sciences, University of Bath, Bath, BA2 7AY

DR J COMYN, Institute of Polymer Technology and Materials Engineering, University of Technology, Loughborough, LE11 3TU

DR B C COPE, De Montfort University, Leicester, PO Box 143, Leicester, LE1 9BN

DR A D CROCOMBE, Department of Mechanical Engineering, University of Surrey, Guildford, GU2 5XH

DR P CULLEN, Loctite (Ireland) Ltd, New Business Development – International, Tallaght Business Park, Whitestown, Dublin 24, Ireland

DR G J CURTIS, Building 521, AEA Harwell, Didcot, Oxfordshire

MR G R DURTNAL, Section Leader, Polymers Division, Royal Arsenal East, Woolwich, London, SE18 6TD

DR J R G EVANS, Department of Materials Technology, Brunel University, Uxbridge, UB8 3PH

DR C A FINCH, Pentafin Associates, 18–20 West End, Weston Turville, Aylesbury, HP22 5TT

MISS M GIRARDI, The Welding Institute, Abington Hall, Abington, Cambridgeshire, CB1 6AL

DR R GREEF, Department of Chemistry, University of Southampton, Southampton, SO9 5NH

DR J GUTHRIE, Loctite (Ireland) Ltd, New Business Development – International, Tallaght Business Park, Whitestown, Dublin 24, Ireland

DR M HADDON, Ciba-Geigy, Duxford, Cambs, CB2 4QA

DR F R JONES, Department of Engineering Materials, The University, Sheffield, S10 2TZ

DR K KENDALL, Advanced Materials, ICI plc, PO Box 11, The Heath, Runcorn, WA7 4QE

PROFESSOR A J KINLOCH, Department of Mechanical Engineering, Imperial College, London SW7 2BX

DR GRAHAM LAKE, MRPRA, Brickendonbury, Hertford, SG13 8NL

MR P LEWIS, MRPRA, Brickendonbury, Hertford, SG13 8NL

MR J LINDSAY, Peradin Ltd, Freshford Mill, Freshford, Bath

PROFESSOR R LINFORD, De Montfort University, Leicester, PO Box 143, Leicester, LE1 9BH

MR G BERNARD LOWE, Morton International Ltd, Specialty Chemical Group, Polymer/Europe, University of Warwick Science Park, Sir William Lyons Road, Coventry CV4 7EZ

DR A MADDISON, 27 Adonis Close, Perrycrofts, Tamworth, Staffordshire, B79 8TY

DR F R MARTIN, Loctite (Ireland) Ltd, New Business Development – International, Tallaght Business Park, Whitestown, Dublin 24, Ireland

DR D P MELODY, Loctite (Ireland) Ltd, New Business Development – International, Tallaght Business Park, Whitestown, Dublin 24, Ireland

DR D E PACKHAM, School of Materials Science, University of Bath, Bath BA2 7AY

DR J F PADDAY, Emerging Technologies – Research Division, Kodak Ltd, Headstone Drive, Harrow, HA1 4TY

DR B M PARKER, Materials and Structures Department, Defence Research Agency, RAE Farnborough, Hampshire, GU14 6TD

MR G C PARKER, Swift Adhesives Ltd, St George's House, Church Street, Twickenham, TW1 3NE

MR B H PAXTON, BRE, Garston, Watford, WD2 7JR

MR J PRITCHARD, 2 Durley Drive, Sutton Coldfield, Birmingham

MR J L PROSSER, 14 Woburn Court, Stanmore Road, Richmond, TW9 2DD

MR H REITER, School of Materials Science, University of Bath, Bath, BA2 7AY

DR A D ROBERTS, Deputy Director, MRPRA, Brickendonbury, Hertford, SG13 8NL

MR D F G RODWELL, BRE, Garston, Watford, WD2 7JR

DR A J G SAGAR, Business Manager – Materials Science, BTTG, Shirley House, 856 Wilmslow Road, Didsbury, Manchester M20 8RX

DR S J SHAW, Materials and Structures Department, Defence Research Agency, RAE Farnborough, Hampshire, GU14 6TD

DR M E R SHANAHAN, École Nationale Supérieure des Mines de Paris, Centre des Matériaux, BP 87, 91003 Evry Cedex, France

DR M G SOMEKH, Department of Electrical and Electronic Engineering, University of Nottingham, NG7 2RD

MR T J SMITH, Ciba-Geigy, Duxford, Cambs, CB2 4QA

MR A J SPARKES, 32 Woodland Way, Stevenage, Herts, SG2 8BT

DR D A TOD, Defence Research Agency, Fort Halstead, Kent, TN14 1BP

MR R F TOUT, FIRA, Maxwell Road, Stevenage, Herts, SG1 2EW

DR S T TREDWELL, BP Chemicals Ltd, Barry Business Group, Sully, South Glamorgan, CF6 2YU

MR D C WAIGHT, Raphoe, 20 Warfield Road, Bracknell, Berkshire, RG12 2JY

MR P WALKER, AWE, Building SB43, Aldermaston, Berkshire

MR C WATSON, Technical Director, Loctite UK, Watchmead, Welwyn Garden City, Hertfordshire AL7 1JB

DR J F WATTS, Department of Materials Science and Engineering, The University of Surrey, Guildford, GU2 5XH

A

Abrasion treatment

J F WATTS

In the removal of large quantities of extraneous material from a metallic surface prior to painting or adhesive bonding it is often appropriate to consider some form of abrasive treatment. In general such treatments can only be employed successfully to remove brittle materials such as native oxides (e.g. mill scale) and attempts to remove large quantities of energy-absorbing material such as previous coats of paint are not successful. Although abrasion cleaning includes all processes that involve the passage of an abrasive medium over the surface to be cleaned, the most widely employed, particularly where consistency of finish and high throughput are required, is abrasive blast cleaning. Other methods, either carried out by hand or machine, include wire brushing and sanding, buffing and so on.

Any abrasive treatment has a twofold purpose: to remove rust, mill scale and other surface detritus, but also to provide an appropriate surface profile to which the organic phase can be applied. Whether blast cleaning or some other abrasive method is used the cleanliness of the abrasive medium is of prime importance. That is to say if the wire brush, or the abrasive in the blast-cleaning process, has been used to clean an oily substrate it will exhibit a memory effect and the oil will contaminate the abrasive and transfer the organic material to the next surface to be cleaned. As far as efficiency of abrasion treatment is concerned it is one of the most effective ways of producing a chemically clean surface,[1] in terms of the residual level of hydrocarbons and associated material.

Any form of abrasion treatment necessary involves the removal of material and the exposure of new surface to the atmosphere. The geometry and size of the abrasive medium will affect the surface profile as illustrated by the examples of Fig. 1, which compares the surface topography of cold rolled steel obtained by emery abrasion (120 grit) and with that generated

1

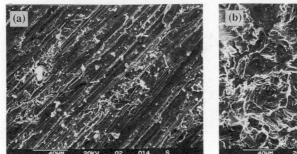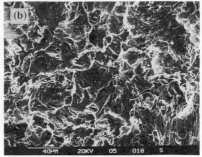

Fig. 1. Surface morphology produced by (a) emery abrasion and (b) grit-
blasting (scanning electron micrographs)

by blasting with chilled iron grit (G12) (see **Roughness of surfaces**). In
the case of blast cleaning the particles may be either inorganic (silica,
mineral slags) or metallic (invariably iron, although zinc is used to confer
temporary corrosion resistance in marine environments), and the shape
may vary from spherical to particular size fractions of the crushed source
material.[2] In the case of spherical abrasive particles a peening action at
the surface is achieved, and although this produces some surface cleaning
the overriding effect is to produce compressive stresses in the surface
region. This is particularly important if the article is to be subjected to
mechanical constraints, as it is known to improve fatigue resistance, in
addition to the important criterion of paint adhesion.

The increase in adhesion of an organic system to an inorganic surface
cleaned by an abrasion method is exclusively a result of the increased
surface area, and therefore increased interfacial-contact that is generated
by these methods.[3]

Although blast cleaning is the most effective method, the performance
of the coated substrate depends on the type of abrasive employed. Sand
and other non-metallic abrasives are somewhat better than metallic shot
or grit, and metallic grit appears to give better adhesion than a shot-peened
surface.[4] Although this subject has yet to be exhaustively investigated it
seems that abrasive shape and fracture characteristics are the important
factors in obtaining a well-cleaned substrate that will yield good durability
when the organic system is applied.

References

1. J E Castle, J F Watts in *Corrosion Control by Organic Coatings*, ed.
   H Leidheiser, NACE, Houston, 1981, pp. 78–87.
2. C G Munger, *Corrosion Prevention by Organic Coatings*, NACE, Houston,
   1984, pp. 229–33.

3. J F Watts, J E Castle, *J. Mater. Sci.*, **19**, 2259–72 (1984).
4. L K Schwab, R W Drisko in *Corrosion Control by Organic Coatings*, ed. H Leidheiser, NACE, Houston, 1981, pp. 222–6.

Accelerated ageing

A MADDISON

Many environmental factors affect durability, in particular, stress, temperature and exposure to fluids. The generation of meaningful data upon which confidence in the long-term durability of candidate bonding systems may be based is therefore extremely important.

Almost universally implicated in the degradation of adhesive bonds, the effects of water have been extensively studied (see **Durability of coatings in water**). The rate at which adhesive bonds lost strength in the presence of water depends principally upon the following:

1. The rate at which moisture enters the bond line by diffusion through the adhesive;
2. Diffusion rates through permeable adherends;
3. Transport of water along the adhesive/adherend or adhesive/carrier interfaces and any cracks present.

Losses in joint strength may then occur as a result of: changes in the properties of the adhesive, e.g. by plasticization, hydrolysis or cracking, and interfacial instability, e.g. by oxide hydration or adhesive displacement.

The concept of a critical water concentration below which bond strength is not affected has been advanced,[1] but in some operating environments the progressive accumulation of water in the bond line is inevitable.

Measurement of the sorption characteristics[2] of the adhesive and its mechanical properties as a function of water uptake provides in principle a basis for the prediction of retained bond strength. In reality, however, irreversible interfacial changes may predominate, a problem most effectively countered by adherend surface treatment.

Much routine testing is done using lap shear joints exposed to controlled laboratory conditions, and withdrawn periodically for residual strength measurement. Some standard test environments, traditionally associated with corrosion testing, e.g. those defined in ASTM–B117 (salt spray) and BS 3900 Part F2 (condensing humidity), find application in bond durability assessment. More extreme conditions are occasionally used, e.g. in a cataplasma test which specifies exposure to 100% relative humidity (RH) at 70 °C followed by cooling to −20 °C. In the field of adhesion, the value of many such accelerated tests is questionable. It is essential that test conditions relevant to the intended application are chosen and that temperatures should approximate to service limits and never approach the **Glass transition temperature** of the adhesive.

Injudicious increases in test solution concentration, while possibly accelerating generalized adherend corrosion, may inhibit bond degradation processes by osmotic effects. Greater sensitivity to realistic test conditions may be obtained by adopting modified specimen geometries which saturate more rapidly.[3]

While test strategies based on the exposure of unstressed specimens are undoubtedly useful for screening purposes, more realistic test procedures demand the simultaneous application of stress.

Boeing have developed a **Wedge test**, a simplified crack propagation technique for surface treatment quality-control purposes. Empirical correlations with aircraft experience have been obtained and the test is claimed to reproduce the modes of failure observed in service. Often yielding results in a few hours, it can be useful for general pretreatment comparisons, provided a high-modulus adhesive and unyielding adherends are used. In this latter respect modification to specimen geometry may be required in order to maintain stress at the crack tip.

More quantitative results may be obtained by applying loads to conventional joint geometries. For example, stress **Humidity** tests using short diffusion path joints may readily discriminate surface treatment effects.[4] Again, care must be exercised in the choice of conditions, for example excessive stress levels may induce creep failure rather than reproduce the interfacial effects observed in service.

For high-performance bonding systems very long periods of exposure to outdoor conditions may be necessary in order to observe the influence of surface treatment and adhesive type on durability.[5] Recent work based on the exposure of sensitive specimen geometries to real operating conditions has demonstrated that the discrimination of surface treatment effects may be achieved in acceptably short times.

The correlation of accelerated test results and the prediction of service performance is particularly difficult when operating conditions are complex, i.e. involve combinations of temperature, humidity and stress cycles. The early implementation of durability trials in real or simulated service environments is therefore highly desirable.

A number of relevant test methods are listed in Appendix 2.

References

1. A J Kinloch, *J. Adhesion*, **10**, 193 (1979).
2. J Comyn, *Developments in Adhesives*, *2*, ed. A J Kinloch, Applied Science Publishers, London, 1981.
3. D J Arrowsmith, A Maddison, *Int. J. Adhesion and Adhesives*, **7**, 1 (1987).

4. P A Fay, A Maddison, *Int. J. Adhesion and Adhesives*, **10**, 3 (1990).
5. J L Cotter, *Developments in Adhesives*, *2*, ed. A J Kinloch, Applied Science Publishers, London, 1981.

Acid–base interactions

K W ALLEN and J R G EVANS

A little over 20 years ago it was recognized that interactions across an interface (whether between two liquids or between a liquid and a solid) arose from several different types of force, and it was suggested that the total interaction was the result of a summation of the parts. Thus Fowkes (see ref. 1 and citations therein) expressed the work of adhesion W_A (see **Contact angles and interfacial tension**) as

$$W_A = W_A^d + W_A^h + W_A^{ab} + W_A^p + W_A^i + \cdots \qquad [1]$$

where the superscripts are d for London **Dispersion forces** (q.v.), h for **Hydrogen bonding** (q.v.), ab for acid–base interactions, p for Keesom dipole–dipole interactions, and i for Debye-induced dipole attractions (**Polar forces**).

For instance where only dispersion forces were significant, he used a geometric mean relationship involving work of cohesion terms (W_c) for the two phases:

$$W_A = (W_{c_1}^d W_{c_2}^d)^{1/2} \qquad [2]$$

This proved useful, and could be theoretically justified (see **Good–Girifalco interaction parameter**). This led others to introduce an expression

$$W_A = W_A^d + W_A^p \qquad [3]$$

where now the superscript p represents all the non-dispersion interactions. The problems arose when geometric mean expressions were also used with these 'polar' contributions:

$$W_A^p = (W_{c_1}^p W_{c_2}^p)^{1/2} \qquad [4]$$

so

$$W_A = (W_{c_1}^d W_{c_2}^d)^{1/2} + (W_{c_1}^p W_{c_2}^p)^{1/2} \qquad [5]$$

The second term in Eqn 5 assumes that the polar interaction between dissimilar materials can be deduced from the internal polar interactions within those materials. This is not necessarily so. Ethers, tertiary amines and esters, having limited internal polarity, nevertheless present strong interactions with dissimilar species.

As long ago as 1968 Bolger and Michaels[2] suggested that the important non-dispersion interactions could be expressed in terms of acid–base theory (see **Acids**). They used the concept of Brönsted acids, but Fowkes generalized the approach in terms of Lewis theory. Lewis acids are defined as electron acceptors and Lewis bases are electron donors so that the slight electronic charge disparity within a dipole can be considered to confer acidic or basic properties (see also **Rubber adhesion**). Fowkes[1] then recast Eqn 1 as

$$W_A = W_A^d + W_A^{ab} + W_A^c + W_A^e \qquad [6]$$

where the superscripts are dispersion, acid–base covalent and electrostatic respectively; W_A^d is given by the first term on the right-hand side of Eqn 5. The covalent interaction in adhesive joints is rare, but may give rise to substantial values of W_A (up to 500 mJ m^{-2}). Here W_A^e is generally regarded as low (~ 1 mJ m^{-2}).

The value of W_A^{ab} can be expressed in terms of the molar enthalpy ΔH^{ab} for reaction between an acid and base:

$$W_A^{ab} = \frac{\Delta H^{ab}}{N} \cdot S \cdot k \qquad [7]$$

where S is the number of acid–base surface sites per unit area, N the Avogadro number and k a term, near unity, which converts enthalpy to free energy.

Fowkes applied the concepts of Drago[3] to evaluate ΔH^{ab}:

$$-\Delta H^{ab} = C^a C^b + E^a E^b \qquad [8]$$

where C and E are constants related to the tendency of the acid or base to form covalent or ionic bonds respectively and have units (J mol^{-1})$^{1/2}$. Values for C and E were obtained from enthalpy of mixing or spectral parameter shifts and are tabulated by Drago et al. using iodine as a reference acid with $C = E = 1$.

Approximate values for polymers can frequently be deduced from Drago's data for homologues. The values of C and E for substrates can be obtained from enthalpies of wetting, from the measurement of frequency shifts caused by adsorption of characterized acids or bases on the substrate or by inverse gas chromatography in which the substrate in powder form occupies the column. Several attempts have been made to correlate the properties of filled polymers with the acidic or basic nature of the filler and matrix.[4] In general, elastic modulus, failure stress and elongation at break increase and the water permeability decreases when an acidic polymer is combined with a basic substrate or vice versa. However, C and E values are available for only a few systems of interest in adhesion at present and this restricts the widespread application of the acid–base interaction theory.

Some authors, for example Huntsberger,[5] have challenged the validity of the assumptions upon which this whole thesis has been based arguing for the importance of polar (Keesom) contributions.

Thus Eqn 1 may be rewritten as

$$W_A = (W_{c_1}^d W_{c_2}^d)^{1/2} + (W_{c_1}^p W_{c_2}^p)^{1/2} - \frac{kS}{N}(C^a C^b + E^a E^b) \qquad [9]$$

Some tend to neglect the last (acid–base) term, others (especially Fowkes) minimize the second (polar component) term: the whole issue is not yet resolved. This is an area of current controversy.

References

1. F M Fowkes et al., *Mater. Sci. Eng.*, **53**, 125 (1982).
2. J C Bolger, A S Michaels in *Interface Conversion for Polymer Coatings*, ed. R Weiss and G D Cheevers, Elsevier, New York, 1968, Ch. 1.
3. R S Drago et al., *J. Amer. Chem. Soc.*, **93**, 6014 (1971), **99**, 3203 (1977).
4. J A Manson, *Pure Appl. Chem.*, **57**, 1667 (1985).
5. J R Huntsberger, *J. Adhesion*, **12**, 3 (1981).

Acids

D E PACKHAM

Science commonly takes words from everyday language, and gives them a special meaning. With the passage of time meaning of a particular scientific term may well change: it is quite possible that several different meanings will be current at the same time, being used in different contexts.

The word 'acid' is a good example. A discussion of its meaning is relevant in this book because **Acid–base interactions** play a significant role in considerations of the theory of interfacial tension and of adhesion.

The scientific concepts of acids and the complementary compounds 'bases' have developed over several centuries with the development of chemistry.[1,2] Here it is only necessary to start with the idea associated with Liebig (1838) and Arrhenius (1887) which is still familiar from introductory courses in chemistry. This defines an acid as a compound containing hydrogen which can be replaced by a metal to form a salt. In ionic terms this amounts to defining an acid as a species which gives hydrogen ions (protons): a base is associated with hydroxyl ions. A typical acid–base reaction can be represented as

$$\text{HCl} + \text{NaOH} \rightarrow \text{NaCl} + \text{H}_2\text{O} \qquad [i]$$
$$\text{acid} \qquad \text{base} \qquad \text{salt}$$

or

$$H^+ + OH^- \rightarrow H_2O \qquad\qquad [ii]$$

The presence of water as a solvent is implicit in these reactions, indeed the hydrogen ion in reaction ii should be shown hydrated:

$$H_3O^+ + OH^- \rightarrow 2H_2O \qquad\qquad [iii]$$

Brönsted acids

In 1923 Brönsted extended the concept of an acid as a species which produces hydrogen ions to put forward a broader definition of an acid as a proton donor. A base by analogy was a proton acceptor (see also **Hydrogen bonding**). The reaction of a conventional acid with water becomes an acid–base reaction:

$$\underset{\text{acid}'}{HCl} + \underset{\text{base}''}{H_2O} \rightleftharpoons \underset{\text{acid}''}{H_3O^+} + \underset{\text{base}'}{Cl^-} \qquad\qquad [iv]$$

and

$$CH_3COOH + H_2O \rightleftharpoons H_3O^+ + CH_3COO^- \qquad [v]$$

Two points should be noted. Water accepts a proton so is acting as a Brönsted base. The 'first' acid (acid$'$) gives rise to a species, e.g. Cl^- which can act as a base – a conjugate base (base$'$). Similarly there is a conjugate acid, H_3O^+ (acid$''$) associated with the base water. Here the chloride ion and acetate ion, which would not traditionally be regarded as bases are so regarded because they are capable of accepting protons (for example in the reverse of reactions iv and v). The Brönsted definitions freed the concept of a base from being associated with the hydroxyl ion.

It is important to remember that the justification of definitions such as these in science is not a matter of 'right' or 'wrong': it is a matter of whether the definitions are 'useful' in the sense of aiding clear thinking and being helpful in the classification of reactions. That the Brönsted definitions have been much used for nearly 70 years suggests that they satisfy these criteria.

Lewis acids

In 1923, and more prominently in 1938, Lewis extended the concept of an acid still further, defining an acid as an acceptor of electron pairs (and a base as an electron pair donor). An electron is in a sense the 'opposite' of a proton, so these definitions can be seen as a development of Brönsted's. Brönsted enabled the concept of 'base' to be freed from exclusive attachment to the hydroxyl ion: Lewis went further in also broadening the concept of 'acidity' beyond hydrogen-containing species.

In conventional neutralization of an acid in aqueous solution the OH base donates a pair of non-bonding electrons, so these are still seen as acid–base reactions. The Lewis definition extends the concept to reactions with a formal similarity, which would not previously have been considered as acid–base reactions.

Consider reactions vi–ix.

$$H^+ + OH^- = H_2O \qquad [vi]$$

or

$$H_3O^+ + OH^- = 2H_2O \qquad [vii]$$

$$Ag^+ + Cl^- = AgCl\downarrow \qquad [viii]$$

$$M^+ + O{=}C\diagup^R_{\diagdown R} = \overset{+}{M}O{=}C\diagup^R_{\diagdown R} \qquad [ix]$$

acceptor   donor
(acid)   (base)

Reactions vi and vii are the traditional neutralization of an acid by an hydroxyl ion: here the donation of a non-bonding electron pair from the oxygen to the proton should be noted. Reaction viii can be considered analogous, a non-bonding electron pair is donated from the chlorine to the silver ion.

Reaction ix represents the reaction between a carbonyl compound, perhaps part of a polymer chain, and a cation on a substrate surface; this is the kind of reaction which might occur at some interfaces in adhesive bonds. The oxygen atom of the carbonyl group donates a non-bonding electron pair, this time into a vacant orbital in the metal ion, so again it is a Lewis acid–base reaction.

Conclusions

Acid–base reactions, especially donor–acceptor interactions occur widely and some knowledge of them is necessary to follow contemporary debates on theories of surface interactions relevant to adhesion. Both Finston[1] and Jensen[2] give good historical surveys of the development of the acid–base concept, and detailed critiques of Brönsted and Lewis theories.

References

1. H L Finston, A C Rychtman, *A New View of Current Acid–Base Theories*, John Wiley, New York, 1982.
2. W B Jensen, *The Lewis Acid–Base Concepts*, Wiley–Interscience, New York, 1980.

Acoustic microscopy

M G SOMEKH

Instrumentation

The acoustic microscope is essentially a refinement of more conventional methods of ultrasonic **Non-destructive testing**. The instrument differs mainly by virtue of the higher frequency of ultrasonic wave used. These are typically from 100 MHz up to over 2 GHz (although systems operating at 50 MHz may often be justifiably called acoustic microscopes).

The heart of the acoustic microscope is the lens. Figure 2 shows a diagram of a typical lens. The transducer is usually made of zinc oxide which is sputtered on to the sapphire buffer rod. At the end of the rod there is a spherical indentation which is filled with fluid (almost invariably water). When the transducer is excited with electrical energy an acoustic wave is generated which travels down the rod until it meets the fluid-filled spherical cavity. Since the velocity of sound in the buffer rod is considerably greater (typically $11\,000\,\mathrm{m\,s^{-1}}$) than the velocity in the fluid (sound velocity in water $= 1500\,\mathrm{m\,s^{-1}}$) the fluid-filled indentation acts like a converging lens so that the parallel beam in the buffer rod is focused to a tight spot on or below the sample surface. The energy reflected back from the sample travels back through the fluid and is detected by the transducer. In order to separate the transmitted signal from the much smaller reflected signal the sound is pulsed so the transmitted signals are separated in time from the reflected signals.

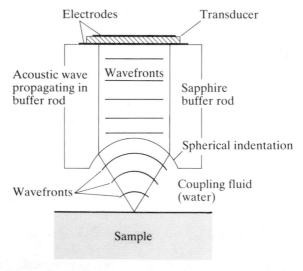

Fig. 2. Schematic diagram of an acoustic lens focused on sample surface

An image is produced by mechanically scanning the sample and building up the image point by point. The magnitude of the reflected pulse is usually used to modulate the image brightness.

The acoustic microscope can produce images with comparable resolution to those obtained with **Optical microscopy** (at 2 GHz the wavelength of sound in water is 0.75 μm, which is similar to near-infra-red radiation). The advantage over the optical microscope, however, lies in the fact that the ultrasound propagates through the material under test. The microscope can thus image through optically opaque materials and also gives quantitative information concerning the mechanical properties of the sample.

Modes of operation

Figure 2 shows the ultrasound beam focused on to the sample surface. In this focal position the images differ little from those obtained in the optical microscope. There are two main modes of operation which both involve moving the sample surface towards the lens; these are depicted schematically in Figs 3(a) and (b).

Figure 3(a) shows subsurface imaging where the acoustic beam is focused on to the interface between two layers. The beam is focused at the interface so that the defects in the bond layer can be imaged and quantified. Subsurface imaging is usually performed with a narrow angle

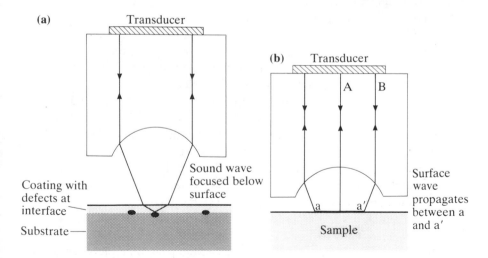

Fig. 3. (a) Schematic diagram of an acoustic lens focused at the interface between coating and substrate; (b) schematic diagram of an acoustic lens operating in surface wave mode

lens which does not excite surface waves (see next paragraph). Lateral resolution in this imaging mode is typically tens of micrometres.

Figure 3(b) shows a schematic diagram of the so-called $V(z)$ mode. This involves splitting the incident sound into two significant components. One is directly reflected from the sample surface (denoted A on Fig. 3(b)). The ray path B on Fig. 3(b) involves a ray of sound which is converted to a surface or Rayleigh wave at the sample–fluid interface. This wave travels along the surface of the sample continuously shedding its energy into the fluid; most of this energy does not return to the transducer. The energy that reradiated from position a' appears to come from the focus of the lens and will return to the transducer normally. The microscope output can thus be thought to arise primarily from ray paths A and B. As the lens to sample separation is changed the relative phase between the two beams varies and the received output undergoes a succession of period oscillations. If appropriate signal processing is applied the period of the oscillation can be used to determine the surface wave velocity to around one part in 10^4.

Accurate knowledge of the surface wave velocity is particularly useful since this can be related to coating thickness and interface contact conditions on the micrometre scale.

Applications to adhesion

The subsurface imaging mode depicted in Fig. 3(a) can be very effective for examining the bonding between two layers. In practice the layer should be fairly thick in order to separate the reflection from the top surface from the interface reflection. Unless special precautions are taken the coating should typically be greater than 150 μm thick for examination by a 100 MHz lens. Local delaminations at the bond layer can often be detected by a phase reversal on reflection at the interface.

The $V(z)$ mode can be used for quantitative measurements, but there is one important difference from the mode described above in that the method is *not* applicable when the layer thickness is much greater than the surface acoustic wavelength in the material. This limitation arises because the surface wave decays in depth and will not penetrate as far as the interface for thick layers. The method is thus really only applicable for thin films. At the very simplest level the method can be used for determination of the film thickness by measuring the difference in velocity of the layered material from that of the substrate. Similarly the microscope can be used to image areas of delamination and differing degrees of adhesion because these can be detected by changes in the surface wave velocity and hence the microscope output. Impressive images of such delaminations have been obtained at high frequencies by Bray et al.[1] More quantitative work[2] has correlated surface wave velocity on thin films to the fabrication procedure; the results can be explained by changes in the presumed conditions at the interface.

The acoustic microscope in both modes has considerable potential in the study of adhesion where the mechanical properties can be measured non-destructively and quantitatively. Clearly more work is required particularly correlating the results with destructive methods such as the **Scratch test** and **Peel tests**. A final note of caution should be added that most low-power ultrasound methods measure the elastic properties of the interface which is often but not always related to the mechanical strength.

References

1. R C Bray, C F Quate, J Calhoun, R Koch, *Thin Solid Films*, **74**, 295 (1980).
2. R C Addison, M Somekh, G A D Briggs, *Proc. IEEE Ultrasonics Symposium*, **2**, 775 (1986).

Acrylic adhesives

F R MARTIN

Introduction

Acrylic adhesives are a versatile class of reactive adhesives that have in recent years gained wide acceptance in industry as bonding agents in the assembly of components. They are also to be found in the consumer market, where the industrially important benefits of rapid bonding, strength and 'ease of use' can also be appreciated.

Setting mechanism

Acrylic adhesives are applied to the substrates to be bonded as a liquid mixture of unreacted methacrylate (or acrylate) monomers, polymers to thicken or toughen, reactive resins for strength, adhesion promoters and polymerization agents. Polymerization or curing is started by the formation, by various methods, of free radicals; the polymerization then proceeds very rapidly by sequential addition to form the solid adhesive in the bond line (see **Addition polymerization**).

The versatility of acrylic adhesives comes from the large number of different monomers available, which give flexible and tough or hard and rigid adhesives when cured, depending on the potential **Glass transition temperature** or on the functionality of the monomers (see **Toughened acrylic adhesives**).

Table 1. Performance* of acrylic adhesives

	Fast adhesive	Tough adhesive
Viscosity	10–20 000 mPa	20–80 000
Speed	30–60 s	60–120
(time to fixture)		
Strength		
Tensile shear	20 N mm^{-2}	15–25
(ASTM D1002-64)		
GB steel		
Toughness		
T-peel (ASTM D1876-69T)	2 N mm^{-1}	3–4
GB aluminium		
Recommended substrates	Metals, ferrites	Metals, ferrites, glass, plastics, wood

* Data are taken from the technical literature and are believed to be typical.[1]

Curing or setting of the adhesive can be achieved by irradiation by ultraviolet (UV) light (through glass) (see **Radiation-cured adhesives**), by the use of an activator on one or both substrates, by mixing two components either before applying the adhesive to the bond area or by mixing within the bond line as the components are assembled, or by heat. In each case free radicals are formed by the reaction of an initiator in the adhesive with a second agent, such as UV light, heat or a chemical activator. The polymerization is rapid, with strong bonds being formed in as little as 10 s. More information on the chemistry and variety of acrylic adhesive systems is given in ref. 1. Some performance data are shown in Table 1, which highlights the strength and speed of bonding obtainable.

Applications

Acrylic adhesives can be formulated to have low, controllable viscosity, which makes possible rapid, precise dispensing to parts to be bonded. In addition, rapid curing or setting, without long periods in ovens, makes these adhesives particularly useful for the assembly of various components in automatic or semi-automatic production lines (see **Engineering design with adhesives**).

Examples of these are the bonding of decorative glass figures, jewellery, sporting equipment, toys, loudspeakers (both ferrite/metal and paper and plastic cones) and small motors. Because the acrylic adhesives bond a wide range of substrates (metals, plastics, wood, glass) they can be used to bond different components and have found wide use in many industries.

Table 2. Comparative benefits of industrial adhesives (epoxies, cyanoacrylates, acrylics)

	Epoxies		Cyanoacrylates	Acrylics
	2-part mixed	1-part heat cured		
Bond strength				
Metals/ferrites	+	+ +	+	+ +
Plastics	0	—	+ +	0
Toughness	—	+ +	0	+ +
Temperature/humidity	+	+ +	0	+
Speed of bonding	0	—	+ +	+
Ease of use				
Automatic application	—	—	+ +	+ +

Note: + + great benefit, + significant benefit, 0 no benefit, — significant disadvantage.

Advantages/disadvantages of acrylic adhesives

Table 2 compares acrylic adhesives to two other important classes of adhesive used widely in industrial assembly operations, **Epoxide adhesives** and **Cyanoacrylate adhesives**. The speed of bonding and ease of use of cyanoacrylates makes them ideal where toughness and high-temperature resistance are not critical. Epoxies provide good bond strengths to many substrates, but are slower curing and require either precise mixing or a heat cure. Acrylic adhesives can be seen to complement epoxies and cyanoacrylates, providing better temperature resistance and toughness than cyanoacrylates and better speed and ease of use than epoxies, while giving good bonding performance.

Other articles of relevance are **Cyanoacrylate adhesives, Epoxide adhesives** and **Anaerobic adhesives**.

References

1. F R Martin, Acrylic adhesives, in *Structural Adhesives – Developments in Resins and Primers*, ed. A J Kinloch, Applied Science Publishers, Barking, 1986, Ch. 2.
2. D J Stamper, Toughened acrylic and epoxy adhesives, in *Synthetic Adhesives and Sealants*, Vol. 16 of Critical Reports in Applied Chemistry, Society of Chemical Industry, London, 1987, pp. 59–88.
3. C W Boeder, Anaerobic and structural acrylic adhesives, in *Structural Adhesives – Chemistry and Technology*, ed. S R Hartshorn, Plenum Press, New York, 1985, Ch. 5.

Addition polymerization

J COMYN

Monomers containing rings or double bonds can be polymerized by addition polymerization. This is a chain reaction involving the sequential steps of initiation, propagation and termination.[1] The term 'chain polymerization' is often preferred to 'addition polymerization'. Initiation is the process by which active centres are formed. These may be free radicals, anions, cations or, in the case of Ziegler–Natta polymerization, a coordinated anion. The free radical addition polymerization of a vinyl monomer is illustrated below.

Initiation

Free radicals can be produced by the thermal or photochemical decomposition of an azo or peroxy compound; examples given below are benzoyl peroxide and persulphate ion.

$$Ph—CO—O—O—CO—Ph \overset{heat/UV}{=} 2Ph—CO—O^{\cdot}$$

$$S_2O_8^{2-} \overset{heat}{=} 2SO_4^{\cdot-}$$

by redox decomposition of a peroxide by transition metal ions, e.g.

$$Fe^{2+} + H_2O_2 = Fe^{3+} + HO^{\cdot} + HO^{-}$$

or by direct photo-initiation using ionizing radiation. The primary radicals then attack the monomer thus,

$$R^{\cdot} + CH_2{=}CH = R—CH_2—CH^{\cdot}$$
$$\phantom{R^{\cdot} + CH_2{=}}\underset{R}{|}\underset{R}{|}$$

Propagation

Many monomer molecules are now added to the radical just formed to produce a long-chain macroradical. Radical lifetimes are typically a few seconds, and during this time thousands of monomer units can be added. The propagation reaction can be symbolized by a monomer molecule adding to the monomer radical unit at the end of a polymer chain. Here $P(n)$ symbolizes a polymer chain with n monomer units.

$$P(n)—CH_2—CH^{\cdot} + CH_2{=}CH = P(n+1)—CH_2—CH^{\cdot}$$
$$\underset{R}{|}\phantom{ + CH_2{=}}\underset{R}{|}\underset{R}{|}$$

Termination

In termination two macroradicals react, either by recombination or disproportionation.

$$2P(n)-CH_2-CH^{\cdot}$$
$$| $$
$$R$$

$$= P(n)-CH_2-CH-CH-CH_2-P(n) \qquad \text{recombination}$$
$$| \qquad |$$
$$R \qquad R$$

$$= P(n)-CH{=}CHR + P(n)-CH_2CH_2R \qquad \text{disproportionation}$$

In addition to initiation, propagation and termination, further processes which can occur are chain transfer, inhibition and retardation.[1]

High-mass macromolecules are produced from the start of addition polymerization, and the molar mass (molecular weight) and its distribution are controlled by the rates of initiation and propagation, the mode and rate of termination, and by chain transfer reactions.

Addition polymerizations can be carried out in bulk, solution or by suspension or emulsion polymerization with water as the medium. Polymer emulsions can be used directly as adhesives, paints and binders (see **Emulsion and dispersion adhesives**).

Addition polymerization and adhesives

There are several polymers which form the basis of adhesives made by addition polymerization. These include the carbon-chain synthetic rubbers, polyacrylates and polymethacrylates, **Ethylene–vinyl acetate copolymers** and emulsion polymers.

Copolymerization, i.e. with two or more monomers,[1] is a useful way of obtaining materials with novel properties, and in the case of **Ethylene–vinyl acetate copolymers** the effects of the vinyl acetate units is to reduce the crystallinity of polyethylene, and to introduce polar groups which improve adhesion.

Some adhesives cure by addition polymerization, notably the **Cyano-acrylate adhesives**,[2] **Alkyl-2-cyanoacrylates** and **Acrylic adhesives**.[3] Cyanoacrylate monomers undergo rapid anionic addition polymerization when the adhesive bond is closed. Hydroxyl groups in adsorbed water are generally thought to be the initiators, and the rapidity of anionic polymerization is due to two electron-withdrawing groups ($-CN$ and $-COOR$) which stabilize the propagating anion. The initiation step is

$$\qquad\qquad\qquad CN \qquad\qquad\qquad CN$$
$$\qquad\qquad\qquad | \qquad\qquad\qquad\quad |$$
$$HO^- + CH_2{=}C = HO-CH_2-C-$$
$$\qquad\qquad\qquad | \qquad\qquad\qquad\quad |$$
$$\qquad\qquad\qquad COOR \qquad\qquad\quad COOR$$

Reactive acrylic adhesives generally consist of a solution of a toughening rubber in partly polymerized monomer. The remaining monomer is polymerized in a free radical addition polymerization; redox initiation involves a peroxide which is dissolved in the adhesive, and a metal salt which can be applied in a solution to one of the surfaces. Acrylic cements consist of a partly polymerized acrylic monomer containing an initiator. Cure is established by the thermal or UV decomposition of the initiator (see **Radiation-cured adhesives**).

When mixed with acid anhydrides or primary or secondary amines, **Epoxide adhesives** cure by **Condensation polymerization**. However, when the hardener is a tertiary amine or a Lewis acid (see **Acids**) such as BF_3 cure is by a ring opening ionic addition polymerization.[4]

References

1. F W Billmeyer, *Textbook of Polymer Science*, 3rd edn, John Wiley, New York, 1984.
2. G H Millet in *Structural Adhesives, Chemistry and Technology*, ed. S R Hartshorn, Plenum Press, New York, 1986, Ch. 6.
3. C W Boeder in *Structural Adhesives, Chemistry and Technology*, ed. S R Hartshorn, Plenum Press, New York, 1986, Ch. 5.
4. Y Tanaka, R S Bauer in *Epoxy Resins, Chemistry and Technology*, 2nd edn, ed. C A May, Marcel Dekker, New York, 1988, Ch. 3.

Adhesion

D E PACKHAM

The chief aim of this entry is to show the various ways in which the term 'adhesion' is employed in a technical sense, drawing attention to confusion which can occur because of its use in different contexts.

Etymology

'Adhesion' seems to have come into the English language in the seventeenth century. It is interesting that one of the earliest examples (1661) quoted by the *Oxford Dictionary* is its use in a scientific context by Robert Boyle.[1]

The word 'adhesion' comes from the Latin *adhaerere* to stick to.[2] Its use by Lucretius of iron sticking to a magnet[3] anticipates the present technological application of the word. *Adhaerere* itself is a compound of *ad* (to) + *haerere*, where *haerere* also means to stick. Cicero uses the expression *haerere in equo* literally 'to stick to a horse' to refer to keeping a firm seat.[3] (It follows from this that the word 'abhesion' sometimes encountered, supposedly meaning 'no adhesion' or 'release' is etymological nonsense and should be abandoned.)

Adhesion as a phenomenon

The use of 'adhesion' to describe the commonplace phenomenon or state where two bodies are stuck together is relatively straightforward. It is unlikely to lead to any serious confusion in a practical context, despite worries that have been expressed about the circularity of such a definition.[4]

It is perhaps worth pointing out that some are not content with so simple a concept, but insist on building theoretical models into a definition of adhesion. ASTM D 907 defines 'adhesion' as 'the state in which two surfaces are held together by interfacial forces which may consist of valence forces or interlocking forces or both'. This means that we must know about the interface on the atomic scale before we can know whether we have an example of the phenomenon of adhesion; it subjugates the readily observed phenomenon to a particular scientific theory and scientific theories of their nature are subject to revision.

The magnitude of adhesion

When we ask how large the adhesion is, we enter a realm where the term is used in a number of quite different ways.

'Fundamental' adhesion In one use, adhesion refers to the forces between atoms at the interface. This is sometimes called 'true adhesion' or 'fundamental adhesion'. Here, of course, the concept is necessarily tied to one of the **Theories of adhesion** and to a particular model for the interface concerned.

Many different measures may be used to specify this 'fundamental' adhesion. It may be expressed in terms of forces or in terms of energies. Again, depending on the context, these may be forces or energies of attachment or else of detachment. Sometimes values of fundamental adhesion can be calculated from a theoretical model (see **Electrostatic theory of adhesion, Good–Girifalco interaction parameter**); occasionally they may be deduced from experimental measurement (see **Adhesion – fundamental and practical**); for many practical adhesive bonds they are not available by either route.

The work of adhesion (see **Wetting and work of adhesion, Contact angles and interfacial tension**) is a simple example of a concept which could be regarded as a 'fundamental' measure of adhesion between two surfaces.

Practical adhesion Destructive **Tests of adhesion** are also said to measure the adhesion of a joint or coating. 'Adhesion' here refers to the number that results from **Shear test** (q.v.), **Tensile tests** or whatever. It can be a peel strength or a **Fracture mechanics** parameter. These measures are sometimes referred to as the 'practical adhesion' of a particular joint: they more or less satisfactorily answer the question, 'How strong is the joint?'

As explained in specialized articles on testing, the numerical result obtained in a particular type of test usually depends on the operational variables – dimensions of the test piece, rate of loading and so forth. Thus there is not one, but many, values of practical adhesion for a particular joint.

The adhesion recorded reflects the 'fundamental' adhesion at the interface but also the mechanical response of the adhesive, substrate and interfacial region (see **Rheological theory**). Equation 7 in the article on **Peel tests** shows how the 'practical' adhesion (peel strength) may be related to the 'fundamental adhesion' (work of adhesion) for that type of test.

The mode of failure in a test may be adhesive, cohesive or mixed (see **Locus of failure**). Sometimes authors talk as if a test only measured 'adhesion' in the first of these cases. Such a distinction does not seem helpful as none directly measures 'fundamental' adhesion and all give a number which reflects indirectly the interfacial forces.

Conclusion

Good[4] has lamented the lack of unanimity on the meaning of the term 'adhesion', and has suggested definitions which would reduce the ambiguity when it is used in a scientific context. This article has been concerned with the situation as it is rather than as, perhaps, it should be. It has tried to indicate the depth of meaning of the term as it is actually used in the technical literature.

References

1. J A Simpson, E S C Weiner, *Oxford English Dictionary*, 2nd edn, Clarendon Press, Oxford, 1989.
2. C T Onions, *The Oxford Dictionary on English Etymology*, Clarendon Press, Oxford, 1966.
3. C T Lewis, C. Short, *A Latin Dictionary*, Clarendon Press, Oxford, 1879.
4. R J Good, *Treatise on Adhesion and Adhesives*, vol. 5, ed. R L Patrick, Marcel Dekker, New York, 1981, p. 293.

Adhesion– fundamental and practical

D E PACKHAM

As discussed in the article **Adhesion**, the term is used both for the fundamental atomic and molecular forces responsible for holding the two phases together and for the quantities such as peel strengths that come from destructive **Tests of adhesion**. Over the past 30 or so years general understanding of the relation between 'fundamental' and 'practical' adhesion has developed considerably.

At one time many thought, or wrote as if they thought, that adhesion tests more or less gave a direct indication of the fundamental forces at the interface. This led to a reaction where it was argued that 'those ... interested in the fundamental interfacial forces responsible for adhesion can get little, if any, information from destructive tests'.[1] The reaction was based on the observation that the results of tests of adhesion depend on the test parameters, for instance the values obtained with most adhesives depend upon the test rate because of the influence of the **Viscoelasticity** of the adhesive on the test result. Thus destructive tests were considered simply to reflect the overall rheology of the joint. Bikerman was a forceful advocate of this position – see **Rheological theory**.

We have now come to a position where most recognize that both joint rheology and fundamental interfacial forces contribute to the measures given by destructive tests, and both determine the loads under which the joint will fail in service. The relation between the three – fundamental forces, rheology and test results – is understood in broad principle, and in a few cases has been elucidated in quantitative detail. This will now be discussed.

A relatively simple way to an understanding is via an energy-balance approach to the theory of peel adhesion which is essentially an application of the first law of thermodynamics. This was developed in the article on **Peel tests**. For ease of illustration, consider the 90° peel of an inextensible strip. Combining Eqns 4, 5 and 7 in the article cited gives peel energy P as

$$P = F/b = W_A \text{ [or } W_C] + \psi_{plast} + \psi_{v/e} + \psi_{bend} + \cdots \qquad [1]$$

This relates the test result (the measured peel load F) to a measure of intermolecular forces (the work of adhesion W_A or of cohesion W_C, see **Wetting and spreading**) and to the rheology via various terms such as energy dissipated in plastic deformation ψ_{max}, in viscoelastic loss $\psi_{v/e}$, in bending ψ_{bend}. Here b is the width of the peeled strip.

A similar approach is possible for other tests which measure in some form an energy of fracture. Tests based on the principles of **Fracture mechanics** can give a measure of G_C the critical energy release rate, which is essentially an energy of fracture under precisely defined conditions. Here

G_C can be expressed as a sum of interfacial energy terms and other terms for bulk energy dissipation in a way analogous to peel energy P in Eqn 1. (The energy-balance approach is essentially one of fracture mechanics.)

Many tests, for example **Tensile tests** and **Shear tests**, measure a critical stress rather than an energy of fracture. Good has pointed out that this critical stress will also depend on the fracture energy.[2] He adapted the Griffith–Irwin theory of fracture to apply to adhesive bonds. According to this theory,[2] the fracture stress σ_f of material of modulus E will be given by

$$\sigma_f = k\sqrt{(EG/l)} \qquad\qquad [2]$$

where k is a constant, l the length of the critical crack and G the work of fracture per unit area. Like peel energy P in Eqn 1, G can be expressed as the sum of the various energy-dissipation terms involved in the fracture. (The first law of thermodynamics again.) So G may be written as

$$G = W_A \text{ [or } W_C] + \psi \qquad\qquad [3]$$

Depending on the **Locus of failure** either the work of adhesion W_A or the work of cohesion W_C will be appropriate; ψ represents all the other energy-dissipation terms.

Good argues that G and E will be semi-local properties which will vary as one passes from one phase bonded through the interface to the second phase. The way in which the product EG varies throughout the bond (and the lengths and positions of potentially critical flaws[2]) will determine the mode of failure, interfacial or cohesive within the bulk or close to the interface. Failure will occur where the term (EG/l) is at a minimum, at the stress given by Eqn 2.

In some uncomplicated examples it has been possible to analyse the results of adhesion tests to obtain numerical values associated with interfacial forces. An example of this was the work of E H Andrews and A J Kinloch[3] who measured the adhesion of SBR (styrene–butadiene rubber) to different polymeric substrates over a range of temperatures and test rates. Three types of test including a 90° peel test were used. The results were analysed to evaluate the fracture energy per unit area G. (For the peel test G was P, the peel energy.)

The polymer SBR is mechanically simple, and it was found that the results for G for a given substrate could be superimposed using the Williams, Landel and Ferry (WLF) technique (see **Viscoelasticity – time–temperature superposition**). The master curves for each substrate were close to being parallel to each other and to a similar curve for cohesive fracture of the SBR (Fig. 4).

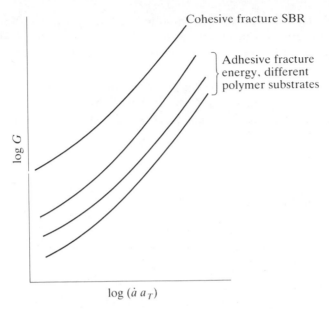

Cohesive fracture SBR

Adhesive fracture
energy, different
polymer substrates

log G

log ($\dot{a}\, a_T$)

Fig. 4. Master curve of fracture energy G against reduced test rate ($\dot{a}a_T$) for SBR and for adhesion bonds between SBR and various polymer substrates. After Andrews and Kinloch[3]

The parallel log/log curves imply some mathematical relationship of the form

$$G = G_0\phi(\dot{a}, T) \qquad [4]$$

where $\phi(\dot{a}, T)$ is a rate ($\dot{a}$) and temperature-dependent viscoelastic factor, and G_0 some intrinsic failure energy characteristic of the substrate and independent of rate and temperature. As a value of G_0 could be obtained for cohesive tearing of the SBR itself, values of G_0 could be deduced from the adhesion measurements for each substrate.

The work of adhesion for the various SBR/substrate combinations were evaluated from contact angle measurements (see **Surface energy components**), and were compared with the values of G_0. For bonds where adhesive failure occurred, there was a good agreement between W_A and G_0. Where the **Locus of failure** was cohesive, G_0 could be related to the cohesive failure energy of the materials concerned.

This work shows how the type and magnitude of interfacial forces can be deduced from the results of adhesion tests. For this simple polymer the adhesive fracture energy is made up of an intrinsic term G_0 related to the interfacial forces, and a term related to the viscoelastic energy losses during testing. The viscoelastic loss can be expressed as an additive term (cf. Eqns 1 and 3) or, for this system, as a multiplicative factor, Eqn 4.

Although for SBR at all normal rates and temperatures of test the viscoelastic term ψ is much greater than the interfacial term W, W exerts a profound influence on the measured adhesion because of the multiplicative relationship.

In summary, the relationship between fundamental and practical adhesion is now understood in broad terms. Information about fundamental interfacial forces can be obtained from destructive tests (*pace* de Bruyne).

References

1. N A de Bruyne in *Adhesion and Cohesion*, ed. P Weiss, Elsevier, Amsterdam, 1962, p. 46.
2. R J Good, *J. Adhesion* **4**, 133 (1972).
3. A J Kinloch, *Adhesion and Adhesives: Science and Technology*, Chapman and Hall, London, p. 314.

Adhesion in medicine

M E R SHANAHAN

Although adhesion intervenes in many phenomena occurring in the industrial world, sight must not be lost of its importance in the natural, biological context and in particular the role it plays in medical applications. It would seem that the vast range of natural phenomena in which adhesion is fundamental has only fairly recently been appreciated. As far as medicine is concerned, research is being actively pursued even if understanding is still at a fairly rudimentary level in many domains. Clearly, enormous problems exist due to the sheer complexity of the systems to be examined. In addition, since virtually all the cases of interest are in an essentially aqueous environment, there are considerable difficulties in direct observation and modelling is not easy. **Polar forces** and **Hydrogen bonding** will be predominant. Nature nevertheless uses adhesion, and polymer science to great benefit! The coagulation of blood at a wound, although a complex process, is a perfect example. Platelets in the blood are attracted to the site of the wound and an enzyme called thrombin is produced from prothrombin. Thrombin acts as a catalyst by converting the protein fibrinogen (from the blood plasma) into fibrin, an insoluble, polymerized protein. This fibrous material constitutes one of the essential materials of the clotting process. The conversion from fibrinogen into fibrin is therefore a reaction of natural polymerization with thrombin as the 'curing' agent and as such the system can be compared not unreasonably to a two-component adhesive!

Although the areas of medicine in which adhesion is important must be very extensive, there are basically two scales to be considered.

Firstly there is the macroscopic aspect which concerns essentially the manufacture of prostheses in their various forms (joints, 'pace-makers', etc.). Problems to be overcome here involve the creation of suitable and resistant bonds to organic elements existing in the body, such as bone or muscle, and the need for the synthetic components to be biocompatible. Not only must adhesive bonds be established in an aqueous environment, conditions which are plainly avoided when possible in an industrial context where the choice usually exists, but also body fluids are potentially capable of reducing adhesion over a long period both on simple thermodynamic grounds and for reasons of degradation (comparable to environmental stress cracking and stress corrosion). The thermodynamic argument is simple and based on Dupré's equation. If substance A (say the naturally occurring material) adheres to substance B (the prosthesis), their thermodynamic work of adhesion (in air) is given by

$$W = \gamma_A + \gamma_B - \gamma_{AB} \qquad [1]$$

where γ_A and γ_B represent the surface free energies of A and B and γ_{AB} is their mutual interfacial free energy. However, if separation is apt to occur in a liquid (body fluid) L, the work of adhesion is modified to W_L:

$$W_L = \gamma_{AL} + \gamma_{BL} - \gamma_{AB} \qquad [2]$$

where γ_{AL} and γ_{BL} are the interfacial free energies of A and B with L. If $\gamma_{AL} + \gamma_{BL} < \gamma_A + \gamma_B$, the presence of L will reduce the propensity for adhesion of the system (see **Durability**).

The second main aspect of **Adhesion in medicine** corresponds to the cellular level. A few examples will be given. In the development of a cancer, one major problem is that the patient is often only aware of the primary tumour when it has evolved sufficiently for metastases to have been released. These malignant cells are free to migrate within the body, often in the blood circulation, and create secondary tumours elsewhere in the body leading to a generalized illness.[1] Were we to be able to increase the adhesion of metastases to their source, many treatments of cancer could be limited to a specific area of the body. By contrast, consider a case where adhesion is undesirable. Heart disease if often related to the blocking of arteries by matter, in particular thrombi, or locally produced blood clots, adhering to the walls of the blood vessels. Reduction of this restriction to the circulation would clearly be beneficial since the essential flow of blood to the heart muscle would then be less impeded. An important step during the ingestion of foreign particles by phagocytes[2] is the establishment of bonds between the immunological effector and its target.

The engulfment of bacteria by macrophages is almost certainly related to surface free energy and thus adhesion-controlled processes.[3]

For these, and other reasons, various researchers have looked at the complexities of cell adhesion.[4,5] The reader is referred to the review by Bongrand and Bell[5] and the references therein for a more complete appraisal of this complex field.

References

1. M McCutcheon, D R Coman, F B Moor, *Cancer*, **1**, 460 (1948).
2. C Capo, P Bongrand, A M Benoliel, R Depieds, *Immunology*, **35**, 177 (1978).
3. A W Neumann, *Advances in Colloid and Interface Sci.*, **4**, 105 (1974).
4. C J van Oss, *Cell Biophysics*, **14** 1 (1989).
5. P Bongrand, G I Bell in *Cell Surface Dynamics: Concepts and Models*, ed. A S Perelson, C Delisi, F W Wiegel, Marcel Dekker, New York, 1984, Ch. 14.

Adhesion under ultra-high vacuum

R G LINFORD

Truly clean metallic surfaces stick together by themselves, without the need for any adhesive and, for dissimilar materials, the adhesive bonding that results is stronger than the cohesive strength of the weaker of the two materials involved. Truly clean surfaces, free from surface oxides and/or absorbates, can only be maintained under ultra-high vacuum (UHV) conditions, i.e. at pressures below 10^{-8} Pa (10^{-10} torr; 10^{-13} bar). This is because the lifetime of a clean surface τ in seconds is approximately related to the pressure p in pascals by

$$\tau = \frac{10^{-4}}{p}$$

Ultra-high vacuum technology involves sophisticated pumps free from backstreaming; turbomolecular systems or combinations of sorption cryopumps, ion pumps and titanium sublimation pumps are often used. Stainless steel vessels are preferred to glass because of the permeability of the latter to atmospheric helium. Soft metal 'O' ring (In or Au), or OFHC copper gaskets and knife-edge seals are used to join vacuum vessel components. Conventional vacuum greases have too high a vapour pressure to be used under UHV conditions. Truly clean surfaces are produced from as-received materials by a series of conventional cleaning procedures (degreasing, abrasion, etc.), followed by argon-ion bombardment (which can be thought of as atomic scale sandblasting) under UHV conditions. Surface oxidation or reduction under controlled gas atmospheres may also be required. The cleanliness of the surface is best monitored by **Auger electron spectroscopy**.

The contact between clean surfaces is limited by topography, as discussed in **Engineering surfaces of metals** and **Roughness of surfaces**. The initial points of contact are between *asperities* which *plastically deform* to provide adhesive contact over more than 1% of the surface.

Metal-to-metal adhesion was a major study in certain areas of the early US space programme because the atmospheric pressure on the moon is about the same as under UHV. For example, the type of astronaut suit designed in the early 1960s had elbow joints that cold-welded and seized under simulated lunar environments, causing discomfort and lack of operational manouevrability! The prototype 'lunar buggy' developed at General Motors in California involved coiled-coil Ti wire tyres to provide some springing and load cushioning; and the chafing and abrasion of touching coils under UHV conditions led to metal-to-metal adhesion and a very bumpy ride. A similar 'surface self-cleaning' and subsequent cold welding at moving joints was anticipated for reactor machinery in high-temperature, helium atmosphere nuclear reactor systems that were under study by the Central Electricity Generating Board (CEGB) and other power utilities in the 1970s. One possible route to the avoidance of cold welding that excited interest at that time, especially by Don Buckley and his associates at NASA Lewis Research Centre in the USA and Roger Linford at Berkeley Nuclear Laboratories in the UK, was to use the lattice mismatch between metals of differing crystal structure. The clean surface adhesion problem is less pronounced for oxides, ceramics and other non-metallic solid materials than for metals.

Adhesive classification

B C COPE

The primary purposes of an adhesive may be said to be to wet and spread on the surfaces of the adherends and to fill the gap between them, and subsequently to form a relatively permanent, coherent bond. The former action must necessarily demand that the adhesive is applied in the liquid phase, whereas the final bond depends on the adhesive being a solid. The two most widely used taxonomic schemes for adhesives address either the way in which this phase change takes place or the chemical nature of the adhesive itself; although other schemes, such as a classification by adherend or end-use industry are also current.

The final bond produced by the vast majority of adhesives consists

Table 3. Classification of adhesives by setting mode

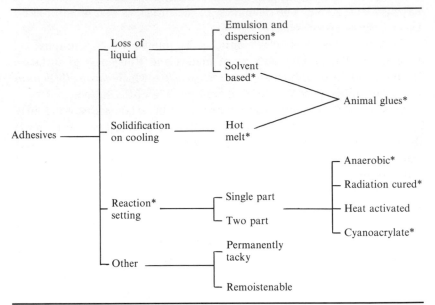

* There are specialized entries with these titles.

largely of a solid organic polymer. The fluidity necessary at the time of application may be attained by various means that thus form the basis of the system of classification, see Table 3.

In the case of **Solvent-based adhesives** the polymer or blend of polymers that will eventually form the bond is dissolved in a carrier solvent. Sometimes, as in many materials for packaging applications, the solvent is water, but since the bonds thus produced will be moisture susceptible the use of organic solvents is more usual. An added advantage of the use of organic solvents is that their lower latent heat of evaporation ensures faster drying. Against this such solvents are more expensive than water, are usually inflammable and are sometimes irritant or toxic. Growing concern about release of solvents to the atmosphere has combined with **Health and safety** considerations and ever rising fire insurance premiums to accelerate a movement away from organic solvent-based systems.

Whatever the solvent system, polymers produce highly viscous solutions at quite low concentrations thus limiting the weight of adhesive coating that may economically be applied in a single operation, and increasing the bulk and cost of an adhesive system containing a given quantity of solid adhesive. High shrinkage can also be a problem.

When a solvent-based system is applied to a porous or permeable substrate the solvent may pass out of the bond through the adherends. When the adherends are relatively impermeable a common practice is to coat both mating surfaces and allow them to become substantially dry before bringing them into contact while the adhesive retains some degree of **Tack**. This technique is familiar to many through its application to the repair of punctures in bicycle tyres or the lamination of decorative laminates to various substrates.

Polymer dispersions (in non-solvent liquids, usually water), have much lower viscosities than solutions with the same polymer content. So great is the disparity in viscosities that a 50% solids dispersion may have a considerably lower viscosity than a 10% solution. Such dispersion or suspension adhesives, often called 'emulsion adhesives' or even more loosely 'water-based adhesives', combine the advantages of a high solids content with low inflammability and toxicity hazard but suffer from long drying times (see **Emulsion and dispersion adhesives**). Their use on non-porous substrates is restricted in a similar way to that of solution adhesives.

There exist two obvious ways of increasing the active solids content of an adhesive effectively to 100%: use of **Hot melt adhesives** and **Reaction setting adhesives**. The main constituent of hot melts is a thermoplastic polymer that may be blended with thermoplastic modifiers and extenders and inert fillers to create a system that is a load-bearing solid at the service temperature, but a mobile liquid at the (higher) application temperature. Polyolefins, ionomers, polyesters and polyamides are among the polymer types that have been used as bases for **Hot melt adhesives**.

The **Animal glues and technical gelatins**, applied hot and previously popular for wood bonding and book binding, are very concentrated aqueous solutions of gelatin. Their hardening is a combination of solution and hot melt mechanism in that solidification on chilling is augmented by diffusion of moisture into the porous substrate.

The second means of transforming a liquid adhesive entirely into a solid without the loss of a solvent or dispersion medium is to produce solidification by a chemical change rather than a physical one. Such reactive adhesives may be single-part materials that generally require heating or exposure to electron beam or UV radiation to perform the reaction and which may be solids, liquids or pastes, or two-part systems where the reactants are stored separately and mixed only shortly before application (see **Radiation-cured adhesives**). The former class are exemplified by the fusible, but ultimately reactive, epoxide film adhesives and the latter by the two-pack **Epoxide adhesives** and **Polyurethane adhesives** and by the **Toughened acrylic adhesives** that cure by a free-radical **Addition polymerization** mechanism.

Table 4. Classification of adhesives by chemical nature

Thermosets (a)	Elastomers (b)
Urea formaldehyde	Nitrile rubber
Phenol formaldehyde*	Butyl rubber
Melamine formaldehyde	Polychloroprene
Resorcinol formaldehyde	Polyisobutene
Furane resins	Styrene–butadiene rubber
Polyurethanes*	Silicone rubber
Epoxides*	Reclaimed rubber
Polyesters	Natural rubber
Anaerobics	Cyclized natural rubber
Polybenzimidazoles*	Chlorinated natural rubber
Polyimide*	Others
Others	

Thermoplastics (c)	Natural products and modified natural polymers
Acrylics*	Animal glue*
Cyanoacrylates*	Fish glue
EVA*	Casein
Polythenes	Bitumens
Polyesters	Dextrins
Polyamides	Starches
Polyvinyl acetate	Gums
Polyvinyl alcohol*	Cellulose esters
Polyvinyl chloride	Cellulose ethers
Polyvinyl butyral	Others
Polyvinyl ethers	
Polyether etherketone*	
Polyimide*	
Polyphenylquinoxaline*	
Others	

* There are specialized articles with these titles. See also
(a) **Reaction setting adhesives**, (b) **Rubber based adhesives**
and (c) **Hot melt adhesives**.

Two types that may be considered to be special cases of single-component reactive systems are the **Anaerobic adhesives** and **Cyano-acrylate adhesives**. These are similar in that they are single-pack liquids or blends of liquids that are stable on storage, but which polymerize and solidify at ambient temperature when spread as a thin film between adherends. The cure mechanisms differ in that the anaerobics depend on the fact that oxygen inhibits their otherwise rapid polymerization, and so exclusion of oxygen, as in a joint, permits polymerization that may be catalysed by species on the adherent surfaces, whereas the cyanoacrylates depend on catalysis by hydroxy groups on the adherends.

Permanently tacky or **Pressure-sensitive adhesives** are those which may be applied as dispersions, solutions or hot melts and which convert to a rubbery, tacky film of relatively low adhesive strength and rather higher cohesive strength. They may be used to produce bonds that are permanent, but not creep resistant, and also temporary or serial temporary bonds. Such materials are often used on flexible tapes.

Remoistenable adhesives are water-based compositions that dry to a non-tacky film that may have its adhesive power regenerated by the absorption of water. Envelope gums are a familiar example.

The classification of adhesives by hardening mechanism (Table 3) is probably the most informative, but a classification based on the chemical type of the main polymer present is often used. Such a system might order matters first by the class of polymer, thermoset, thermoplastic or elastomeric, often with a separate class for natural polymers, and then subdivide these classes according to the chemical structure. Table 4 shows how this may be done.

Select references

W C Wake, *Synthetic Adhesives and Sealants*, John Wiley, Chichester, 1987.
W C Wake, *Adhesion and the Formulation of Adhesives*, 2nd edn, Applied Science Publishers, London, 1982.
J Shields, *Adhesives Handbook*, 3rd edn, Butterworths, London, 1984.
I Skeist, *Handbook of Adhesives*, Van Nostrand-Reinhold, New York, 1977.
S C Termin, Adhesive compositions, in *Encyclopedia of Polymer Science and Engineering*, Vol. 1, Wiley Interscience, 1985, pp. 547–77.

Adhesives for textile fibre bonding

A J G SAGAR

A general discussion of **Adhesives in the textile industry** is given in an article of that title.

Non-woven materials

These are produced by converting a loose web or fleece of fibres into an integral structure. To stabilize the construction of most non-wovens an adhesive is applied either as a polymer dispersion/latex (binder), or incorporated in the fleece as a thermoplastic (fusible) material.

Latex adhesives, of which there are many types (nitrile rubbers (NBR), SBR, **Ethylene-vinyl acetate copolymers**, acrylics, polyvinylidene chloride, etc.) should ideally be cross-linkable (for wash and dry-clean resistance), preferably with a low cure (reaction) temperature (see **Reaction setting adhesives**), as some fibres, such as polypropylene, may be damaged by high temperatures. The binder (adhesive) type can markedly affect physical properties and performance; the amount of hard and soft polymer in the binder controls the softness of the finished product. Most binders impart adequate dry-state adhesion and so the choice of adhesive is usually governed by secondary requirements such as method and conditions of application and costs.

Synthetic rubbers (SBR, NBR) are used extensively where soft fabrics having some degree of stretch are required, whereas acrylic copolymers are widely used for less extensible fabrics. Acrylics are the best all-round binders, having good lightfastness in particular, and good wet strength. Nitrile rubbers (NBR) impart resistance to dry-cleaning solvents and to oils.

Thermoplastic fusible adhesives (see **Hot melt adhesives**) may be used in the production of adhesive-bonded non-wovens, as powders, fibres or sheet films. The formation of strong bonds between the fibres occurs as the softened thermoplastic adhesive material resolidifies on cooling. The amount of adhesive incorporated in the web governs the strength of the fabric.

The use of bicomponent fibres, in which each individual fibre is made up of two types of polymer, lying either side by side or in a sheath-core formation, may be regarded as a special case of adhesive bonding with a thermoplastic material. One polymer, with a lower melting-point than the other, melts and forms the bond while the second remains solid and maintains the fibrous network.

Flocking

This is a means of attaching a fibre pile to a base fabric or paper; the flock is attached using an adhesive. The adhesives may be water-based (synthetic latices), **Solvent-based adhesives** or solvent-free (see **Adhesive classification**); the preferred ones are acrylic latices, polyurethanes, and PVC **Plastisols**. On curing, to form a cohesive solid, acrylic and polyurethane adhesives yield bonds that are resistant to dry-cleaning and laundering; PVC adhesives are only resistant to water-based cleaning agents as solvents extract plasticizer from PVC. The high gel temperature required for PVC (170–180 °C) may restrict the choice of base material.

Sizing

Size, which is used to treat warp yarns prior to weaving, commonly consists of an adhesive and an emulsified lubricant. The function of the adhesive is to bind together the fibres or filaments of a yarn, thus supplementing the cohesive effect of twist, and to form a film on the surface of the threads as protection against the abrasive action of the loom. The requirements of the adhesive are that it must be water dispersible or soluble, form a film, have good adhesion, allow separation of each warp yarn, be flexible and finally must be easily removed from the woven material during desizing.

Adhesives used in size formulations include starches (natural such as potato, maize or sago, or chemically modified, e.g. starch ethers), natural gums (guar, carob or locust bean flour), gelatin, **Polyvinyl alcohol**, polyacrylic acid, vinyl and acrylic resins and sodium carboxymethyl cellulose (SCMC).

Tyres, belts and hose

Rubber products such as tyres, belts and hose rely on reinforcement by textiles to achieve the required physical properties. To effect reinforcement, textile and rubber must be adequately bonded together, and to promote adhesion there is a range of treatments to suit most fibre–rubber systems. The adhesion-promoting material (dip) is usually a terpolymer latex of butadiene–styrene–vinyl pyridine (or a blend of SBR and vinyl pyridine) which bonds well to the fibres, together with a resorcinol formaldehyde precondensate which, on curing, bonds well to rubber; a three-dimensional resin network is formed.

Adhesives in textile coating, laminating and printing are discussed in a separate article.

Select references

W C Wake, D B Wooton (eds), *Textile Reinforcement of Elastomers*, Applied Science Publishers, London, 1982.
D B Wooton in *Developments in Adhesives – 1*, ed. W C Wake, Applied Science Publishers, London, 1977.
J A Radley (ed.), *Industrial Uses of Starch and its Derivatives*, Applied Science Publishers, London, 1976.

Adhesives in textile coating, laminating and printing

A J G SAGAR

Adhesives for textile fibre bonding are discussed elsewhere; a more general article considers **Adhesives in the textile industry**.

Coated fabrics

Adhesives in relation to coated fabrics can be thought of as either the tie-coat to bond the surface coating to the fabric, or the surface coating itself, directly bonded to the fabric. The types of polymers used in coating fabrics are rubbers, PVC, polyurethane, acrylics and silicones; the coatings are usually cross-linked to increase durability but some topcoats can be thermoplastic, for example polyurethanes, to facilitate thermal bonding of seams in production of garments or other articles.

Lamination

Lamination (sticking together of layers) of fabric, foam, and film, in various combinations, can be performed using latex- and solvent-based or fusible adhesives (see **Adhesive classification**). In addition, adhesives are sometimes applied from solution or latex for subsequent reactivation by heat during lamination. Acrylic dispersions are the main type of latex-based adhesives used in the laminating industry; butadiene rubbers and polyurethanes are also utilized in smaller quantities. The latex-based adhesives are used for aesthetic reasons, and also on heat-sensitive or open-structured fabrics. The solvent-based adhesives, e.g. rubbers and polyurethanes, are used in similar applications to the latex adhesives.

Thermal bonding using fusible adhesives (see **Hot melt adhesives**) can be achieved in three ways. Where one of the layers is a thermoplastic material or can be softened and tackified by heat, the use of a separate adhesive is not necessary. A special type of this sort of fusing is flame lamination using polyurethane foam. Besides mechanical bonding of the softened foam to the second substrate, some chemical bonding may be possible due to decomposition of the foam to give reactive chemical groups. Flame lamination is not a suitable method for use on heat-sensitive or open fabrics, but on other fabrics is a simple and cheap method of laminating; another plus is there are no solvents or water to remove, but this must be set against the production of toxic fumes during the flaming process.

The second method of thermal bonding utilizes nets and webs of thermoplastic material; this is now challenging flame lamination. Coextruded nets, where one polymer acts as a cheap carrier for a lower-melting, more expensive polymer, are also available.

The third thermal bonding method involves the application of powders to one substrate in a separate application. The powders can be applied as a compounded paste by rotary screen printing, as powders from hot engraved rolls (powder point process), or by scatter coating followed by sintering (heating below the polymer's melting-point with application of light pressure). The scatter-coating method is probably the most versatile and widely used.

Whichever method is used to get adhesive on to a fabric, the distribution of adhesive usually has to be discontinuous to retain the flexibility, handle and other aesthetics of the bonded assembly.

A special example of lamination, and probably the most important use of adhesives in the assembly of garments, is fusible interlinings. These are used in garment production for collars, cuffs, jacket fronts, etc. so they retain their shape after repeated laundering or dry-cleaning. The performance of interlining-stabilized fabrics very much depends on the properties of the adhesive; the desirable properties are low cost, strong bond at low add-on (weight of adhesive used), minimum strike-back or strike-through and stability to dry-cleaning and/or laundering.

The adhesive coating materials, applied to the base cloth to form the fusible interlining, are thermoplastic materials which melt and flow in the fusing press and so form the bond between interlining and face fabric. The materials most commonly used as adhesives have already been described (under **Hot melt adhesives**), i.e. polyethylene, copolyamide, copolyester, plasticized PVC, cellulose acetate and polyvinyl acetate.

Seaming

In the seaming of articles, use of hot melts is limited to relatively simple operations such as joining of straight edges. The thermoplastic material can be incorporated into the fabric or there is a wide range of films, tapes, nets and even coated threads which can be sandwiched between layers of non-thermoplastic material.

Use of **Solvent-based adhesives** in seaming stems from the pioneering work of Charles Macintosh in the last century; seam bonding with rubber solutions was a method used in the original waterproof coats carrying his name. Developments of the rubber-bonded seams are used in the manufacture of items such as life-jackets and life-rafts. The normally vulcanized (cross-linked) polymers used are natural rubber, butyl rubber, neoprene and polyurethane (see **Rubber-based adhesives**).

Carpets

Adhesives are used in tufted carpets to anchor and stabilize the yarns; they can also be used to attach a foam backing or a laminated secondary backing. Latices predominate with SBR (cross-linked) being the most widely used; EVA latex (see **Ethylene–vinyl acetate copolymers**) is cheaper but has lower bond strength and is water-sensitive. Other adhesives used are PVC **Plastisols** in carpet tiles, and hot melts and two-component polyurethanes in top-quality products (because hot melts and polyurethanes are relatively expensive).

Pigment printing

This technique, the surface application of insoluble colourants, has several potential advantages over dyes which react with the fibre; their colour is fully developed at the time of application, as compared with reactive dyes which only develop full colour after complete fixation, and the colour of the pigment is intrinsic to that pigment and will not change appreciably on different fibre substrates. The drawback is that, as the colour is inert, it needs an adhesive to hold it in place and prevent loss through abrasion, washing, or dry-cleaning.

The binder, therefore, must have the following properties: good adhesion to fibre and pigment in the dry state, and in water and solvent; a soft, non-tacky handle; elasticity; lightfastness, clarity and no colour; compatibility with, and stability towards, other ingredients in the mix. Some styrene butadiene copolymer rubbers and vinyl latices are used successfully but the most suited are the acrylics. The polymers are usually either self-cross-linking (with heat) or incorporate a cross-linkable resin such as melamine–formaldehyde.

Select references

L M Smith in *Developments in Adhesives – 1*, ed. W C Wake, Applied Science Publishers, London, 1977.
G R Lomax, *Textiles*, **14** (1), 2–8; **14** (2), 47–56 (1985).

Adhesives in the textile industry

A J G SAGAR

This article gives a survey of the types of adhesive used; more specialized aspects are covered in **Adhesives for textile fibre bonding** and **Adhesives in textile coating, laminating and printing**.

Adhesives are widely used in the textile industry as shown in Table 5. If sizes and tyre cord dips are excluded, approximately 65% are used in carpet backing (mostly styrene–butadiene rubbers). Other major uses of adhesives are in non-wovens as binders (15%), in fabric backing (8%), in flocking and laminating (8%) and in fusible interlinings (2%).

Excluding warp yarn sizes for weaving, which are still dominated by starch and cellulose derivatives, most applications use synthetic polymers.

Table 5. Applications for textile adhesives

Type of use	Textile application
Bonding layers of materials together	Lamination, fusible interlinings, seam and hem sealing, labels, bookbinding, surgical dressings
Bonding fibres together	Joining yarns, warp sizing for weaving, twistless spinning, formation of bonded-fibre fabrics (non-wovens)
Application of protective layers	Coated fabrics, carpet backing
Application of decorative finishes	Flock and pigment printing, motifs
Adhesion promotion	Tyre cord and belt fabric dips

Table 6. Classes of polymers used as textile adhesives

Natural products
Starches, modified starches, cellulose derivatives, natural gums, natural rubber

Synthetic polymers
Polyethylene, polypropylene, polyvinyl chloride, polyvinyl acetate and alcohol, **Ethylene–vinyl acetate copolymers**, polyacrylates, polyamide, polyester, **Polyurethane adhesives**, polychloroprene (neoprene), butadiene rubber, nitrile rubbers

The quantities used are relatively small amounts compared with, say, the construction and the **Packaging industry** so there is less incentive for manufacturers to develop products specifically for textiles. Nevertheless a good variety of products are made, which are well able to meet the often stringent requirements in the textile industry (Table 6).

If a polymer is to be used as an adhesive, several properties have to be considered, strength, elongation, flexibility, melt viscosity (for **Hot melt adhesives**), solubility or swelling characteristics in aqueous or dry-cleaning solvents, resistance to heat, light and hydrolysis and resistance to microbiological degradation.

These types of adhesives can be broadly classified by their physical form of application and/or the way they are activated to form a bond with the material to which they are applied, i.e. as emulsion polymers, solution polymers, hot melts and reactive polymers (see **Adhesive classification**). Each of these will now be considered in turn.

Emulsion polymers

These products consist of a dispersion of very small particles of polymer in an aqueous medium (see **Emulsion and dispersion adhesives**). Their major advantage, apart from being based on water rather than organic solvents, is that high-molecular-weight products can be made and dispersed, to give high-solids-content emulsions (up to 65%) which have low viscosities (water-thin). Solution-based polymers of the same molecular weight and solids content would be unworkable because of their very high viscosities. The main disadvantage is that the water medium can cause swelling, and hence curl, of water-sensitive substrates such as fabrics. Emulsion adhesives include natural rubber latex and polymers and copolymers of acrylic esters and acids, vinyl and vinylidene chloride, vinyl acetate and chloroprene.

Solution polymers

Water can be used as the medium for a solvent-based adhesive when the polymeric material is water-soluble. An example of such an adhesive would be a yarn-sizing formulation based on modified cellulose, **Polyvinyl alcohol** or polyacrylic acid. Generally, however, solvents are organic compounds, for example hydrocarbons (hexane, toluene), ketones (butan-2-one), and esters (ethyl acetate), see **Solvent-based adhesives**.

Usually only rubbery polymers are considered for textile applications; however, for example in tyre cord and vee belt and transmission belt fabric dips stiffer (resinous) polymers can be used.

Hot melts

The **Hot melt adhesives** suitable for textile use are copolyamides, copolyesters, **Polyurethanes**, acrylics, polyethylenes, polypropylene, ethylene copolymers (mainly with vinyl acetate and vinyl alcohol), polyvinyl acetate and blends with **Phenolic resins**, plasticized vinyl acetate/PVC copolymers and plasticized cellulose acetate. They are used extensively for fusible interlinings, application of motifs and decorations, in seam and hem sealing, for fabric lamination (including wall coverings) and in carpet backing and bonding; in addition, fusible bicomponent fibres (two polymers in one fibre) in non-woven melded fabrics are, in effect, hot melt adhesives; when an assembly partially or wholly composed of these fibres is heated, the contact points of the fibres fuse together, thereby giving structural coherence to the fabric. In selecting hot melts, serviceability in use must be balanced with ease of application.

The melting-point needs to be above service temperatures (NB medical textiles are often steam-sterilizable at 130 °C for several minutes), but too high a melting-point may affect adversely the dimensional stability, colour and handle of the fabric.

If the melt viscosity is too high the polymer will not flow into the substrate and an inadequate bond results: if too low, strike-through or strike-back can occur, resulting in a stiff substrate and a bond starved of adhesive. Thermal stability must be adequate should the polymer be kept molten in the application for long periods. The physical form is important; for example, a discontinuous coating is preferred in some applications to give products with a soft handle, and the ability to form reticulated films is therefore a consideration.

Reactive adhesive compositions

Many of the emulsion- and solution-based adhesives, already described, can be cross-linked, and can therefore be termed **Reaction setting adhesives**.

There are other systems, where a prepolymer of low molecular weight reacts with a second component to give a three-dimensional structure and a very durable polymer. Examples include urea and melamine resins used with formaldehyde in fabric finishing, resorcinol/formaldehyde systems used in cord and fabric dips and low-molecular-weight silicones which react on the substrate to give durable water-repellent finishes.

Further details appear in other articles, **Emulsion and dispersion adhesives, Solvent-based adhesives, Adhesive classification, Hot melt adhesives, Rubber-based adhesives**.

Select references

W C Wake (ed.), *Synthetic Adhesives and Sealants*, John Wiley, Chichester, 1987.
K O Calvert (ed.), *Polymer Latices and their Applications*, Applied Science Publishers, London, 1982.

Adsorption theory of adhesion

K W ALLEN

The title, adsorption theory, is a reflection of the extent to which there are very close parallels between the ultimate forces involved in adhesion and those involved in the adsorption of gases and vapours on to solid materials. It has been well known for 70 years or more that, in the latter case, the phenomena can be divided into two separate groupings: physical adsorption (or physisorption) and chemical adsorption (or chemisorption).[1] It is now clear that a similar distinction has to be made in the discussion of adhesive bonding. However, in this case forces of the type which have been described as 'physical' are always involved and forces of the type described as 'chemical' sometimes augment them.

Table 7. Valence forces

Bond type	Bond energy $(kJ\,mol^{-1})$	Equilibrium length (Å)
Primary		
Ionic	600–1200	2–4
Covalent	60–800	0.7–3
Secondary		
Hydrogen	~ 50	3
Dipole interactions	~ 20	4*
London dispersion	~ 40	$< 10^*$

* Dipole interactions (also sometimes called **Polar forces**) and **Dispersion forces** are not infrequently referred to collectively as van der Waals forces.

The forces involved are those which act between the atoms and molecules in the structure of matter and are commonly known as 'valence forces'.[2] They are summarized with some of their major characteristics in Table 7.

It will be immediately obvious that these forces and bonds are all of exceedingly short range. However, London dispersion forces are especially important because they are universally present since they arise from and depend solely upon the presence of nuclei and electrons. The other types of force can only act when appropriate chemical groupings occur. Dispersion forces can exert a significant influence at greater distances (a few tens of Ångström units) than primary chemical bonds because of the additive effects of all the atoms in a body.

The adsorption theory of adhesion attributes adhesive strength to the action of London dispersion forces combined, in many instances, with dipole or **Polar forces** or **Primary bonding at the interface**. It is because these forces are all of very short range that adhesives all have to pass through a liquid stage so that they may achieve the necessary close intimate contact and interaction (wetting and spreading). Thus the whole topic of the energies of interaction of solid surfaces with liquids (surface thermodynamics) is important and has been studied extensively in this context.

A comparison of different theories is given under **Theories of adhesion**.

References

1. S Glasstone in *Textbook of Physical Chemistry*, 2nd edn, Macmillan, London, 1946, Ch. 14.

2. R J Good in *Treatise on Adhesion and Adhesives*, Vol. 1, ed. R L Patrick, Marcel Dekker, New York, 1967, Ch. 2.
   M Rigby et al. in *Forces Between Molecules*, M Rigby, E B Smith, W A Wakeham and G C Maitland, Clarendon Press, Oxford, 1986, Ch. 1.

Aerospace applications

S TREDWELL

Aircraft built of plywood had been in use in the First World War; however, it was the Second World War which added impetus to the use of adhesives in aircraft.

The first of such **Structural adhesives** was developed in the UK during the Second World War for assembling De Havilland aircraft. It was named the Redux adhesive system by its inventors at Aero Research Ltd. Although the initial technique was crude it was the starting-point from which all such modern structural adhesive systems have evolved. The Redux system was the first to demonstrate the basic principle of toughening a thermosetting resin with a high-molecular-weight linear polymer. Such phenolic-cured adhesives (see **Phenolic adhesives**) had, and continue to have, one major performance advantage, i.e. bond durability, that is still not matched by the newest **Epoxide adhesive** systems which are now in use for aircraft assembly.[1]

Even these systems, however, are likely to be superseded to some extent by the polyimides (see **Polyimide adhesives**). The aerospace industry, perhaps the strongest force behind the development of improved adhesives, has been evaluating polyimide adhesives for service up to 300 °C. A good illustration, of the size and scope of the technology involved, has been presented by Albericci.[2]

Most recently, some bonded aluminium in aircraft, particularly bonded aluminium honeycomb, has been replaced by bonded fibre composite structure (see **Fibre composites – introduction**) because of the inherent improvements in specific strength and specific modulus. These non-metallic composite structures represent an extension of adhesive bonding technology in that a resin system is used to bond fibres and fibre layers together in much the same way as an adhesive system is used to bond metallic details together. Concurrent with the evolution of bonding applications has been an evolution in the materials themselves and the process used to treat and bond metallic details. The evolution of these processes and new materials has been traced by Arnold.[3]

Current applications[4] of structural adhesive bonding in aircraft, include the following:

1. *Bonded stiffeners.* Various types of stiffeners, including ⌐⌐, ⌐, ⊥, ⊏ sections and corrugated backing are bonded to fuselage and wing skin panels.
2. *Bonded laminates.* Doubles are bonded to skin panels to reinforce areas where holes are required for mechanical fasteners or where large openings are required for doors, windows and other access holes. Multiple reinforcement is sometimes used around large openings. Multiple or bonded laminates in the form of tapered or flat plates or built-up shapes, are used for other applications, for example root ends of lower wing skins.
3. *Bonded honeycomb sandwich structures.* Thin skins are bonded to aluminium alloy or Nomex honeycomb and to edge closure members. Bonded honeycomb sandwich structures are used for many applications, e.g. flaps, elevators, rudders, tailerons, vertical fins, doors.
4. *Bonded joints.* Various joints are adhesive bonded, including skin panel splices (with or without mechanical fasteners) and closure joints at the edge of honeycomb sandwich structures.
5. *Bonded helicopter rotor blades.* A typical blade consists of metallic or composite skins bonded to metal spars or ribs, to form the leading edge, and to a honeycomb core to form the trailing edge. Tapered bonded laminates are used in the construction of root ends of rotor blades.

Improved **Fatigue** performance and weight saving are the main advantages of adhesive bonding compared to other joining methods. Such aircraft structures may be exposed to a wide range of environmental conditions. The temperatures encountered in service may influence the selection of adhesives for specific aircraft applications. Several different types of structural adhesive are available including phenolic, polyamide-epoxy and epoxy-phenolic adhesives.

Bonded assemblies are complex structures with many individual constituents and interfaces.[5] Most metal bond service disbonds involve interfacial fractures, especially at the metal-oxide/primer and metal/metal-oxide interfaces. This type of disbond is characteristically an environmental effect. Research has established that the durability of adhesive bonded joints, in aerospace applications, depends on many factors including environment, applied stress, corrosion adhesive type, primer type substrate and substrate pretreatment (see also **Durability – fundamentals** and **Honeycomb structures**).

References
1. R L Patrick (ed.), *Structural Adhesives with Emphasis on Aerospace Applications*, Vol. 4, Marcel Dekker, New York, 1976, Ch. 4.

2. P Albericci in *Durability of Structural Adhesives*, ed. A J Kinloch, Applied Science Publishers, London, 1983, Ch. 8.
3. D B Arnold in *Developments in Adhesives – 2*, ed. A J Kinloch, Applied Science Publishers, London, 1981, Ch. 6.
4. P Poole, *Inst. Metall.* (course vol.), Series 3, **18**, 202 (1981).
5. J C McMillan in *Developments in Adhesives – 2*, ed. A J Kinloch, Applied Science Publishers, London, 1981, Ch. 7.

Alkyl-2-cyanoacrylates

J GUTHRIE

Basic materials

Cyanoacrylate adhesives were introduced in 1958 by Eastman Kodak. They are based on alkyl-2-cyanoacrylate monomers.[1]

$$CH_2 = C \diagup{CN} \diagdown{C=O} \diagup{RO}$$

Cyanoacrylates are rapidly polymerized by nucleophilic species to give linear polymers with high molecular weights. Unlike other adhesives cyanoacrylates consist of pure monomer in admixture with low levels of stabilizers, accelerators and modifiers. The patent literature abounds with references to many different monomers, but only a restricted group are used in commercial products. The structures of the most common monomers are shown below.

Commercially important esters

Name	R
Methyl	CH_3-
Ethyl	CH_3CH_2-
n-Butyl	$CH_3(CH_2)_3-$
Allyl	$CH_2=CH \cdot CH_2-$
β-Methoxyethyl	$CH_3OCH_2CH_2-$
β-Ethoxyethyl	$CH_3CH_2OCH_2CH_2-$

Monomer synthesis

Alkyl-2-cyanoacrylates can be prepared by several synthetic procedures. The only method of importance involves the Knoevenagel condensation of an alkylcyanoacetate with formaldehyde. As this is a base-catalysed reaction, the monomer is rapidly polymerized to give a low-molecular-weight poly(alkyl-2-cyanoacrylate). The resulting polymer is retro-polymerized by heating under controlled conditions to yield monomeric cyanoacrylate (Scheme 1).

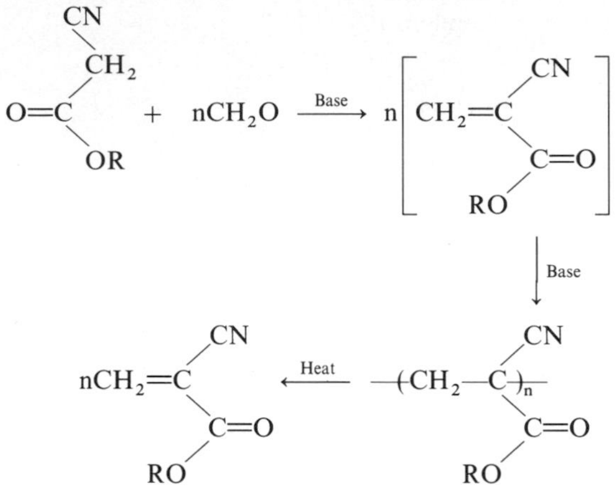

$$
\begin{array}{c}
\overset{\displaystyle CN}{\underset{\displaystyle OR}{O=C{-}CH_2}} \quad + \quad nCH_2O \xrightarrow{\text{Base}} \quad n\left[\ CH_2=C\overset{\displaystyle CN}{\underset{\displaystyle RO}{C=O}}\ \right]
\end{array}
$$

$$\Big\downarrow \text{Base}$$

$$
nCH_2=C\overset{\displaystyle CN}{\underset{\displaystyle RO}{C=O}} \xleftarrow{\ \text{Heat}\ } \ -\!(CH_2-C)_n-\ \overset{\displaystyle CN}{\underset{\displaystyle RO}{C=O}}
$$

Scheme 1

Depolymerization is carried out under vacuum in the presence of an acid such as sulphur dioxide. The monomer which distils from the reaction mixture is collected in a vessel containing radical and anionic polymerization inhibitors.

Polymerization of alkyl-2-cyanoacrylates

Cyanoacrylates can be polymerized by free radical and ionic initiators (see **Addition polymerization**). In adhesive applications ionic polymerization is by far the most important mode of chain growth. It is their marked susceptibility to initiation by anions and nucleophiles that is responsible for their usefulness as adhesives. The cyanoacrylate π-electron system is under the influence of two strongly electron-attracting groups. This results in a reduced electron density on the β-carbon and an enhanced susceptibility to nucleophilic attack (Scheme 2):

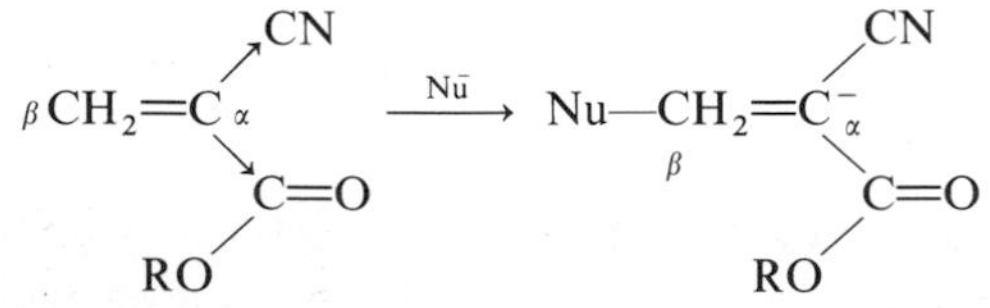

$$
\beta\,CH_2{=}\overset{\displaystyle CN}{\underset{\displaystyle RO}{C_\alpha\ C=O}} \xrightarrow{\ N\bar{u}\ } Nu{-}\underset{\beta}{CH_2}{=}\overset{\displaystyle CN}{\underset{\displaystyle RO}{C_\alpha^-\ C=O}}
$$

Scheme 2

The carbanion formed at the α-carbon is stabilized by delocalization (II and III):

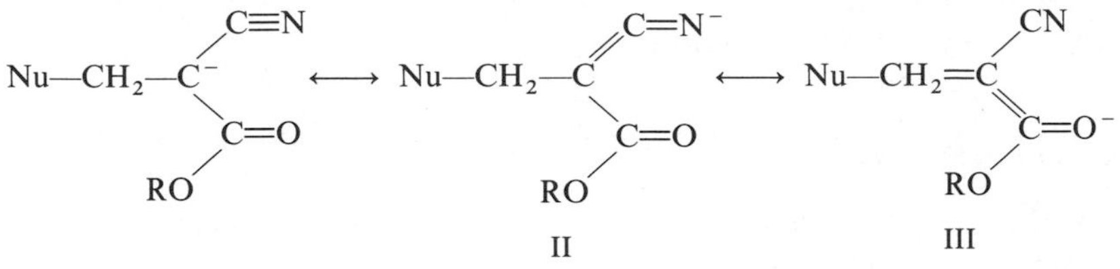

$$\text{Nu—CH}_2\text{—C}^- \overset{\text{C}\equiv\text{N}}{\underset{\text{C=O}}{\diagup}}_{\text{RO}} \longleftrightarrow \text{Nu—CH}_2\text{—C} \overset{\text{C=N}^-}{\underset{\text{C=O}}{\diagup}}_{\text{RO}} \longleftrightarrow \text{Nu—CH}_2\text{=C} \overset{\text{CN}}{\underset{\text{C=O}^-}{\diagup}}_{\text{RO}}$$

II   III

The combination of a highly electrophilic β-carbon, a stable carbanion and an unhindered β-carbon confer on alkyl-2-cyanoacrylates their unique reactivity.

When used as adhesives, polymerization or curing is brought about by nucleophilic species present on the substrate. The nucleophile can be an ion or a neutral molecule (amine). If the initiating species is an ion the polymerization proceeds by an anionic mechanism. If a neutral molecule is involved, the reaction is referred to as a zwitterionic polymerization (Scheme 3).

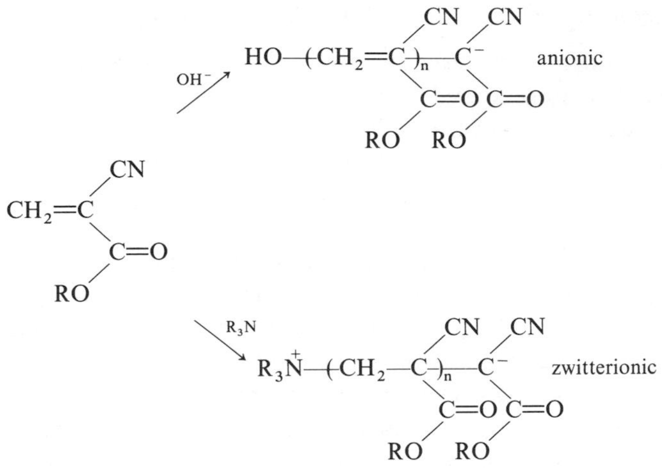

Scheme 3

The nature of the polymerization is presumably dependent on the substrates involved. In any case both are extremely rapid with overall rates much greater than radical polymerizations. Ionic polymerizations are normally sensitive to termination reactions, but in the case of cyanoacrylates only strong acids are capable of terminating the growing chain. Some chain-transfer reactions are believed to occur due to the presence of carboxylic acids formed by monomer hydrolysis. A proton is transferred to the growing carbonium to give a 'dead' polymer chain. The weak acid anion is capable of acting as an initiator and a new active centre is generated by reaction with monomer.

Select references

H W Coover, J M McIntire in *Handbook of Adhesives*, 2nd edn, ed. I Skeist, Van Nostrand Reinhold, New York, 1977, p. 569.
G H Millet in *Structural Adhesives, Chemistry and Technology*, ed. S R Hartshorn, Plenum Press, New York, 1986, Ch. 6.
D C Pepper, *J. Polym. Sci., Polym. Symp.*, **62**, 65 (1978).

Anaerobic adhesives

D P MELODY

Basic chemistry and properties of the uncured adhesives

Anaerobic adhesives and sealants are based on acrylic (usually *meth*acrylic) functional monomers and cure by a redox initiated free radical polymerization (see **Addition polymerization**). They are so named due to the characteristic of requiring a relatively air-free condition to allow curing. Hence anaerobic adhesives are very suitable for the bonding and sealing of close-fitting metal components.

Anaerobic adhesive formulations vary substantially depending upon the properties sought in the cured and uncured composition. Typical formulations contain (meth)acrylate monomer(s), modifiers, initiator(s), accelerator(s) of redox-initiated free radical polymerization and stabilizers/inhibitors. Methacrylate monomers are preferred to acrylate as these latter have a greater tendency to cause skin irritation (see **Health and safety**).

While dimethacrylates of polyalkoxy diols, such as triethyleneglycol dimethacrylate, are commonly employed, advanced resins carrying suitable functionality have been used in many recent anaerobic adhesives to expand the range of properties available in the cured products.

Modifiers can include dyes, thickeners, gelling agents, fillers and lubricity controllers and the like as agents that influence the properties of the uncured composition. Plasticizers, reinforcing and/or toughening agents, adhesion promoters and thermally polymerizable prepolymers may be added as 'modifiers' of the cured adhesives.

The initiator is usually a hydroperoxide.

Accelerators, which interact with initiator(s) speed up the curing reaction. Many conventional free radical polymerization stabilizers and inhibitors can be used. However, such additives are minimized to give a high level of reactivity. Oxygen in air is often enlisted as a fugitive inhibitor by using partially filled, low-density polythene bottles (or tubes) as packages.

Anaerobic adhesives differ from related **Acrylic adhesives** by being very finely balanced in terms of stability and reactivity.

Properties of cured anaerobic adhesives

Properties of cured anaerobic adhesives are related to the formulation chosen for a given application area. For example anaerobic threadlocking formulations cure to very hard materials for studlocking applications and to relatively soft solids for locking precision screws. Usually, cured anaerobic products are highly cross-linked and form strongly adhesive but somewhat brittle solids. They are resistant to water and solvents and perform well under extremes of temperature ($-50\,°C$ to $+150\,°C$).

Due to their curing properties, anaerobic adhesives remain uncured outside the **Joint assembly**: this allows removal with solvent. Some formulations now include sensitizers which allow drying of the excess with UV light.

Applications

The major application areas for anaerobic adhesives are as follows: locking and vibration proofing threaded assemblies; sealing threaded joints and pipe fittings; sealing porous and flanged metal structures; strengthening cylindrical fits; structural bonding of rigid metal and ceramic components.

Adhesives for threadlocking provide resistance to vibration and maintain the tension achieved in threaded joints. Adhesives for thread sealing generally contain fillers to give suitable viscosity characteristics and some physical sealing in the uncured form. When cured, they give good durability (see **Durability – fundamentals**) with the ability to disassemble later. Anaerobic porosity sealants are used effectively to ensure sealing of welds and coatings.

Anaerobic sealants are particularly effective as form-in-place gaskets for rigid flange structures. The 'anaerobic gasket' replaces gasket inventories, but more importantly, it eliminates gasket yielding (with loss in structural rigidity) by allowing metal–metal contact in the clamped assembly.

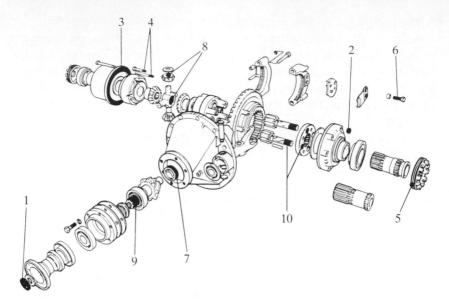

Fig. 5. Applications for anaerobic adhesives in an automotive final drive

Traditional methods of cylindrical fitting parts such as gears, bushings and bearings have involved interference fitting. The expense and risks of defect in this methodology can now be eliminated by use of anaerobic adhesives. The combination of anaerobic adhesives and 'interference' fitting' can be used to give outstanding performance (see **Joint design – cylindrical joints**).

Given the surface initiation, central to the function of anaerobic structural adhesives, they work best in face-to-face bonding of small flat components. Whereas traditional anaerobic adhesives cure to rigid solids, which have high shear and tensile strengths but low cleavage and peel strengths, newer anaerobic structural adhesives offer good impact, fair peel strengths and more tolerance for surface contamination. In bonding, as in all anaerobic applications, use of a redox activator can be used to increase cure speed and to facilitate cure through larger gaps ($\geqslant 0.2$ mm).

Figure 5 shows a series of uses for anaerobic adhesives in the final drive of an automobile as examples of application areas (see also **Automotive applications**).

Aspects of this subject are also dealt with in other articles notably **Acrylic adhesives, Durability – fundamentals, Engineering practice, Joint design**. In the article on **Toughened acrylic adhesives** some properties are compared with those of **Epoxide adhesives** and anaerobic adhesives.

Some test methods are given in Appendix 1.

Select references

C W Boeder in *Structural Adhesives, Chemistry and Technology*, ed. S R Hartshorn, Plenum Press, New York, 1986.

G S Haviland, *Machinery Adhesives for Locking, Retaining and Sealing*, Marcel Dekker, New York, 1986.

Animal glues and technical gelatins

C A FINCH

Animal glues are probably the oldest type of adhesives: they were used to bond the wood of the coffins of Egyptian mummies. Several animal proteins have been used as the raw materials for adhesives, including blood proteins, caseins (the protein of milk), albumin (from egg white) and collagen products from skin, leather and, especially, animal bone. Over the centuries animal glues have been used extensively, as concentrated aqueous solutions, to make furniture, wooden weapons (such as longbows and crossbows), to bind the inflammable heads of matches, and to coat paper, especially abrasive papers ('sandpaper' or 'glasspaper') and high-quality rag papers for banknotes and security documents. Some animal glues are still employed for these purposes, and also in craft woodworking, for making musical instruments, such as violins, which are conventionally made using a high-quality glue derived from rabbit skins.

Technical gelatin is a convenient term for gelatin and animal glues used for non-edible purposes. Chemically, it is, like pure edible gelatin, a mixture of gelling proteins obtained by partial hydrolysis, by thermal or chemical means, of waste collagen protein from animal connective tissue, skin and bone. The chief features of the amino-acid composition of animal glues are those of collagen – an unusually high content of glycine and of the imino acids, proline and hydroxyproline (which affect gelling properties due to inter-chain **Hydrogen bonding**), with very small amounts of the aromatic and sulphur-containing amino acids. There is very little difference in amino-acid composition between animal glues from different types of raw material (bone or skin), or from different animals. However, non-gelling fish glues, from cold-blooded organisms, have a low hydroxyproline content. In all animal glue production the degreased raw material is subjected to the basic reaction of the hydrolysis of collagen, a multiple helical chain protein: the rate of this hydrolysis increases with temperature and with stronger acid or alkaline conditions. Hide glues are usually made by an acid process. The production conditions are designed to break down the collagen but retain large molecules in the resulting soluble proteins. The resulting dilute protein solutions are concentrated by evaporation, then gelled by cooling. The resulting gels are chopped into small pieces, and then dried by passage through a series of warm, dehumidified chambers.

Production of animal glue is considered as environmentally 'friendly', since it involves conversion of unpleasant waste into useful products, but economic operation depends on satisfactory disposal of residues: some of these are used as slow-release nitrogenous fertilizers. The mineral residue from bone glue production ('bone meal') by thermal methods is used as a phosphate supplement in animal feed and fertilizers: it is also calcined at 1000–1200 °C to produce calcined bone, used in the manufacture of bone china to provide strength and translucency. Animal glues and technical gelatins contain added preservatives and a proportion of non-protein materials, usually mucopolysaccharides and soluble inorganic salts. These impurities differ with the source of raw material.

The physical and chemical properties of animal glues depend on the nature of the raw material and on the method of processing. The 'quality' depends on the gelling point, and the viscosity, measured at standard concentration (6.67 wt% for technical gelatin; 12.5 wt% for animal glue). The 'jelly strength' (or 'bloom strength') is measured using a 'gelometer' (a type of penetrometer) by determining the force (in grams) needed to depress the surface of a standard gel, conditioned at 10°, by a flat plunger 12.5 mm in diameter, to a depth of 4 mm. Viscosity is measured with a calibrated Ostwald viscometer at 60°: results are normally reported in millipoise (mP), centistokes (cP), or millipascal seconds (mPa s). The test methods are detailed in well-established national and international standards, which also describe the measurement of pH, foam characteristics, grease content and melting and setting points (see Select references).

Only small amounts of technical gelatin are used in adhesives in industrialized countries, mainly in specialist applications, and to a limited extent in craft woodworking, since the concentrated technical gelatin solutions are viscoelastic and have the property of **Tack**, which allows the temporary 'grab' between the surfaces being joined, so that they can be adjusted to their exact position before the final joint is allowed to harden and become permanent. Animal glue bonds between suitable wood surfaces can have tensile strengths comparable to those of high-performance synthetic adhesives. Other craft operations using animal glues (as warm aqueous solutions, often formulated with polyhydroxy compounds as humectants, soluble salts (usually magnesium sulphate), gel depressants and preservatives) include bookbinding (which involves a long-life paperboard-to-book cloth or leather bond), and other paper-to-paper applications, such as remoistenable gummed paper tape, and decorative paper box making. Significant amounts of low-quality animal glue are also manufactured in developing countries for general and domestic use.

There are several specialist uses of technical gelatin: one of the most important includes the binder component of the head composition mixture of matches. The gelatin quality controls the ignition performance of the match, by affecting the microfoam properties of the match head, and the adhesion of the head to the match splint. Technical gelatin is used (among other polymers) in coated abrasive paper ('sandpaper') as the adhesive between the paper substrate and the abrasive component (of glass powder). The glue binder is dried by passing the paper web through a heated oven, and treated with a formaldehyde-releasing compound, which cross-links the dried film, so improving adhesion and reducing moisture sensitivity and increasing the working life of the product. Technical gelatin is also used as size coating for high quality rag-based paper used for security documents such as banknotes. The coating provides good moisture and abrasion resistance, and good adhesion to the special inks used in the banknote design.

Select references

E Sauer, *Tierische Leime und Gelatine*, Springer-Verlag, Berlin, 1958. This describes the technology of skin-based and bone-based glues, and also casein adhesives.

A G Ward, A Court (eds), *Science and Technology of Gelatin*, Academic Press, London, 1977. This volume gives the definitive account of the classical nature, properties and manufacturing technology of edible and pharmaceutical gelatin, including technical gelatin, based on the many research reports of the Gelatine and Glue Research Association.

British Standard 647: 1981, *Methods for Sampling and Testing Glues (Bones, Skin and Fish Glues)* (under revision, 1990). German Standard DIN 53260: 1963 (rev.), *Pruefung von Glutinleimen*. (These standards are incorporated into International Standard IS 9665, to be published in 1992, and CEN Standard 29.665 in 1993.) The older standards describe several chemical test methods not included in the International Standards.

J Alleavitch, W A Turner, C A Finch, Gelatin, in *Ullmann's Encyclopedia of Industrial Chemistry*, 5th edn, Vol A12, VCH Verlagsgesellschaft, Weinheim, 1989.

Anodizing

A MADDISON

Anodizing has been used as a **Pretreatment for metals prior to bonding** both in practical processes and fundamental scientific studies where it has played a role in the development of the **Mechanical theory**. It is applied most commonly to aluminium.

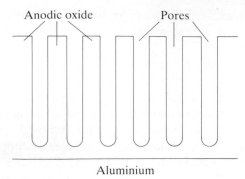

Fig. 6. Schematic representation of porous anodic oxide layer on aluminium

The basic principle of anoziding, namely that the component to be treated is made the anode in an electrolytic cell, may be applied using a great variety of conditions. The morphology and chemistry of these electrochemically generated surfaces is influenced by a number of factors, principally cell voltage, the solubility of the film in the electrolyte, component and electrolyte composition and the pretreatment employed. In anodizing processes considerable scope therefore exists for the optimization of surface properties for specific applications.

The anodic film formed on aluminium in aggressive electrolytes such as phosphoric, sulphuric and chromic acids is porous. The classical description of the film is of a regular hexagonal array of cylindrical pores penetrating normal to the surface almost to the base metal (Fig. 6). The actual structure revealed by **Electron microscopy** varies greatly with electrolyte and anodizing conditions.[1-3] Some details are given in **Mechanical theory**: see also Fig. 38 in **Electron microscopy**.

Most commercial anodizing of aluminium is performed in sulphuric acid electrolytes to yield oxide films which, of the order of tens of micrometres thick, provide good corrosion and abrasion resistance, in addition to a useful dye-absorption capability. Other anodizing processes utilize chromic, oxalic, nitric or phosphoric acids, or mixed electrolytes.

Relatively few anodic treatments find application in structural adhesive bonding where durability (see **Durability – fundamentals**) in hostile service conditions is of paramount importance. Only two processes are routinely practised in the aircraft industry.

The long-established DEF 151 Type 2 method uses a dilute chromic acid solution operated at approximately 40 °C and a programmed voltage sequence peaking at 50 V d.c. A dense oxide structure a few micrometres thick is formed.

Anodizing for 20 min at 10 V d.c. in a 10% solution of phosphoric acid at room temperature (i.e. to Boeing Aircraft Specification BAC 5555) generates a surface structure on aluminium alloy to which highly durable bonds may be obtained.

Typical procedures for chromic and phosphoric acid anodizing are given in **Pretreatment of aluminium**, and the durability (see **Durability – fundamentals**) of bonds to the surfaces formed is discussed. An example of a profile of elemental composition in depth for a phosphoric acid anodized film is shown in **Auger electron spectroscopy**.

The mechanisms by which stable bonds to these anodized surfaces are achieved have been extensively studied.[3] In the BAC 5555 process a whiskered porous oxide structure less than 1 μm thick is developed over a thin barrier layer. The incorporation of phosphate ions in cell walls inhibits oxide hydration and contributes to greater bond durability.

For bonding applications the performance of sulphuric acid and chromic acid anodized surfaces may be improved by controlled oxide dissolution in phosphoric acid to yield morphologies more readily penetrated by adhesives and primers.

In contrast to the above slow batch methods a.c. anodizing processes have been developed for the treatment of aluminium at the coil stage, using either sulphuric acid[4] or phosphoric acid electrolytes.

Anodic treatment may also be considered when bonding other metals. Examples include the anodizing of magnesium in ammonium bifluoride solution, the anodic etching of stainless steel in nitric acid and anodizing of titanium in sodium hydroxide or chromic acid[3,4] (see **Pretreatment of titanium**). Alternative anodizing of copper can produce a **Microfibrous surface** (see **Pretreatment of copper**).

References

1. S Wernick, R Pinner, *Surface Treatment and Finishing of Aluminium and its Alloys*, 5th edn, Vols 1 and 2, Finishing Publishers, Stevenage, 1987.
2. V F Henley, *Anodic Oxidation of Aluminium and its Alloys*, Pergamon, Oxford, 1982.
3. J D Venables, *J. Mat. Sci.*, **19**, 2431 (1984).
4. J A Filby, J P Wightman, *J. Adhesion*, **28**, 1 (1989).

Auger electron spectroscopy

J F WATTS

Auger electron spectroscopy (AES)[1,2] is one of a family of surface analytical techniques that have assumed increasing importance in materials science investigations over the last decade or so. The other two

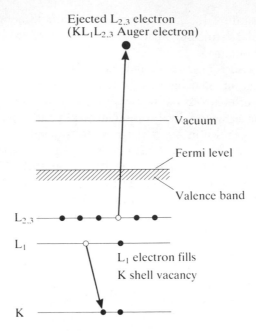

Ejected $L_{2,3}$ electron
($KL_1L_{2,3}$ Auger electron)

Vacuum

Fermi level

Valence band

$L_{2,3}$

L_1

L_1 electron fills
K shell vacancy

K

Fig. 7. Schematic representation of the process of Auger electron emission.
The initial core (K shell) vacancy is created by a finely focused
electron beam

are **X-ray photoelectron spectroscopy** (XPS or ESCA)[1,2] and **Secondary
ion mass spectrometry** (SIMS);[3] indeed AES has many features in common
with the former technique and for this reason they are often to be found
on the same analytical system.

The basis of AES is the relaxation of an excited atom or ion following
the emission of a core level electron, i.e. the generation of a core level
vacancy. The process is illustrated in Fig. 7 which shows the three processes
involved in the generation of a $KL_1L_{2,3}$ Auger electron; the initial process
is the ejection of a K electron; this will usually be achieved by electron
beam irradiation although both X-ray photons or ions may also be
employed, the vacancy in the K shell which results is filled by the
promotion of an electron from the L_1 level, and in order to conform with
conservation of energy principles another electron must be ejected, in this
case a $L_{2,3}$ electron. This electron is termed the $KL_1L_{2,3}$ Auger electron,
and the kinetic energy it possesses as it leaves the atom or ion is
approximately the difference of the three energy levels involved:

$$E_{KL_1L_{2,3}} = E_K - E_{L_1} - E_{L_{2,3}}$$

In most cases the kinetic energy of the outgoing Auger electron is in the

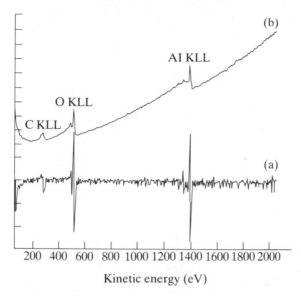

Fig. 8. Auger electron spectra of an oxidized aluminium surface recorded (a) in the differential, and (b) in the pulse-counting modes of operation

range 50–1500 eV and it is this relatively modest energy that gives the technique its surface specificity, as the distance an electron will travel in a solid varies as $E^{-1/2}$. This gives an analysis depth of 2–5 nm depending on experimental conditions, in contrast to characteristic X-ray analysis in the scanning electron microscope/energy dispersive analysis of X-rays (SEM/EDX) it is the energy of the outgoing particles rather than the primary beam that determines the analysis depth of the technique.

The emitted electron spectrum can be presented in two possible ways, the differential spectrum of Fig. 8(a) or the pulse counted (direct) spectrum of Fig. 8(b). The reasons for the choice of spectrum type are many and varied, and are often determined by the spectrometer design, but the former mode is often used for point analysis at spatial resolutions of $>1\ \mu$m whereas the latter is invariably employed for high spatial resolution ($<0.1\ \mu$m) scanning Auger microscopy (SAM). In SAM it is possible to obtain chemical maps as one does in **Electron probe microanalysis** (EPMA), the difference being that the image reflects surface (3–5 nm) rather than bulk (1 μm) composition.

The spatial resolution of AES can be used to provide either point analyses or chemical maps of a material surface. The spectra of Fig. 9 are taken from an emery-abraded mild steel surface (see **Abrasion treatment**) which is known to have a low level of carbon contamination; however,

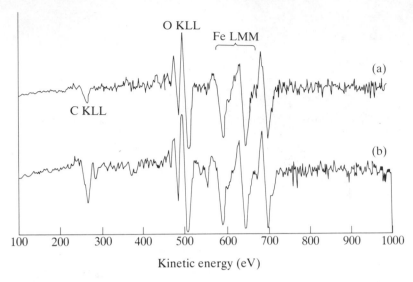

Fig. 9.   Auger spectra from an emery abraded mild steel surface, illustrating
the difference in carbon concentration between (a) peak and
(b) valley positions on the surface profile

the intensity of the C_{KLL} transition is clearly more intense in the
spectrum recorded from the valley of the surface profile than that recorded
from the peak. Such variations in contamination on steel surfaces have
been shown by SAM to interfere with the production of a uniform
Conversion coating or other metal pretreatment process[4] (see **Pretreatment
of steel**).

The SAM images of Fig. 10 are from the failure surface of a chromated
aluminium substrate to which an organic coating had been applied (see
Pretreatment of aluminium, Locus of failure). **X-ray photoelectron
spectroscopy** analysis indicated a very thin film of polymer remaining on
the substrate after failure, and the C_{KLL} Auger image clearly shows the
layer to be non-uniform. The Al_{KLL} and the Cr_{LMM} images are
complementary, indicating that the chromate coverage is incomplete, and
the islands of polymer remaining on the surface are associated with the
fragmented chromate layer. This implies that the adhesion of the polymer
is better for the **Conversion coating** than the metallic substrate. The
secondary electron image shows surface texture which is, by and large,
irrelevant to the chemical information concerning the failure mode.

Although AES is essentially a surface technique it is possible to combine
it with inert gas ion sputtering to remove material and thus produce a
depth profile many nanometres into the sample. In general the sputtering

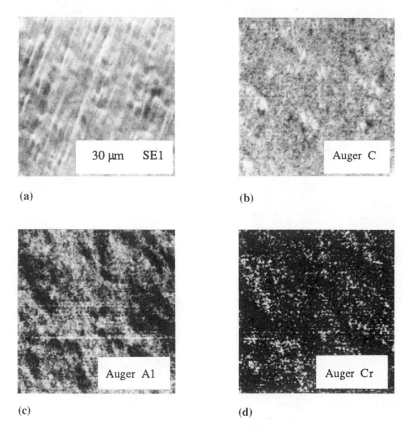

(a) (b) (c) (d)

30 μm SE1

Auger C

Auger Al

Auger Cr

Fig. 10. Scanning Auger microscopy of an aluminium substrate following failure of a polymeric coating. (a) Secondary electron image; (b) carbon Auger image; (c) aluminium Auger image; (d) chromium Auger image

depth will not exceed 1 μm, but the depth profile will have a depth resolution of 1–3 nm. For depths greater than 1 μm some form of mechanical removal of material will be required. Such depth profiles are particularly useful in adhesion studies for the determination of segregated or impurity layers at surfaces, and the estimation of the thickness of **Conversion coatings**. Figure 11 (overleaf) shows such a depth profile for phosphoric acid anodized (see **Anodizing**) aluminium alloy substrate.

References

1. D Briggs, M P Seah, *Practical Surface Analysis*, Vol. 1, *Auger and X-ray Photoelectron Spectroscopy*, John Wiley, Chichester, 1990.

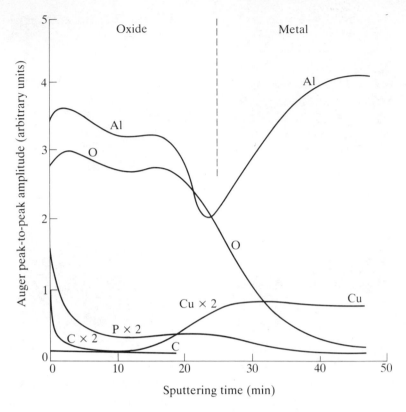

Fig. 11. Auger depth profile of an anodized aluminium substrate. The oxide is approximately 300 nm thick

2. J F Watts, *An Introduction to Surface Analysis by Electron Spectroscopy*, RMS Handbook No. 22, Oxford University Press, 1990.
3. J C Vickerman, A. Brown, N Reed, *Secondary Ion Mass Spectrometry: Principles and Applications*, Oxford University Press, 1989.
4. R A Iezzi, H Leidheiser, *Corrosion*, **37**, 28–38 (1981).

Autohesion

J COMYN

Autohesion or self-sticking is the phenomenon by which a measurable adhesive bond is formed, when the surfaces of two polymers are placed in contact. A strict definition would limit the phenomenon to the surfaces of two identical polymers, but the term is used in a wider sense than this.

The phenomenon only occurs with uncross-linked or lightly cross-linked polymers which are above their **Glass transition temperatures**; this is because the development of an adhesive bond requires molecular mobility. The phenomenon occurs in the practical situations of bonding with contact adhesives and the **Solvent welding** of thermoplastics.

There are two opposing views on the mechanism of autohesion, the contact theory of Anand and the **Diffusion theory** of Voyutskii. The authors have summarized their positions and criticized each other.[1,2]

Contact theory

The view of the contact theory is that the molecules of one polymer make intimate contact with those of the other; van der Waals forces between the polymers give the adhesive bond its strength. The opposing force for the making of contact is the viscosity of the polymer, such that the process will be slowed down if the viscosity is high (see **Adsorption theory**).

Interfacial diffusion and mixing

Voyutskii's proposal[3] is that interdiffusion occurs between the chains of each polymer, so that the interface eventually disappears; this process is illustrated in Fig. 12. Interdiffusion leads to mixing of the two polymers, but as polymers which are chemically very similar often do not mix, the applicability of the theory is limited (see **Compatibility**). There are, however, a number of instances where interfacial diffusion has been demonstrated experimentally between differing polymers; they include the following: polyvinylchloride and polycaprolactone; polyvinylchloride and polymethylmethacrylate; polyvinylchloride and styrene-acrylonitrile copolymer; polyvinylidene fluoride and polymethylmethacrylate.

The solubility parameter (see **Compatibility**) is a simple, but not particularly reliable guide to the possibility of mixing: the smaller the difference in solubility parameters between two polymers the more likely they are to be miscible. Forbes and McLeod[4] measured the **Tack** between a number of different rubbers, taken in pairs. They compared the adhesive tack between two rubbers $[X - Y]$ with the mean of the autohesive tack $[(X - X + Y - Y)/2]$. The adhesive tack was always smaller than the mean autohesion, but difference decreased as the difference in solubility parameter between the two rubbers decreased (Fig. 13). This suggests that with significant interdiffusion between polymers there is high adhesion and vice versa (see **Diffusion theory**).

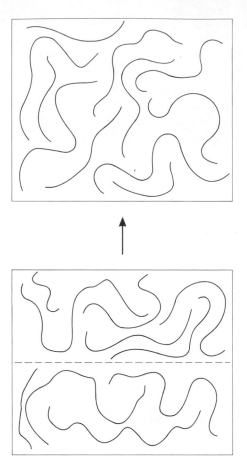

Fig. 12.   Diffusion across the interface between two polymers in contact

Development of bond strength and polymer self-diffusion coefficients

If autohesion is controlled by self-diffusion of the polymer, then the rate at which bond strength increases will depend on the self-diffusion coefficient, and strength should eventually reach the cohesive strength of the rubber. The time required for this varies considerably; Forbes and McLeod[4] give values of 3–5 min for butyl rubber, 4–5 h for natural rubber and 14–17 h for styrene–butadiene rubber. Many studies have shown that bond strength does increase with contact time, some information from Skewis[5] for rubber–rubber and rubber–glass being shown in Fig. 14. Diffusion will not occur between rubbers and glass, and this may be the reason for the low bond strengths. Higher levels of adhesion develop

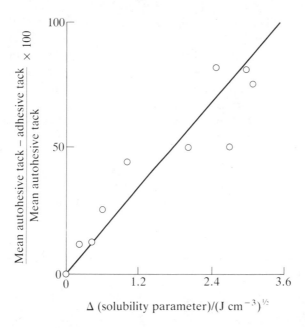

Fig. 13. Effect of differences in polymer solubility on adhesive tack between pairs of rubbers. The abscissa represents the normalized difference between the mean of the autohesive tack $[(X - X + Y - Y)/2]$ and the adhesive tack $[X - Y]$

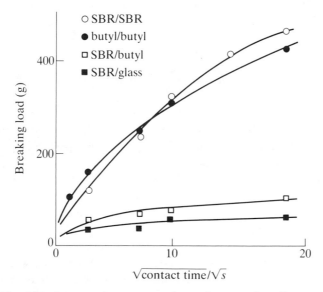

Fig. 14. Adhesion at various contact times of styrene–butadiene and butyl rubbers. After Skewis[5]

between like rubber pairs, and this was attributed by Skewis to diffusion across the interface.

Other evidence put forward in support of the diffusion theory includes the observation that both the rate of bond formation and self-diffusion coefficient decrease as the molar mass (molecular weight) of the polymer increases.

Activation energies have been used in the contest between the two theories. By measuring the rate of bond development between poly-isobutenes, Voyutskii obtained an activation energy for autohesion of 11.7 kJ mol^{-1}; however, as this is much less than the activation energy for the self-diffusion of polyisobutene (23 kJ mol^{-1}) it seems to give poor support for this mechanism. Self-diffusion coefficients and viscosities of polymers are dependent on the same molecular and intermolecular parameters; the consequence of this is that the activation energies for the two processes are the same, and these parameters are therefore neutral in competition between the two theories.

References

1. S S Voyutskii, *J. Adhesion*, **3**, 69 (1971).
2. J N Anand, *J. Adhesion*, **5**, 265 (1973).
3. S S Voyutskii, *Autohesion and Adhesion of High Polymers*, Interscience, New York, 1963.
4. W G Forbes, L A McLeod, *Trans. IRI*, **34**, 154 (1958).
5. J D Skewis, *Rubber Chem. Tech.*, **39**, 217 (1966).

Automotive applications

A MADDISON

In recent years the range of adhesive materials used in automotive manufacture has expanded to include polyurethanes, plastisols, phenolics, hot melts, anaerobics, cyanoacrylates, toughened acrylics and epoxies (see **Structural** and **Hot melt adhesives**). Selection criteria are based principally upon the nature of the adherends, the mechanical properties required under service conditions and application and curing characteristics.

Compliant, gap-filling materials are effective in direct glazing and panel-stiffening roles in which the service stresses are low, but the evaluation of high-performance **Toughened adhesives** now challenges traditional techniques in the construction of safety-critical assemblies. In body-shell manufacture there are a number of advantages of **Structural adhesive** bonding. Reduction of stress concentrations means that thinner-gauge materials can be used. Dissimilar materials can be joined, and seam sealants are eliminated. There are improvements in **Fatigue** and **Impact resistance**. All of this gives new freedom in design and the scope for improved appearance of the vehicle (see **Joint assembly**).

For jigging purposes, and to provide handling strength prior to adhesive cure, supplementary mechanical fasteners are desirable. A further consideration is the improved peel resistance and impact performance which may be realized by strategic mechanical fastening. In this respect spot welding remains the favoured method.

British Leyland's energy conservation vehicle (ECV) programme, which culminated in the construction of a series of lightweight, energy-efficient vehicles[1] exemplified the radical exploitation of structural adhesive bonding. A crucial programme objective addressed the requirement for long-term bond durability, and in this respect coil pretreatments were developed for aluminium and evaluated in laboratory and complete vehicle studies.

In contrast to the aerospace industry (see **Aerospace applications**), conditions prevailing in automobile manufacture dictate that pretreatments (see **Pretreatment of aluminium**) should have short process times and be tolerant of production variables. The surfaces produced should retain adequate bonding properties during extended storage under poorly controlled conditions. The surface may need to be sufficiently conductive to permit weld-bonding techniques. Pretreatments applied at the coil stage must be compatible with mill oils and press lubricants, and withstand mechanical deformation in the press.

The above approach has not, to date, been extended to include steel, although considerable effort is expended in its pretreatment prior to painting. The current practice is to weld bond oil steel using single-part epoxy adhesives curing during the first (electrocoat) paint stage.

It has been shown, however, that the **Durability** of such bonds is questionable, particularly when stressed.[2] Reduced dependence on mechanical fasteners will require the development of production-viable treatments for steel or the introduction of coated variants.

Further examples are given in **Anaerobic adhesives, Industrial applications of adhesives** and **Joint design – cylindrical joints**.

References

1. D Kewley, Aluminium alloy body structures for future vehicles, *Proc. Instn. Mech. Engrs.*, **201** (D2), 129–34 (1987).
2. P A Fay and A Maddison, Durability of adhesively bonded steel under salt spray and hydrothermal stress conditions, *Int. J. Adhesion and Adhesives*, **10** (3), 179 (1990).

Autophoretic primers

J L PROSSER

These primers are a recent development in waterborne coatings, combining pretreatment and priming. They comprise aqueous dispersions of resins, usually acrylics, containing acids and oxidizing agents; they deposit firmly adherent corrosion-resistant films on metals by dipping. The acid is usually hydrofluoric and the oxidizing agent hydrogen peroxide, so that a carefully controlled plant must be used. It is postulated that the hydrofluoric acid attacks steel, liberating ferrous ions; these ions are oxidized to ferric by the H_2O_2, promoting coagulation of resin with deposition of a gelatinous film on to the metal, which has been cleaned and etched by the acid. Finally, the coated metal is stoved to drive off water and to compact the film. The choice of resin is important, the wrong choice can result in poor adhesion. The technique is still novel, although it has already been exploited industrially.

Select reference

UK Patent 1 130 687.

B

Blister test

A J KINLOCH

Introduction

The blister test is a method for determining the adhesive fracture energy of a joint or coating. The common form of the pressurized blister test specimen basically consists of pressurizing the adhesive, or coating layer, so that it forms a blister which at a critical pressure begins to debond from the rigid substrate to which it is bonded (see **Fracture mechanics**). A useful account of the development of the test to 1977 is given by Anderson et al.;[1] more up-to-date summaries have been provided by Kinloch,[2] Gent and Lewandowski[3] and Briscoe and Panesar;[4] original accounts of recent developments of the test are given in ref. 5.

The blister test specimen was originally suggested by Danneberg, developed by Malyshev and Salganik and subsequently modified by Anderson et al.[1] and Andrews and Stevenson.[2] A good feature of the test is that the pressurizing medium may be an inert one, such as dry nitrogen, or an active hostile environment such as water. The test has been adapted to study the effectiveness of **Release** agents.[4] A problem with the blister test when using very rigid and brittle adhesives or coatings is that of inserting a sharp crack at the interface. A debond is often inserted using a release film or release paint, but this may not be sufficiently sharp for these materials.

Theoretical aspects

To determine the value of the adhesive fracture energy one has to decide the mode of deformation of the pressurized layer. In the case of a relatively thin blister the mode of deformation is considered to be mainly that of tensile deformation of the blister, and the blister is then modelled as an elastic membrane.[3,4] Alternatively, in the case of a relatively thick blister, the pressurized layer is considered to deform mainly by bending, and this is modelled as an elastic circular plate with a built-in edge constraint.[2] A further contribution to the stored elastic energy which is available to assist growth of a debond arises from an internal stress inherent in the test specimen. Such stresses may be introduced during manufacture of the specimen. The contribution of residual stresses to the blister mechanics has been considered in detail by Allen and Senturia.[5]

Recent developments

Recently, there have been reported several developments of the common blister test method. Allen and Senturia[5] have developed the 'island-blister' test and Fernando and Kinloch[6] have developed the 'inverted-blister' test method. Both of these test methods are designed to enable the adhesive fracture energy to be measured for a coating of thin adhered film where the coating or film has insufficient strength to resist the pressure needed for debonding if the standard blister test was employed.

References

1. G P Anderson, S J Bennett, K L DeVries, *Analysis and Testing of Adhesive Bonds*, Academic Press, New York, 1977.
2. A J Kinloch, *Adhesion and Adhesives*, Chapman and Hall, London, 1987.
3. A N Gent, L H Lewandowski, *J. Appl. Pol. Sci.*, **33**, 1567 (1987).
4. B J Briscoe, S S Panesar, *J. Appl. Phys.*, **D19**, 841 (1986).
5. M G Allen, S D Senturia, *J. Adhesion*, **25**, 303 (1988).
6. M Fernando, A J Kinloch, *Int. J. Adh. Adh.*, **10**, 69 (1990).

C

Cathodic disbondment

J F WATTS

Cathodic disbondment (or delamination) occurs when failure of an organic coating is associated with a cathodic potential on the metal substrate, a phenomenon most widely considered with respect to iron-based alloys. Such a potential may emanate from the electrolytic cell that exists at the surface of a corroding metal or it may have been applied by way of an impressed potential to ensure the entire metal structure is cathodically protected.

In the former category we are concerned with localized damage to a paint film (see **Paint service properties and adhesion**) exemplified by a stone chip on the body of a motor car. The exposed metal surface in the case of steel undergoes the anodic reaction which will eventually lead to a characteristic rust deposit on the surface:

$$Fe \rightarrow Fe^{2+} + 2e^-$$

The electrons produced by this reaction are consumed in the cathodic reaction which occurs adjacent to the anodic area, and may involve both oxygen and water:

$$H_2O + \tfrac{1}{2}O_2 + 2e^- \rightarrow 2OH^-$$

Thus the cathodic reduction of water and oxygen leads to the production of hydroxyl ions which in turn leads to an increase in the pH of the electrolyte in the environs of the coating–substrate interface. It is this alkalinity which is responsible for the rapid failure of coatings or adhesive joints and has given rise to the term 'cathodic disbondment'. The case of an entire structure being polarized cathodically is frequently encountered when cathodic protection is used to prevent corrosion of massive steel structures such as ships or pipelines.[1] The cathodic reaction will occur at any regions of exposed metal bringing about rapid failure of the coating.

(a)

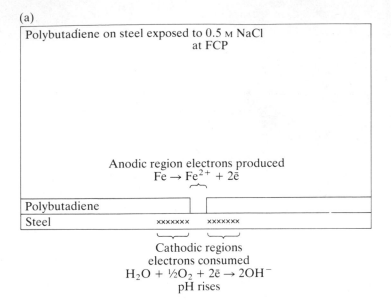

Polybutadiene on steel exposed to 0.5 M NaCl at FCP

Anodic region electrons produced
$$Fe \rightarrow Fe^{2+} + 2\bar{e}$$

Polybutadiene

Steel

×××××××   ×××××××

Cathodic regions
electrons consumed
$$H_2O + \frac{1}{2}O_2 + 2\bar{e} \rightarrow 2OH^-$$
pH rises

(b)

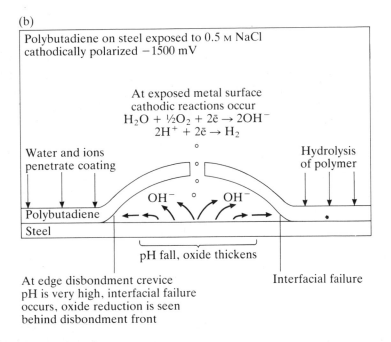

Polybutadiene on steel exposed to 0.5 M NaCl
cathodically polarized −1500 mV

At exposed metal surface
cathodic reactions occur
$$H_2O + \frac{1}{2}O_2 + 2\bar{e} \rightarrow 2OH^-$$
$$2H^+ + 2\bar{e} \rightarrow H_2$$

Water and ions
penetrate coating

Hydrolysis
of polymer

OH^- OH^-

Polybutadiene

Steel

pH fall, oxide thickens

At edge disbondment crevice
pH is very high, interfacial failure
occurs, oxide reduction is seen
behind disbondment front

Interfacial failure

Fig. 15. Electrochemical reactions involved in cathodic disbondment: (a) in a system at rest potential; (b) in a system that is cathodically polarized (i.e. cathodically protected by an impressed current)

The manner in which these two situations can lead to cathodic disbondment is shown schematically in Fig. 15.

There are several ways in which the susceptibility to cathodic disbondment of coatings applied to a steel surface may be assessed. The most widely used in the UK[2] makes use of an impressed cathodic potential of 1.5 V vs saturated calomel electrode as illustrated in Fig. 16, while the test popular in the USA couples the coated panel to a zinc electrode providing a cathodic potential of 1.0 V. In both tests the pre-damaged panel is exposed in an aqueous solution and the results are reported as the extent of disbondment from the exposed metal as a function of time. In practice a set of tests would be terminated at predetermined time intervals (e.g., 1, 3, 6 months) to compare candidate coatings, the system showing the least disbondment being the best.

In this type of test it is essential, when evaluating coatings, that the substrate pretreatment is kept constant, as changes in surface profile will alter the interfacial pathlength and hence the kinetics of failure if interface diffusion is the rate-controlling step.[3]

There is no universal mechanism of cathodic disbondment, and although it is the production of hydroxyl ions that is eventually responsible for the degradation of adhesion this may be achieved in any of three ways depending on polymer, substrate, and exposure conditions.[4] The classical mode of failure by cathodic disbondment is that of interfacial separation, that is to say the interfacial bonding is attacked by the alkali to such an extent that adhesive failure occurs with no traces of polymer remaining on the metal surface. This type of failure is often associated with lateral diffusion of the active species from the exposed metal. If failure does not occur at the interface it may occur within the interfacial region of the

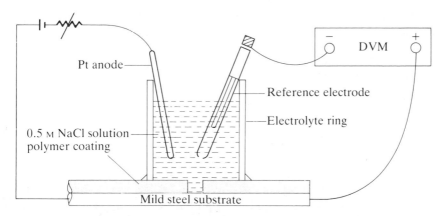

Fig. 16. The standard cathodic disbonding test used in the UK

polymer leaving a very thin (< 5 nm) polymer residue; such a residue can only be detected by surface analysis methods such as **X-ray photoelectron spectroscopy** (XPS), or **Auger electron spectroscopy** (AES). This failure mechanism is often described as alkaline hydrolysis of the polymer and results in the cohesive failure of the polymer very close to the interface as a result of downward diffusion active species through the thickness of the coating and is thus usually confined to thin coatings. The final possibility is the degradation of the substrate oxide, and this phenomenon will occur as a result of the combination of high underfilm pH and very noble electrode potential, the situation that is known to occur in the case of impressed current cathodic protection. However, the likelihood of such an oxide reduction mechanism being the only cause of failure is remote, and although the process may act as an initiator of disbondment[3] or occur in a random manner[5] the kinetics of the process ensure that it will be generally overtaken by either interfacial failure or cohesive failure within the polymer.

With the advent of use of high-performance organic coatings on cathodically protected structures such as gas and oil transmission pipeline, the cathodic disbondment properties of a particular coatings system have become important material parameters and are now to be found in the manufacturer's technical data for such systems. However, the standards employed are many and varied with regard to temperature, potential and surface finish prior to application, and care must be taken in interpreting data from dissimilar tests.

References

1. J Morgan, *Cathodic Protection*, NACE, Houston, 1987.
2. British Gas Specification PS/CW6.
3. J F Watts, J E Castle, *J. Mater. Sci.*, **19**, 2259–72 (1984).
4. J F Watts, *J. Adhes.*, **31**, 73–85 (1989).
5. J F Watts, J E Castle, *J. Mater. Sci.*, **18**, 2987–3003 (1983).

Climbing drum peel test

K B ARMSTRONG

This test is used where the bonded members are not flexible enough to be peeled in a more usual manner (see **Peel tests**). The less rigid of the two members is peeled by winding it around a rigid cylinder – the 'climbing drum' (Fig. 17) – in this way a very high force can be applied to the peel front.

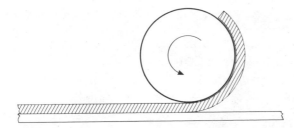

Fig. 17. Schematic representation of the climbing drum peel test

It is a practical test used both for assessing the quality of adhesive bonds between thin skins (usually aluminium alloy) and honeycomb or other core materials for lightweight sandwich panels,[1,2] and also for testing adhesives by peeling a thin sheet from a much thicker one.[3] In the former case its purpose is to peel the skin away from the core and thereby to assess the bond strength/quality. This test is used mainly by the aircraft industry. In the latter case it measures the peel performance of adhesives (see **Tests of adhesion, Honeycomb structures**). The standard test method is ASTM D-1781, which describes the test and the apparatus required. The apparatus consists of a drum with each end of larger diameter than the centre section to which one skin of the honeycomb panel is connected. Flexible bands or bicycle chains are connected to the ends of the drum at one end and to the base of the testing machine at the other. The opposite end of the specimen is held in a grip which is connected to the crosshead of the testing machine. This test is easier to demonstrate than it is to describe. During loading the flexible bands, or chains, unwind from the drum causing it to travel up the specimen while wrapping one skin of the specimen around the centre part of the drum. It is recommended that the specimen be 308 mm (12 in) long by 76 mm (3 in) wide with 38 mm ($1\frac{1}{2}$ in) of skin and honeycomb cut away at one end and 25 mm (1 in) at the other to leave just the skin for attachment to the climbing drum at one end and the testing machine at the other (Fig. 18).

Loading should be carried out in a screw-driven tensile testing machine with the crosshead speed set to 25 mm (1 in) per minute and at least 152 mm (6 in) of the skin/honeycomb bond should be peeled. The machine should have a chart recorder so that the peeling force can be recorded and an average taken for calculation of the peel torque. Testing has shown that the frequency of the oscillations on the chart record can be related to the cell size of the honeycomb used. This test can result in a variety of failure modes and mixtures of them.

With care, this test can also be performed on honeycomb sandwich test pieces having thin Fibreglass, Kevlar or carbon-fibre skins.

Fig. 18. Side view of a climbing drum peel test in which the skin of a honeycomb structure is being peeled away

Failure modes

Various modes of failure may be observed depending on the adhesive, the materials bonded and their surface pretreatments (see **Locus of failure**). For metal–metal peeling very effective surface treatment is necessary to achieve cohesive failure in the adhesive; otherwise failure occurs at, or close to, one substrate surface.

Similarly with aluminium honeycomb structures failure may occur at the skin or the surface of the honeycomb itself, especially if the respective surface preparation is poor (see **Pretreatment of aluminium**).

With appropriate pretreatment, however, failure will occur either by tearing the honeycomb or cohesively in the adhesive fillet. The locus of failure will depend on the strength of the honeycomb itself: this in turn depends upon the thickness and strength of the alloy from which it is made.

'Nomex' honeycomb, made from aromatic nylon fibre in an aromatic polyamide resin, will also fail in the honeycomb, if the adhesive bond is good.

Peel torque and peel strength

According to ASTM D-1781, the results of the test should be expressed as the torque (per unit width) required to peel the strip. This to be calculated from the load and dimensions of the apparatus: a formula is given in the standard. A test run using the drum alone is required to obtain the peel torque needed for the drum to climb in the absence of an adhesive. This may be carried out using a piece of strong fabric so that only the torque to climb the weight of the drum is measured, or a piece of metal skin of the same gauge and type as the test piece can be used. A significant torque can be required to bend the skin. It can be argued that in practice the forces required to separate the skin and honeycomb may include those required to bend the skin. In this case the calibration torque obtained using fabric can be used.

Thus there is a choice as to whether the peel torque calculated included the contribution needed to bend the flexible member. Circumstances will determine which is the more appropriate.

The peel torque is a satisfactory number to quote when comparing like systems with like, but it has no direct fundamental significance. It is straightforward to apply the energy-balance approach to the climbing drum test and to calculate a 'peel strength', i.e. an energy of fracture during peeling per unit area from the peel force recorded and the dimensions of the apparatus (cf. **Peel tests**).

References

1. G C Grimes, The adhesive–honeycomb relationship, *Applied Polymer Symposium*, No. 3, pp. 157–90 (1966).
2. R D Adams, W C Wake, *Structural Adhesive Joints in Engineering*, Applied Science Publishers, London, 1984.
3. R Houwink, G Salomon (eds), *Adhesion and Adhesives*, Vol. 1, *Adhesives*, Elsevier, Amsterdam, 1965.

Coextrusion

R J ASHLEY

Lamination techniques allow the combination of several materials to form a multilayer structure with properties superior to their component layers (see **Laminating**). However, adhesive lamination as a means of producing a multilayer laminate generally involves several machine passes or expensive, large, complex multiple-station laminators. In addition these techniques generally rely on solvent-based materials although some 100% solids systems can be used in limited circumstances. Environmental protection laws have challenged conventional techniques and the coextrusion route to preparation of multilayer structures has gained importance. Coextrusion allows the production of a laminate in a single process at an economic cost. However, the initial equipment is expensive and the process best suited to large-volume outlets where a structure can be produced in continuous operation. The process is generally used for laminates comprising thermoplastic polymers and a wide variety of structures can be made. Very thin layers of expensive polymers can be incorporated effectively. It finds application in the **Packaging industry**.

In the process two or more thermoplastic materials are extruded simultaneously through adjacent die lips and are combined either internally in the die or immediately after leaving the die. Two or more extruder barrels are connected to the multichannel die and the flow rates adjusted to give plies of the film at the desired gauge. Either circular dies for blown coextruded or slot dies for cast laminates may be used.

Adhesion between the plies is partially mechanical since the materials are molten when combined, but the major bond has to be chemical (see **Theories of Adhesion**). Where compatible materials are used such as low density polyethylene or high density polyethylene (LDPE/HDPE) strong bonds can be achieved between the layers without additional means (see **Compatibility**). However, where the wider opportunities to use coextrusion are taken and multilayers conceived with non-compatible materials such as polyamides, polyolefins, polycarbonate, poly(ethylene-vinyl alcohol), adequate adhesion may be difficult to achieve. In these circumstances a specific polymeric layer may be used to act as a compatibilizing agent or **Tie layer**. A very thin layer of the extrudable adhesive has functional groups to enhance the chemical bonding at the interface.

Also see **Extrusion coating**.

Select reference

R J Ashley et al., *Industrial Adhesion Problems*, eds D M Brewis and D Briggs, Orbital Press, Oxford, 1985, Ch. 8.

Compatibility

J COMYN

Compatibility is a measure of the ability of two substances to mix. If two compatible substances are placed in contact then one of the following may take place.

1. The two substances completely mix to form a homogeneous mixture. An example is ethanol and water which are miscible in all proportions.
2. One substance dissolved in the other to form a homogeneous mixture; an example is to place a sheet of rubber in benzene, when the solvent swells and mixes with the polymer.
3. Two separate phases are formed, but each phase is a mixture of two components. Under some conditions phenol and water behave in this manner; one phase is a solution of water in phenol, and the other a solution of phenol in water.

Conditions for compatibility

The condition for compatibility is that the Gibbs free energy for mixing is negative. The Gibbs free energy of mixing ΔG_m is related to the enthalpy (heat) of mixing ΔH_m and entropy of mixing ΔS_m by the following equation, where T is the absolute temperature:

$$\Delta G_m = \Delta H_m - T \Delta S_m \qquad [1]$$

As mixing always increases disorder, ΔS_m is positive so that the $- T \Delta S_m$ term is always negative and therefore favours mixing. Whether mixing actually may occur, therefore, depends on the enthalpy of mixing, and of course the most favourable situation is for this to be large and negative, in other words for a large amount of heat to be evolved on mixing.

It is, however, common where polymers are involved for the enthalpy term to be positive. The enthalpy expression for a 'regular solution' of component 1 in component 2, often used in this context, is

$$\Delta H_m = V\phi_1\phi_2(a_1^{1/2}/b_1 - a_2^{1/2}/b_2)^2 \qquad [2]$$

where a and b are constants from the van der Waals equation, the ϕ's are volume fractions and V is molar volume.

This approach to mixing is very abstract, and it is not possible to take values of the parameters from collections of data and then predict whether mixing may take place. Simpler and more practicable approaches are to compare chemical structures or to use solubility parameters. The first of these depends on the sound principle of 'like generally dissolves like', and it is successful in predicting the solubility of polyvinyl alcohol in water and of polystyrene in benzene, but does not predict the insolubility of cellulose in water. The latter is an interesting case. The similarity of water and cellulose is that they contain hydroxyl groups, but cellulose is prevented from dissolving in water because it is cross-linked by hydrogen bonds. However, this does not mean that they are incompatible; indeed cellulose in the form of cotton absorbs about a quarter of its own weight of water from saturated air.

Solubility parameter δ is a measure of the energy required to separate the molecules of a liquid, and is given by Eqn 3, where ΔU_E is the change in internal energy on evaporation, ΔH_E the enthalpy of evaporation, R the gas constant, T the absolute temperature and V_m the molar volume:

$$\delta = (\Delta U_E/V_m)^{1/2} = ([\Delta H_E - RT]/V_m)^{1/2} \qquad [3]$$

Solubility parameters of liquids can be calculated from experimentally measured values of ΔH_E and V_m. In contrast, the vapour state is inaccessible to polymers, so ΔH_E cannot be measured; they can, however be estimated by one of the group contribution methods,[1] or by measuring the swelling of a polymer in a range of solvents and taking the position that the highest swelling occurs in the solvent most similar to the polymer in the value of its solubility parameter.

Equation 2 can be rewritten as

$$\Delta H_m = V\phi_1\phi_2(\delta_1 - \delta_2)^2 \qquad [4]$$

which shows the significance of solubility parameters: the closer the values for two substances the more likely they are to be compatible. Improvements in prediction can be gained if comparisons of **Hydrogen bonding** and dipole moment are also made.[2]

Some values of δ for a range of solvents are given in Table 8, and some for some common polymers appear in Table 9. In both cases substances are listed in order of increasing solubility parameter.

Compatibility in the context of adhesion

Compatibility is an issue of relevance in a number of aspects of adhesion. Perhaps the most obvious is the selection of solvents in the making of a **Solvent-based adhesive**, or for **Solvent welding** of thermoplastics. In both cases solubility parameters can be employed.

If an adhesive joint has to survive in the presence of a particular liquid, then the solubility parameter can be used to indicate adhesives which would not swell.

The majority of adhesives contain various components added to the base polymer, so the question of compatibility is one of major importance.

Table 8. Solubility parameters of some solvents

Solvent	$\delta/(MPa)^{1/2}$
Aliphatic fluorocarbons	11.3–12.7
Diethyl ether	15.1
Octane	15.6
Dioctyl phthalate	16.2
Tricresyl phosphate	17.2
Toluene	18.2
Ethyl acetate	18.6
Benzene	18.8
Trichloromethane	19.0
Tetrachloroethene	19.0
Acetone	20.3
Dimethylformamide	21.7
Ethanol	26.0
Ethane diol	29.9
Water	47.9

Table 9. Solubility parameters of some polymers

Polymer	$\delta/(MPa)^{1/2}$
Polytetrafluoroethylene	12.7
Polyethylene	16.0–16.4
Natural rubber	16.2–17.0
Polymethylmethacrylate	18.6–19.5
Polyvinyl acetate	19.2
Polyvinyl alcohol	25.8
Cellulose	32.0

It is often obvious that the retention of additives, such as tackifiers in **Pressure-sensitive adhesives**, plasticizers in PVC and antioxidants in many polymers, is essential for making the adhesive and for successful functioning in service. Less obvious is the danger of the migration of additives to the adhesive–substrate interface forming a **Weak boundary layer** to the detriment of adhesion, either during application or service (see **Surface nature of polymers**).

On the other hand, some additives are designed with an incompatibility which causes them to migrate to the interface. Silane **Coupling agents** in thermoset resins and internal **Release** agents are examples. A deeper issue than these however, is compatibility between polymers at an interface. It arises from Voyutskii's **Diffusion theory** or **Autohesion** which depends on the ability of polymer chains at an interface to mix by interdiffusion. Even polymers which are chemically very similar often do not mix. This is because although the entropy of mixing is negative, enthalpies of mixing are often small and positive, so making the Gibbs free energy of mixing positive. The incompatibility of polymers is a fundamental problem in the making of polymer blends.

There are, however, a number of instances where interfacial diffusion has been demonstrated experimentally between differing polymers; they include the following: polyvinylchloride and polycaprolactone; polyvinyl chloride and polymethacrylate; polyvinylchloride and styrene– acrylonitrile copolymer; polyvinylidene fluoride and polymethylmethacrylate.

References

1. E A Grulke in *Polymer Handbook*, 3rd edn, eds J Brandrup, E H Immergut, Wiley–Interscience, New York, 1989, pp. vii/519.
2. K W Harrison, *Adhesion* 3, ed. K W Allen, Applied Science Publishers, London, 1979, Ch. 9.
3. S S Voyutskii, *Autohesion and Adhesion of High Polymers*, Interscience, New York, 1963.

Composite materials

D E PACKHAM

If the term 'composite material' is applied loosely to encompass any material consisting of two or more phases, its scope is very wide. The vast majority of materials come into the category, both those developed technologically and those occurring naturally. Some polymers, many alloys and most ceramics comprise more than one phase, as do concrete and timber, the most widely used constructional materials.

The properties of these materials depend not simply on the properties of the individual phases, but on the adhesion between them. The influence of this interfacial adhesion is multifarious. High adhesion may give rise to high mechanical strength, but lower adhesion in a similar material may give greater fracture energy. If the adhesion between phases can be modified, there is scope for developing materials with a range of properties and applications.

Many of the entries in this book dealing with the science of adhesion and **Theories of adhesion** are relevant to composition materials in general. There are a number of articles which discuss specific aspects of composite materials which come within the scope of the book.

Toughened adhesives, particular **Epoxide adhesives** and **Toughened acrylics**, consist of polymers with a rubbery phase dispersed as small spheres within a more glassy matrix.[1] Appropriate adhesion between the phases is crucial for effective toughening.

Dispersion of inorganic particles gives rise to **Filled polymers**. A wide variety of properties can be achieved depending on the filler and its adhesion to the matrix.[2] The scope of application of rubbers is enormously extended by use of fillers which can give increased strength, stiffness and abrasion resistance to the material (see **Rubber fillers**).

Fibre compositions represent a class of composite material which combine low weight with good mechanical properties.[3,4] Usually a strong and stiff fibre, such as glass or carbon is incorporated in a polymer matrix.

Several articles are included on different aspects of **Fibre–matrix adhesion** in these materials. Special techniques are required for **Fibre composite joining**.

The application of **Coupling agents** to an inorganic surface provides a powerful way of affecting the bonding properties of the surface. This plays an important part both in **Filled polymers** and **Fibre composites**.

Attention is also drawn to articles on textiles and wood, both of which may be regarded as composite materials.

References

1. D R Paul, J W Barlow, H Keskkula in *Encyclopedia of Polymer Science and Engineering*, Vol. 12, ed. H F Mark et al., Wiley–Interscience, New York, 1988, p. 399.
2. L Mascia, *The Role of Additives in Plastics*, Edward Arnold, 1974.
3. B Harris, *Engineering Composite Materials*, Inst. of Metals, 1986.
4. D Hull, *An Introduction to Composite Materials*, Cambridge University Press, 1981.

Condensation polymerization

J COMYN

In condensation polymerization,[1] also known as 'step growth polymerization', monomers react with one another because they contain at least two types of mutually reactive chemical groups, and there are at least two of these groups per monomer molecule. Some common groups and their reactions are given below. In many but not all cases a small molecule such as water is released.

$$-OH + -COOH = -COO- + H_2O$$
alcohol   acid   (poly)ester   water

$$-OH + -NCO = -NHCOO-$$
alcohol   isocyanate   (poly)urethane

$$-NH_2 + -COOH = -CONH- + H_2O$$
amine   acid   (poly)amide   water

A specific example of a condensation polymerization is the reaction between 1,6-diaminohexane and adipic acid to form nylon 66.

$$nNH_2CH_2CH_2CH_2CH_2CH_2CH_2NH_2 + nHOOCCH_2CH_2CH_2CH_2COOH$$

$$= NH_2(CH_2CH_2CH_2CH_2CH_2CH_2NHCOCH_2CH_2CH_2CH_2)_nCOOH + nH_2O$$

Here both monomers each contain two functional groups ($-NH_2$ or

—COOH) and the resulting polymer is linear. In some cases a single monomer which contains two different functional groups will self-condense. 1-Hydroxypropionic acid (lactic acid) is an example of this, but the cyclic dimer which is known as lactone is also formed.

$$nHO—CH—COOH = H(—O—CH—COO—)_n + nH_2O$$
$$\qquad\ \ |\qquad\qquad\qquad\qquad\ \ |$$
$$\qquad\ \ CH\qquad\qquad\qquad\qquad CH_3$$

Polymer

$$2HO—CH—COOH = \qquad\qquad\qquad + 2H_2O$$
$$\qquad\ \ |$$
$$\qquad\ \ CH_3$$

Lactone

To obtain a high molar mass (molecular weight product) the mixture must be stoichiometric; on the other hand controlled deviations from stoichiometry can be used to lower the molar mass. In contrast with **Addition polymerization** where high molar mass molecules are formed right from the start, molar mass increases with the degree of conversion and large molecules are only formed at very high conversion. A consequence of this is that condensation polymers tend to have lower masses and narrower mass distributions than addition polymers.

When a monomer with more than two functional groups is added, branching, and eventually cross-linking will occur.

Adhesives and condensation polymerization

Polyester and polyamide **Hot melt adhesives** are made by condensation polymerization.

There are a number of reactive adhesives that cure by condensation polymerization, and these include **Phenolic adhesives, Epoxide adhesives**, isocyanates, urea and melamine formaldehyde and **Polyimide adhesives**. Condensation polymerization also occurs when silane coupling agents (see **Coupling agents – application**) are applied to surfaces.

Structural **Phenolic adhesives**[2] are based on resoles, which are made by the condensation polymerization of phenol with excess formaldehyde under basic conditions. They contain phenol units bridged by —CH_2— and —CH_2OCH_2— groups, and there are some methylol (—CH_2OH) groups on the rings. A typical structure is

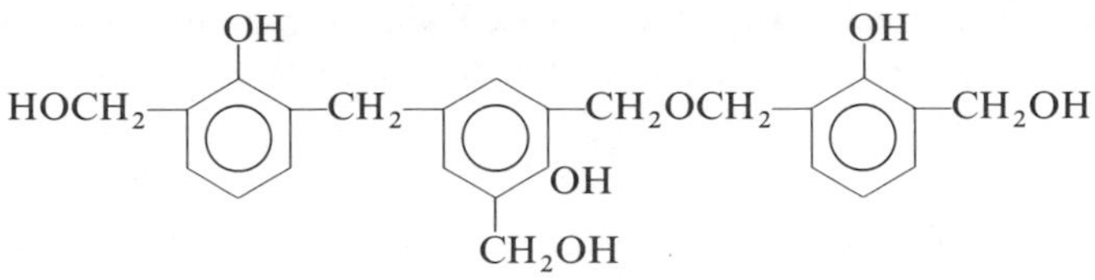

$$HOCH_2 \!-\!\!\bigcirc\!\!-\!CH_2 \!-\!\!\bigcirc\!\!-\!CH_2OCH_2 \!-\!\!\bigcirc\!\!-\!CH_2OH$$

(Structure: three phenol rings, each bearing OH groups. The first ring has OH at top, linked by $-CH_2-$ to the second ring; the second ring has OH at top, CH_2OH at bottom, linked by $-CH_2OCH_2-$ to the third ring which has OH at top and $-CH_2OH$ substituent.)

On heating to 130–160 °C, further condensation polymerization takes place between methylol groups, thus

$$2-CH_2OH = -CH_2OCH_2- + H_2O$$

The products of this reaction are a cross-linked polymer and steam, and to avoid formation of bubbles it is necessary to cure phenolic adhesives in a press.

The reactions of urea or melamine with formaldehyde are the basis of aminoresin adhesives for wood[3] (see **Wood adhesives**).

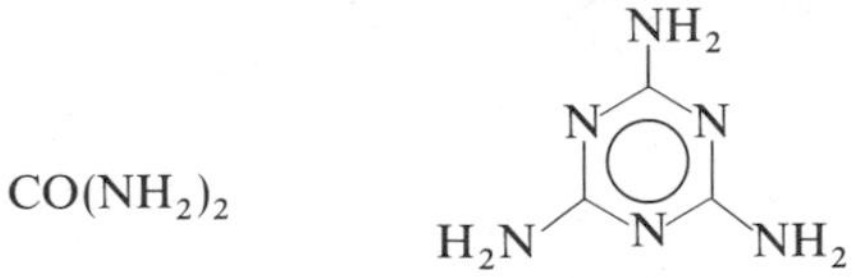

$$CO(NH_2)_2$$

Urea

(Melamine structure: a triazine ring with three NH_2 groups — one at top, H_2N at lower left, NH_2 at lower right — and N atoms in the ring.)

Melamine

In both cases formaldehyde is first added to the amine groups to produce methylols. The $-CH_2OH$ groups later condense in the manner shown above for phenolics.

$$-NH_2 + CH_2O = NHCH_2OH$$

$$-NHCH_2OH + CH_2O = -N(CH_2OH)_2$$

As urea is 4-functional and melamine 6-functional the final products are cross-linked.

One advantage with **Epoxide adhesives**[4] is that no volatiles, which have potential for pollution, are given off during cure. Curing agents which react with epoxides by condensation polymerization include amines, thiols and acid anhydrides, and with the first two of these groups each active hydrogen is capable of reacting with one epoxide in the following manner:

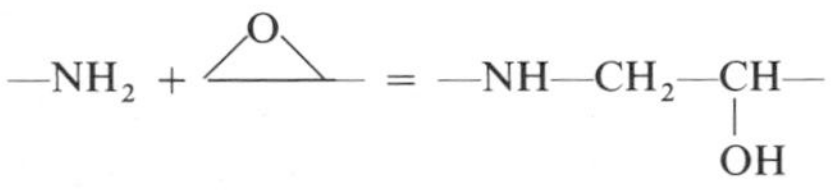

$$-NH_2 + \overset{O}{\triangle} = -NH-CH_2-\underset{\underset{OH}{|}}{CH}-$$

Triethylenetetramine is a commonly used aliphatic amine, and its high functionality (6 amine hydrogens) leads to a high level of cross-linking:

$$NH_2CH_2CH_2NHCH_2CH_2NHCH_2CH_2NH_2$$

Polyimide adhesives are used as high-temperature adhesives,[5] and they are made by reacting a dianhydride with a diamine. In the example shown below pyromellitic anhydride is reacted with 1,4-diaminobenzene. The first step in the reaction gives a polyamic acid which is soluble and fusible, and it is at this stage that it would be applied to an adhesive joint. The second stage takes place during high-temperature cure, and the resulting polyimide is intractable.

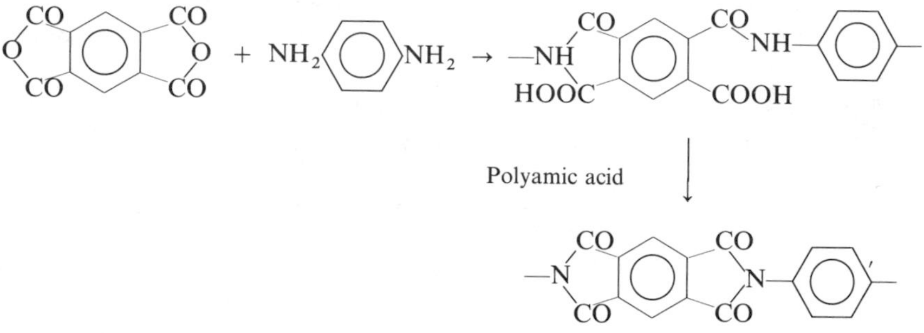

Polyamic acid

Polyimide

References

1. F W Billmeyer, *Textbook of Polymer Science*, 3rd edn, John Wiley, New York, 1984.
2. J Robins in *Structural Adhesives, Chemistry and Technology*, ed. S R Hartshorn, Plenum Press, New York, Ch. 2.
3. A Pizzi in *Wood Adhesives, Chemistry and Technology*, ed. A. Pizzi, Marcel Dekker, New York, 1983, Ch. 2.
4. Y Tanaka, R S Bauer in *Epoxy Resins, Chemistry and Technology* 2nd edn, ed. C A May, Marcel Dekker, New York, 1988, Ch. 3.
5. K P Subrahmanian in *Structural Adhesives, Chemistry and Technology*, ed. S R Hartshorn, Plenum Press, New York, Ch. 7.

Contact angle

J F PADDAY

The nature of the contact angle

The contact angle[1,2] is an angle measured at the triple phase line, between the tangents at the two surfaces of the advancing liquid; see Fig. 145 of

Wetting and spreading for an example. The angle that is measured is a macroscopic quantity usually part of a capillary system which makes the angle measurable without much magnification. The thermodynamic contact angle used in Young's equation [1] (see **Wetting and spreading**) and equations related to it (e.g. Eqns 5–8 in **Wetting and work of adhesion**), is a microscopic quantity which is assumed to equal the measured value:

$$\gamma_{13} - \gamma_{12} = \gamma_{23} \cos \theta \qquad\qquad [1]$$

Unseen surface inhomogeneities very often occur at a microscopic level and cause variations of contact angle according to whether the liquid is advancing θ_a, or receding θ_r, such variations caused by the triple phase line finding contact with different effective surface energies of the solid. Such contact angle hysteresis means that measured values cannot be relied upon unless the cause of such hysteresis is known or determined. Various causes are recognized. The system may not have reached equilibrium, either thermodynamically (the triple phase line may be moving very slowly) or thermally. A rough or porous surface gives rise to hysteresis. Heterogeneity is another cause. This may take the form of contamination with patches of material of differing surface energy or of islands of electrical charge on an insulator. Surface-active contaminants in the wetting liquid may be adsorbed on the surface: such contaminants may be in solution or in the form of microdrops of insoluble surfactant. Finally, if the liquid reacts with the solid surface hysteresis will be observed.

The principle of measuring both advancing and receding angles and using their arithmetic mean as the thermodynamic mean is justified only when the hysteresis is produced by a regular array of inhomogeneities such as with certain types of roughness. Large unexplained differences between advancing and receding contact angles suggest that the system is not properly characterized.

Contact angle at a moving line

Contact angles are formed with wetting systems only. The liquid wetting the solid can be forced to spread by the application of some external force so as to produce 'dynamic wetting', not to be confused with spreading. Further, in many systems the wetting liquid, once spread artificially, can remain in place after the external force is removed, providing the hysteresis is sufficiently large.

Dynamic wetting has been studied extensively[3] and the main feature is that the contact angle changes with the speed at which the wetting line is forced across the solid surface as shown in Fig. 19. Here it is noted that both wetting and dewetting can be forced on a system with a small amount of contact angle hysteresis.

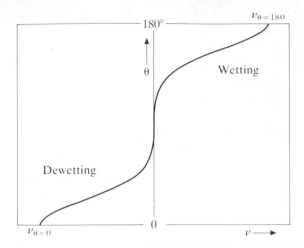

Fig. 19. Contact angle as a function of the speed of wetting line movement
(from T D Blake[3])

At very low wetting speeds the triple phase line moves erratically with
stick–slip movement, thereafter the angle increases continually with speed
for the advancing lie and decreases continually for the receding line until
180° or 0° are reached for each condition respectively. Further speed
increases for either advancing or receding lines results in the break-up of
the line so as to form a serrated or 'saw-tooth' wetting line.

Contact angle and roughness

The effect of roughness of the solid surface is to increase the actual area
of contact between solid and liquid and also that of the unwetted solid
per unit geometric area. The roughness factor, r, is defined as the real
area of the solid divided by this geometrical area described by a boundary.
Wenzel therefore proposed that Young's equation [1] be modified as
follows:[1]

$$r(\dot{\gamma}_{13} - \gamma_{12}) = \gamma_{23} \cos \theta \qquad [2]$$

Hysteresis arises from roughness mainly because the wetting line sees
a different effective macroscopic contact angle arising from the slope at
the position in the groove of the wetting line which is different for
advancing and receding conditions. For each condition the microscopic
contact angle remains the same as seen in Fig. 20.

When the pore size of the roughness approaches molecular dimensions
the roughness factor requires modification to account for the adhesion
forces changing with distance from the solid surface.

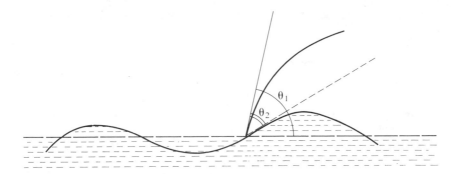

Fig. 20. Macroscopic, θ_1 and microscopic, θ_2, contact angles

Roughness will affect spreading and wetting and so all the equations (1–8) discussed in **Wetting and work of adhesion** require modification by the roughness factor. As spreading coefficients and the corresponding wetting energies can be positive or negative the effect of r is to increase contact angles greater than 90° and decrease those less than 90°.

Contact angle and surface energy of the solid

The spreading coefficient and the wetting energy depend critically on the magnitude of the surface energy of the solid[2] or the array of molecules nearest to the wetting or spreading liquid. Metals, some metal oxides and some non-metal oxides such a silica, possess very high surface energies and thereby encourage spreading because the work of adhesion is large. Platinum metal forms such a surface at which most liquids spread spontaneously at room temperature.

Low-energy solids such as those of polyethylene and polytetrafluoro-ethylene do not encourage spreading and are termed 'hydrophobic'. All these low-energy surfaces possess works of adhesion which are positive but less than the work of cohesion of the liquid, as already noted. When such low-energy solid systems possess very little hysteresis they are described as 'abhesive' because the liquid may be pulled away from the solid with ease (see **Release**).

Surface contamination

The most common cause of contact angle hysteresis arises from heterogeneous contamination of the solid with islands of adsorbed or smeared low-energy impurities. High-energy surfaces attract such contamination more readily than low and often exhibit the surface of low-energy surfaces. Metals also may possess islands of oxidation which induce hysteresis. Low-energy surfaces, on the other hand, can also acquire contamination, more often with surface-active material which is amphipathic in nature.

The effect of surface-active material dissolved in the wetting liquid is to adsorb at the liquid–solid interface and decrease the interfacial tension by a large amount. Such adsorption is usually irreversible in that when the liquid is removed a monolayer of the surfactant remains which further adds to the hysteresis, especially so when the surfactant is an adsorbed polymer or macromolecule. The removal of surface contamination with consequent enhancement of surface energy and lowering of contact angle is a major objective of **Pretreatment** prior to adhesive bonding, painting or printing.

References

1. A W Adamson, *Physical Chemistry of Surfaces*, 3rd edn, John Wiley, New York.
2. W A Zisman in R F Gould, Advances in Chemistry Series 43, *Contact Angle and Wettability*, American Chemical Society, Washington, 1964, pp. 1–51.
3. T D Blake, Wetting kinetics – how do wetting lines move?, *Proc. A.I.Ch.E. International Symposium on the Mechanics of Thin-film Coating*, New Orleans, 1988, Paper 1a.

Contact angles and interfacial tension

D E PACKHAM

It seems inherently reasonable that the energy of an interface (interfacial tension) between phases 1 and 2 should be related to the individual values of **Surface energy**, γ_1 and γ_2: the same molecules are responsible for each. If there actually is an interface between 1 and 2, it is, *ipso facto*, stable so Eqns 1 or 2 must apply

$$\gamma_{12} < \gamma_1 + \gamma_2 \qquad [1]$$

$$\gamma_{12} = \gamma_1 + \gamma_2 - \Delta \qquad [2]$$

where Δ is positive and represents the stability of the interface with respect to separate phases 1 and 2. The value of Δ will depend on the type of bonding across the interface.

For secondary bonds, **Dispersion forces, Polar forces** and **Hydrogen bonds**, various explicit forms of Eqn 2 have been proposed. Their use requires a familiarity with Young's equation and the definition of work of adhesion and the work of cohesion (Fig. 21), discussed in **Wetting and spreading**. It is convenient to quote these relationships here:

$$\gamma_{13} = \gamma_{12} + \gamma_{23} \cos \theta \qquad [3a]$$

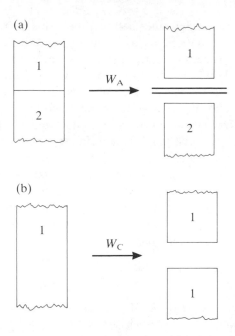

Fig. 21. Representatives of the changes for which (a) work of adhesion, and (b) work of cohesion are the changes in free energy

$$\gamma_{SF} = \gamma_{SL} + \gamma_{LF} \cos \theta \qquad [3b]$$

$$W_A = \gamma_1 + \gamma_2 - \gamma_{12} \qquad [4]$$

$$W_{C_1} = 2\gamma_1 \qquad [5]$$

The subscripts 1 or S, 2 or L and 3 or F refer to the three phases; the letters will be used when it is desirable to refer specifically to solid S, wetting liquid L and fluid (vapour or second liquid) F.

Note that the work of adhesion is defined here as the free energy change required to separate phases 1 and 2 cleanly along the interface and to place each in a separate vacuum enclosure (Fig. 21(a)). It is sometimes defined in a subtly different way, see **Wetting and work of adhesion**, Eqns 1 and 4.

It is important to realize that the work of adhesion and work of cohesion defined by Eqns 4 and 5 refer to the hypothetical changes described: they may not, and usually will not, be closely related to practical measures of fracture energy of an adhesive bond or a single phase material. This is because in the context of practical fracture much larger amounts of work are dissipated by other mechanisms, such as the plastic deformation of the materials around the fracture zone.

There is a significant difference between the terms for the solid surface energy in Young's equation, γ_{13} or γ_{SF} and γ_1 and in Eqn 4 of work of adhesion W_A. This is because in the former case the solid is in equilibrium with phase 3 which may well be air saturated with vapour of liquid 2 (see **Wetting and spreading** especially Fig. 145 therein), but for the latter the solid is *in vacuo*. The difference between these two terms is the spreading pressure, see **Surface characterization by contact angles – polymers**.

Knowledge of interfacial tensions is of interest because of their direct theoretical and (less direct) practical relationship to adhesion (see **Adsorption theory, Peel tests, Adhesion – fundamental and practical**). Three somewhat different approaches to estimating interfacial tensions by Eqn 2 are commonly found in discussions of adhesion. These are discussed in articles on **Good–Girifalco interaction parameter, Surface energy components** and **Acid–base interactions**.

References

1. W A Zisman in *Contact Angle, Wettability and Adhesion*, ed. R F Gould, Advances in Chemistry Series No. 43, American Chemical Society, Washington, 1964, p. 1.
2. B W Cherry, *Polymer Surfaces*, Cambridge University Press, Cambridge, 1981.

Contact angle measurement

J F PADDAY

General

The macroscopic **Contact angle**, θ, is of considerable importance in wetting studies and, in principle is readily measured.[1,2] The experimentalist must recall that it is measured within the advancing liquid phase. In principle, four methods of contact angle measurement are known. They are as follows:

1. Direct geometric measurement;
2. Calculation from measured capillary shape properties of the system;
3. Measurement using a force balance and comparison with the surface tension;
4. Calculation from intermolecular forces.

The first three methods involve determining the macroscopic angle, whereas the fourth involves the microscopic angle. The problems of possible hysteresis are associated with the first three only and demand that both advancing and receding angles be measured several times on the same patch of solid surface. Also, the surface tension of the spreading liquid should be measured both before and after wetting or spreading so as to ensure no contamination of the liquid by interaction with the solid.

Direct geometric measurement

Direct measurement consists of capturing an image of the three surfaces at the triple-phase line of systems shown in Fig. 22.

Very often a reflected image of the liquid–fluid interface is seen at the solid surface and this makes it unnecessary to determine the plane in which the solid surface lies as the angle measured is that between the liquid–liquid surface and its reflected image. Lighting the meniscus at the triple-phase boundary has been solved satisfactorily, the main problem being the interference fringe produced at a curved surface. This problem is particularly difficult when the contact angle is small ($<25°$). Controlled interference methods and Moiré fringes may be used to obtain greater precision.

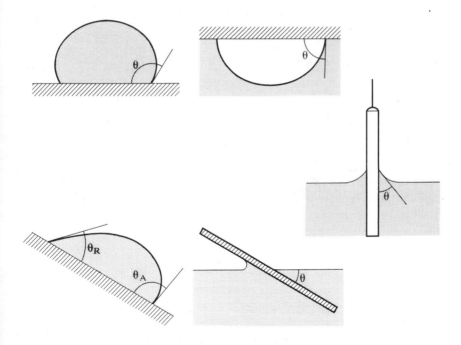

Fig. 22. Methods of measuring the contact angle

Capillary shape methods

The availability of high-speed computing and data processing has led to the development of indirect methods from geometric measurements of capillary shape. However, some of these methods require a knowledge of the capillary constant or the values of the properties that go to make it up.

A well-known example of this method is that of measuring the radius and height of a sessile drop and then with a knowledge of the capillary constant, estimate the best fit of contact angle using the tables of Bashforth and Adam. A simpler method is to determine the spreading of coefficient from the limiting height h of a very large sessile drop of density ρ. The relationship to contact angle is given by

$$S_{C_{123}} = -\rho g h^2/2 = \gamma_{23}(\cos \theta - 1) \qquad [1]$$

See **Wetting and work of adhesion** for the relation between spreading coefficient $S_{C_{123}}$, surface tension γ_{23} and contact angle θ. As such limiting conditions are reached only when the drop is about 1 m diameter it is easier to determine the height of a smaller drop and multiply the numerator by a factor, F, the value of which is determined by the ratio of two linear shape properties such as height and maximum diameter.[1] The contact angle is then derived from the spreading coefficient with a knowledge of the surface tension. When the capillary constant, a, given by

$$a^2 = 2\gamma_{23}/\rho g \qquad [2]$$

is known, the reduced spreading coefficient is given by

$$S_{C_{123}}/\gamma_{23} = -h^2/a^2 = (\cos \theta - 1) \qquad [3]$$

Force balance method

The measurement of the surface tension of a liquid by the Wilhelmy – the force balance method – is widely used and well documented.[1] The same apparatus may be used to measure the wetting energy directly by replacing the Wilhelmy plate ($\theta = 0°$) with a similar plate made of the solid to be wetting. Let the perimeter of the plate along the triple-phase line be L, and the force be F, then

$$F = \gamma_{23}L \cos \theta \qquad [4]$$

Experimentally, the surface tension is measured first using a clean platinum plate. Next, F is determined for the experimental surface with a finite contact angle and finally the surface tension once again. If advancing and receding values are required, a buoyancy correction must be made to compensate for the upthrust as the bottom edge of the plate becomes lower than the free surface of the liquid.

Calculation from surface properties of the system

In principle the method consists of estimating the values of γ_{13}, γ_{12} and γ_{23} of Young's equation which then allows calculation (see **Wetting and work of adhesion**). The problem lies in obtaining a reliable value of the interfacial free energies of the solid surface. Three methods are used:

1. Calculation of γ_{SV} and γ_{SL} from Hamaker constants, assuming dispersion force interaction only;[3,4]
2. Taking γ_{SV} as the **Critical surface tension** as defined by Zisman;[5]
3. Using the Van Laar–Fowkes[5] method for estimating the work of adhesion as the geometric mean of the dispersion force contribution of the solid and of the liquid (see **Surface energy components**)

$$W_a = \sqrt{\gamma_{SV}^d \gamma d_{SL}} \qquad [5]$$

All three approaches involve the assumption that the solid–liquid attraction is derived from dispersion force interaction only. This is only true in a limited number of systems such as with water wetting paraffin or polytetrafluoroethylene. The main advantage of this method is that it provides an estimate of the work of adhesion that is independent of roughness, contamination and surface inhomogeneities that produce hysteresis.

References

1. J F Padday, *Surface and Colloid Science*, Vol. 1, ed. E Matejevic, Wiley–Interscience, New York, 1969, pp. 39–251.
2. A W Neumann and R J Good, *Surface and Colloid Science*, **11**, 31 (1979).
3. J N Israelachvili, *Intermolecular and Surfaces Forces*, Academic Press, London, 1985.
4. A D Zimon, *Adhesion of Dust and Powder*, trans. from Russian by Morton Corn, Plenum Press, New York, 1969.
5. W A Zisman in Advances in Chemistry Series 43, *Contact Angle and Wettability*, American Chemical Society, Washington, 1964, pp. 1–51.

Conversion coating

A MADDISON

Conversion coating processes find extensive application in the **Pretreatment of metals prior to painting** and the development of other important surface properties, e.g. corrosion resistance, lubrication characteristics suitable for tube and wire manufacture and low electrical conductivity for transformer core laminates. Phosphate, chromate and alkaline oxide processes are commercially the most important, compared for example, with fluoride treatments for titanium and zirconium, or oxalate processes for stainless steel.[1]

Phosphates

Iron phosphate and zinc phosphate conversion processes are widely employed in the treatment of steel- and zinc-coated materials. Mechanisms of film formation are complex, but basically involve the anodic dissolution of the metal and the precipitation of insoluble phosphates at cathodic sites on the surface. Polarization by hydrogen evolution reduces reaction rates, and, to counter the effect, practical solutions contain additions of oxidizing agents or heavy metal salts to accelerate processing.

Phosphate coatings may alternatively be produced using solutions containing ammonium, potassium or sodium dihydrogen phosphates, and these types may be preferred in applications where paint adhesion must be retained after mechanical forming.

Zinc phosphate coatings are also widely utilized as paint pretreatments, particularly when service conditions are severe. Coating structure is influenced by previous cleaning stages, though sensitivity to these operations can be reduced by using a pre-dip or virtually eliminated by adopting calcium-modified solutions which incorporate zinc calcium phosphate into the coating. Zinc phosphate treatments are also effective prior to the application of epoxy powder and cathodic electropaint systems. Most iron and zinc phosphate conversion sequences include a final chromate rinse prior to drying.

Mixed material assemblies require compromise formulations. Fluoride-modified alkali metal phosphate processes are used, for example, in the treatment of steel, zinc and aluminium combinations. Phosphate coatings may also be applied to cadmium and magnesium.

Magnesium phosphate processes, which are usually nitrate or nitro-guanidine-accelerated, are particularly recommended when the requirement exists for high corrosion resistance or exceptional cold-forming lubrication characteristics.

Chromates

Chromates constitute a second major category of conversion treatments which includes both acid and alkaline types and a considerable variety of formulations. 'Chromium chromate' films are produced on aluminium surfaces by spraying or immersion in solutions based on chromates or chromic acid and hydrofluoric acid. **X-ray photoelectron spectroscopy** (XPS) studies of a ferricyanide-accelerated chromate coating suggest a structure consistent with that of hydrated chromium oxide.[2] Other chromate processes find application in the treatment of zinc, cadmium and magnesium.

Chromium phosphate coatings are developed on aluminium surfaces using solutions containing fluoride, phosphoric acid and chromic acid or chromate. Coating weight may be controlled over a wide range by varying the fluoride concentration.

Alkaline oxide

Important in the treatment of hot-dip galvanized and electrogalvanized steels the alkaline solution produces zinc oxide coatings containing typically iron, cobalt and nickel ions. Chromium is incorporated into the film during post-treatment.

Coil line pretreatments

The short process times associated with phosphate, chromate and alkaline oxide conversion treatments permit coil line deployment, in common with no-rinse treatments such as Accomet C.

Adhesive bonding

Although some conversion processes are primarily intended to promote the adhesion of paint, the demands of **Structural adhesive** bonding are significantly different. Nevertheless, some may be optimized to yield bonds of adequate strength and durability. If also developed for application at the coil stage,[3] considerable, possibly decisive, manufacturing advantages may be realized.

References

1. D B Freeman, *Phosphating and Metal Pre-treatment*, Woodhead-Faulkner, Cambridge, 1986.
2. J A Treverton and N C Davies, XPS studies of a ferricyanide accelerated chromate paint film on an aluminium surface, *Surface & Interface Analysis*, **3** (1), 194 (1981).
3. British Patent GB 2 139 540A.

Corona discharge treatment

D BRIGGS

Corona discharge treatment is a form of **Plasma treatment**, operated at atmospheric pressure, used widely for enhancing the adhesive characteristics of plastic materials. It is very widely used for treating plastic films (specially polyolefin films) and conical containers (tubs) but also finds application in other areas such as wire coatings and moulded composite parts.

Illustrated in Fig. 23 is the process for film treatment. The film is passed over an earthed metal roller which is covered with an insulating material (the 'dielectric sleeve'). Separated from the film by ~ 1–2 mm is a metal electrode, usually made from aluminium. A high-frequency (typically 10–20 kHz) generator and step-up transformer provides a high-voltage (peak typically 20 kV) to this electrode.

In each half-cycle the applied voltage increases until it exceeds the threshold value for electrical breakdown of the air gap, when the air is ionized and becomes plasma. Thus each cycle involves current flow in each direction. In continuous operation the discharge appears to be a random series of faint sparks (streamers) superimposed on a blue–purple glow. Discrete intense sparks are also often seen originating from localized regions of the electrode, so 'corona' is not an exact description for this discharge.

The ionized air consists of ions, electrons, excited neutrals and photons in the UV-visible region. All of these are energetic enough to break C—C and C—H bonds to form radicals which via subsequent radical oxidation chemistry yield a variety of oxygen functionalities on the surface.

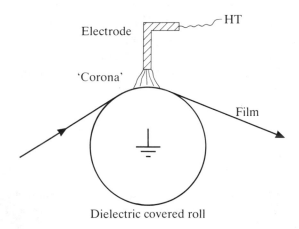

Fig. 23.   Schematic diagram of the electrical discharge treatment process

Taking a polyethylene surface as an example this change in surface chemistry leads to significant increases in **Surface energy** (from 31 up to 50+ mN m^{-2}) and hence **Wettability** and also to a capacity for specific interactions with adhesives, inks, coatings, etc. Particularly important seems to be the **Hydrogen bonding** capacity of enolic —OH groups which can be formed via tautomerization of ketonic groups (—CH$_2$CO—). Both factors are important in improving surface adhesive properties.

The degree of surface change is a function of operating parameters such as the exposure time and the discharge power (and efficiency of coupling power from the generator into the discharge); the latter is itself related to the frequency, discharge gap, electrode cleanliness and so on. Additive bloom to a plastic surface inhibits these changes and requires higher treatment levels to 'burn off' this material so that the underlying surface can be oxidized.

Select reference

D Briggs in *Surface Analysis and Pretreatment of Plastics and Metals*, ed. D M Brewis, Applied Science Publishers, London, 1982, p. 199.

Coupling agents – application

P WALKER

Adhesion promoters or coupling agents as they are sometimes called are compounds used specifically to improve the adhesion characteristics of adhesives and surface coatings and the strength of **Composite materials** by the formation of primary bonds to either the substrate or the applied adhesive/coating, or both, or to the filler or reinforcing material in composite technology, e.g., glass reinforced plastics (see **Coupling agents – chemical aspects, Fibre composites, Fibre matrix adhesion – glass fibre**). In the case of adhesives and surface coatings the emphasis is on improving both the initial adhesion and the adhesion under the most adverse service conditions (see **Durability, Underwater adhesives**). In composite technology the aim is to improve both the initial physical properties and the resistance to water. Water, particularly under saturated conditions is known to cause a dramatic decrease in the bond strength of adhesives[1] and coatings.[2]

A variety of techniques are used to improve initial adhesion including removal of surface contamination (see solvent and vapour **Degreasing**, high temperature), surface profile changes (gritblasting (see **Abrasion**), acid etch) and surface modification (**Conversion coatings, Anodizing,** oxidation). Adhesion promoters are becoming an important addition to these techniques and may be used in conjunction with or as a replacement for these processes.

A relatively large number of compounds from several generic classes have been investigated for use as potential adhesion promoters, the most important of which are silanes, titanates, zirconates and chromium-containing compounds. Other materials include phosphorus-containing compounds; amines, metal salts of organic acids and a variety of organic resins.[3] A comprehensive review by Cassidy and Yager covers the period 1950–71.[4]

In adhesive and surface coating technology adhesion promoters can be used either as pretreatment **Primers** applied from solvent solution to the substrate prior to bonding or coating, or as additives incorporated into the bulk material. In composite technology the fillers, whether chopped fibre, woven or particulate can be pretreated with an adhesion promoter prior to use or the promoter incorporated into the polymer–filler mixture. Whichever method is used extreme care is required to ensure that all relevant factors have been addressed, otherwise no improvement in adhesion will be obtained, or in extreme cases the adhesion may be considerably weaker than in the absence of the adhesion promoter. Factors known to effect bond strength when the pretreatment primer method is used include the pH, concentration and age of the solution, the nature of the substrate and adhesive or coating and the film thickness deposited. When used as an additive it is important that the adhesion promoter is not a co-reactant with the binder otherwise premature cure, instability or promoter depletion will occur on storage. Depletion may also occur by reaction with fillers, solvents or any water present. In the case of silanes which are easily hydrolysed according to

$$R - SiX_3 + 3H_2O \rightarrow R - Si(OH)_3 + 3HX$$

the silanols produced may condense to form higher-molecular-weight oligomers. As it is essential to retain the silane as a monomer or dimer to preserve coupling activity, this condensation is undesirable. Amino-silanes are known to react with oxygenated solvents such as ketones, commonly used as solvents in paints and some adhesives. For example γ-amino-propylmethoxysilane will react with methyl ethylketone to form ketamine plus highly undesirable water according to

$$\begin{array}{c} CH_3 \\ \diagdown \\ C = O + H_2 - N - R \rightarrow \\ \diagup \\ C_2H_5 \end{array} \qquad \begin{array}{c} CH_3 \\ \diagdown \\ C = N - R + H_2O \\ \diagup \\ C_2H_5 \end{array}$$

Using too large an addition may also lead to failure, particularly with titanates. In general the one-pack promoter incorporated method is almost universally desirable; but it should be noted that in addition to improving adhesion a particular promoter may have other effects, frequently beneficial but not invariably so. For example titanates may have a profound effect on the rheology of a paint system by improving dispersion or lowering the viscosity. Increased corrosion resistance may be obtained both from coatings and adhesives.

Adhesion promoters may be of value in a wide range of applications other than those in which improvements in initial and wet adhesion are required. These include the following:

1. Pretreatment primers or additives for sealants and antiglare coatings on glass;
2. As replacements for chromate containing wash primers and conversion coatings for use on aluminium;
3. As pretreatment primers or additives to improve the strength and/or adhesion of chopped glass-fibre sprayed or glass-cloth reinforced coatings and adhesives applied to concrete, plaster, glass and metal substrates.

References

1. D J Falconer, N C MacDonald, P Walker, *Chem. and Ind.*, 1230 (1964).
2. P Walker, *Official Digest*, **37**, 1561 (1965).
3. P Walker in *Surface Coatings – 1*, eds A D Wilson, J W Nicholson, H J Prosser, Elsevier Applied Science Publishers, 1986, Ch. 6.
4. P E Cassidy, B J Yager, *Polym. Technol.*, **1**, 1 (1972).

Coupling agents – chemical aspects

J COMYN

Silanes

Silane coupling agents originated in the 1940s to pretreat glass fibres for use in **Fibre composites**, with a view to improving their water resistance (see **Fibre matrix adhesion – glass fibres**). Their structure is $R—Si(OR')_3$, where R is a functional group which can chemically react with an adhesive, resin or surface coating and R' is usually methyl or ethyl. Some common silane coupling agents are as follows:

$$NH_2CH_2CH_2CH_2Si(OCH_2CH_3)_3$$ 3-aminopropyltriethoxysilane (APES)

$$\overset{\displaystyle O}{\overset{\diagup\diagdown}{CH_2-CH}}-CH_2-O-CH_2CH_2CH_2Si(OCH_3)_3$$

3-glycidoxypropyltrimethoxysilane (GPMS)

$$CH_2{=}\underset{\underset{CH_3}{\big|}}{C}-COOCH_2CH_2CH_2Si(OCH_3)_3$$

3-methacrylpropyltrimethoxysilane (MPMS)

The R groups in these compounds respectively contain amine, epoxide and carbon–carbon double bonds; this makes APES and GPMS suitable for use with **Epoxide adhesives** and resins, while MPMS is suitable for materials which harden by free radical **Addition polymerization** such as reactive **Acrylic adhesives**. They are normally applied to adherends from dilute aqueous solution.

There are several theories as to how the $-Si(OR')_3$ moiety reacts with substrates, but the chemical bonding theory is generally accepted. This involves hydrolysis of $-Si(OR')_3$ to produce a trisilanol $-Si(OH)_3$, followed by condensation copolymerization of trisilanol and surface $-OH$ groups to produce a polysiloxane network which is covalently bonded to the surface; the process is illustrated in Fig. 24. The result is that there is a continuous chain of covalent bonds from the substrate through to the adhesive or resin. The main advantage in using silane coupling agents is to improve the durability of adhesive bonds in the presence of water or water vapour.

In treating glass fibres with silanes we have the inherently favourable situation of two silicon compounds being placed in contact. Silanes are generally effective in improving the adhesion of epoxide and polyurethane paints to a range of metals including aluminium, steel, cadmium, copper and zinc.

There are a significant number of papers in the literature in which spectroscopic techniques have been used to examine the interaction of silanes with silica, glass and metal surfaces; they have recently been reviewed.[1] In particular Koenig and his collaborators have mainly used diffuse reflectance **Fourier transform infrared spectroscopy** (DRIFT) and Boerio has employed reflection–absorption infrared (RAIR) to silanes on polished metal mirrors, but other techniques including nuclear magnetic resonance spectroscopy, static **Secondary ion mass spectrometry** and **Inelastic electron tunnelling spectroscopy** have been used. Some of the important facts which have emerged are as follows. The silane layer forms

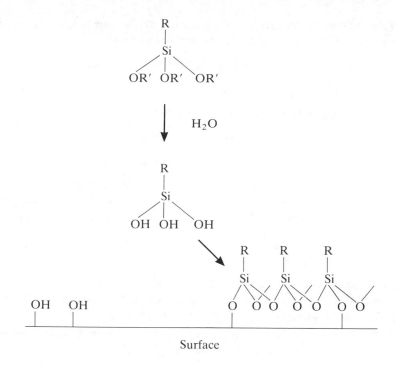

Fig. 24. The hydrolysis of a silane coupling agent to a silane triol, followed by condensation polymerization to give a siloxane network covalently bound to the surface

an interphase with epoxide adhesives such that the concentration of silane decreases gradually as one moves from the surface. There is spectroscopic evidence for the formation of interfacial Si—O—metal bonds with iron, aluminium, titanium and lead. The amine and epoxide functionalities can undergo a range of chemical reactions prior to the application of any adhesive or resin.

Titanates and zirconates

Coupling agents based on titanium and zirconium have been used to improve filler–polymer adhesion and to pretreat aluminium alloy for adhesive bonding.[2] Like the silanes they react with surface hydroxyl groups, but there is no condensation polymerization to produce a polymer network at the interface.

$$ROTi(OR')_3 + \text{Surface–OH} = \text{Surface–OTi}(OR')_3 + ROH$$

Two examples of titanate coupling agents are shown below, zirconate coupling agents have very similar structures to these.

$$(CH_3)_2CH—O—Ti(OCH_2CH_2NHCH_2CH_2NH_2)_3$$

Isopropoxytri(ethylaminoethylamino)titanate

$$(CH_3)_2CH—O—Ti(O—PO—O—PO—OC_8H_{17})_3$$
$$\qquad\qquad\qquad\qquad\quad | \qquad\qquad |$$
$$\qquad\qquad\qquad\qquad OH \qquad\quad OC_8H_{17}$$

Isopropoxytri(dioctylpyrophosphate)titanate

Zirconium propionate is used as an adhesion promoter for printing inks on polyolefins, but zirconium compounds are also used to reduce the stickiness of particles of adhesives in the reclamation of waste paper.

Other coupling agents

Other types of coupling agents include 1,2-diketones for steel,[3] nitrogen heterocyclic compounds for copper,[4] and some cobalt compounds for the adhesion of brass-plated tyre cords to rubber.[5]

Other aspects of coupling agents are described under **Coupling agents – application** and **Fibre matrix adhesion – glass fibre**.

References

1. J Comyn in *Structural Adhesives; Developments in Resins and Primers*, ed. A J Kinloch, Elsevier Applied Science Publishers, 1986, Ch. 8.
2. P D Calvert, R P Lalanandham, D R M Walton, in *Adhesion Aspects of Polymeric Coating*, ed. K L Mittal, Plenum Press, 1983, p. 457.
3. A J Nicola, J P Bell in *Adhesion Aspects of Polymeric Coatings*, ed. K L Mittal, Plenum Press, 1983, p. 443.
4. S Yoshida and H Ishida, *J. Adhesion*, **16**, 217 (1984).
5. W J van Ooij, M E F Biemond, *Rubber Chem. Tech.*, **57**, 688 (1984).

Creep

R D ADAMS

The phenomenon of creep involves the continuous deflection of a material with time under the action of a constant load. All materials experience creep to a greater or lesser extent. For polymers, significant creep only occurs when they are near their **Glass transition temperature**. If an adhesive is exposed to hot and wet conditions, it will absorb moisture and become plasticized: this causes its glass transition temperature to be reduced.

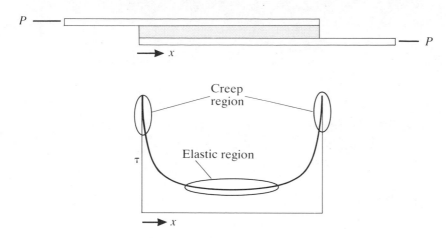

Fig. 25. Variation of shear stress τ with distance x along a lap joint loaded in shear

Creep behaviour is generally non-linear in that 90% of the deflection of a material often occurs in the last 10% of its creep lifetime.

There are few data available for creep of adhesives, and much of what exists is summarized in Adams and Wake.[1]

The general design rule for situations where creep may occur is to use a longer than normal lap-shear joint (see **Shear tests, Stress in joints**). At the joint ends, creep will occur under the high stresses which are there, but the low stress region in the middle of the joint (see Fig. 25) will provide an elastic buffer zone which will tend to hold the joint together.

Appendix 2 gives some standard test methods.

Reference

1. R D Adams, W C Wake, *Structural Adhesive Joints in Engineering*, Elsevier Applied Science Publishers, London, 1984.

Critical surface tension

D E PACKHAM

The concept of critical surface tension of wetting was introduced by Zisman as a method of characterizing the nature of polymer surfaces.[1] It involves **Contact angle measurement** of a series of liquids on the solid surface of interest. Zisman noticed that if the cosine of the contact angle (cos θ) was

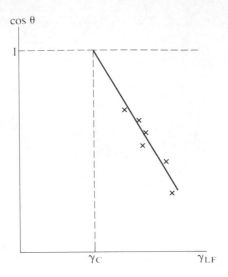

Fig. 26.   Graph of cosine of contact angle θ for a series of liquids of surface tension γ_{LF} on a given solid surface showing evaluation of critical surface tension γ_C

Table 10. Values of critical surface
tension in mJ m^{-2}.[1]

Polymer	γ_C
PTFE	18.5
Polyethylene	31
Polyvinyl alcohol	37
PMMA	39
Nylon 66	46

plotted against the surface tension of the liquid (γ_{LF}) a straight line, or 'narrow rectilinear band', resulted (Fig. 26). The line was extrapolated to $\cos \theta = 1$, and the corresponding value of γ_{LF} was taken as the critical surface tension, γ_C; γ_C then was the surface tension of the liquid that would just spread (see **Wetting and work of adhesion**) on the solid surface. Early work suggested that its value was characteristic of the solid and independent of the liquids used to establish it. Zisman regarded it as an empirical parameter 'whose relative values act as one would expect of the specific surface free energies of the solids'[1] – values in Table 10 illustrate this point. Less cautious authors have called γ_C the surface energy of the solid: this is incorrect.

Refinement of the concept

The availability of a wider range of experimental results, and advances in the theory of polymer wettability lead to a refinement of the concept of critical surface tension. Dann[2] noted a pronounced curvature of some of the $\cos \theta$ vs γ_{LF} plots, and the tendency for different liquids to give somewhat different values of γ_C. He explained these points as follows.

The theoretical relationship between $\cos \theta$ and γ_{LF} for a non-polar solid, derived in **Surface energy components** (Eqn 8) can be written

$$\cos \theta = -1 + \frac{2(\gamma_S \gamma_L^d)^{1/2}}{\gamma_{LF}} \qquad [1]$$

It is clearly not linear. If $\cos \theta$ is made unity in this equation, γ_{LF} becomes (by definition) the critical surface tension, giving

$$\gamma_C = (\gamma_L^d \gamma_S)^{1/2} \qquad [2]$$

Two cases can be distinguished, use of a non-polar liquid and of a polar one. For a non-polar liquid γ_L^d equals the surface tension, which at the condition of $\cos \theta$ of unity, will be the critical surface tension. Thus

$$\gamma_C = \gamma_S \qquad [3]$$

However, for a polar liquid Eqn 2 applies: inspection shows that this predicts a lower value of γ_C when a polar liquid is used.

Further insight comes from consideration of the **Good–Girifalco interaction parameter** theory which gives the following equation:

$$\gamma_S = \gamma_{LF}(1 + \cos \theta)^2/4\phi^2 \qquad [4]$$

When the critical condition of $\cos \theta$ being unity is put in, it can be seen that

$$\gamma_C = \phi^2 \gamma_S \qquad [5]$$

This too predicts that γ_C will depend on the nature of the liquids used, as this will influence the value of ϕ. The critical surface tension will only be the same as the surface energy of the polymer if the interaction parameter is unity: this can occur with a non-polar liquid on a non-polar solid. More usually the value of ϕ is less than one.

Conclusion

Despite these considerations, the critical surface tension is an empirical parameter, often quoted and relatively easy to measure, which helps to characterize the surface of a low-energy solid such as a polymer. Values of this are available from review articles[3] and Bandrup and Immergut's *Polymer Handbook*.[4] A much-simplified practical routine based on the concept is used in ASTM D-2578 to test the surface of plastic film; see **Surface characterization by contact angles – polymers**.

References

1. W A Zisman in *Contact Angle, Wettability and Adhesion*, ed. R F Gould, Advances in Chemistry Series No. 43, American Chemical Society, 1965, p. 1.
2. J R Dann, *J. Colloid Interf. Sci.*, **32**, 302, 321 (1970).
3. D Briggs, D G Rancee, B J Briscoe in *Comprehensive Polymer Science*, Vol. 2, ed. G Allen, Pergamon, Oxford, 1989, p. 715.
4. J Bandrup, E H Immergut, *Polymer Handbook*, 3rd edn, Wiley–Interscience, 1989.

Cyanoacrylate adhesives

J GUTHRIE

Cyanoacrylate adhesives are rapid-curing 'one-part' adhesives based on alkyl-2-cyanoacrylate monomers (I):

$$CH_2{=}C \begin{matrix} \diagup CN \\ \diagdown \\ C{=}O \diagup \\ RO \end{matrix}$$

I

   A brief description of the chemistry of these materials is described in the article entitled **Alkyl-2-cyanoacrylates**. Commercial adhesive products are usually based on the ethyl ester but methyl, n-butyl, allyl β-methoxyethyl and β-ethoxyethyl are also important. Formulated adhesives consist of essentially pure monomer with relatively small amounts of property-modifying additives. The curing reaction is anionic polymerization, initiated by traces of alkaline material present on most substrate surfaces, particularly in conjunction with low levels of surface moisture (see **Addition polymerization**).

Additives

As already stated, formulated cyanoacrylate adhesives contain low levels of additives which improve the performance profile of the product. They can be divided into two general groups: those which modify the polymerization process and those which alter the properties of the final polymer.

Stabilizers The major difficulty associated with the manufacture of cyanoacrylate adhesives is ensuring a balance between stability of the

Table 11. Anionic polymerization inhibitors

Sulphur dioxide	Sulphamides
Sulphur trioxide	Cationic exchange resins
Sulphonic acids	Boric acid chelates
Sulphones	

product and cure speed. This problem has been solved by careful choice of anionic polymerization inhibitors. The materials employed are acidic compounds present at levels between 5 and 100 ppm. Table 11 shows a list of typical additives.

Free radical polymerization inhibitors are also added. These are phenolic compounds such as hydroquinone or hindered phenols.

Accelerators These increase the rate of polymerization. They should not be confused with polymerization initiators as they are not sufficiently nucleophilic to induce polymerization. The compounds described in the literature have one common feature, namely they are all capable of sequestering alkali metal cations. The mechanism by which accelerators function is not clear, but it is believed to involve either increasing ion separation at the growing chain end or activation of anions on the substrate by cation sequestration to give so called 'naked' anions in the liquid adhesive. Examples of compounds used as accelerators are crown ethers (II), polyalkene oxides (III), podands (IV) and calixarenes (V).

$$RO-CH_2CH_2O(CH_2CH_2O)_n-R$$

III

II

IV

X = multifunctional moiety

V

R = alkyl

n = 4, 6, 8

These types of accelerator are particularly effective on porous substrates such as wood and paper.

Adhesion promoters The patent literature describes the use of carboxylic acids and anhydrides as adhesion promoters on metallic substrates. It is assumed that the carboxylic acid group is able to complex with the metal surface and that some degree of copolymerization takes place. There is, however, little or no experimental evidence to substantiate copolymerization. The addition of these acidic materials may also result in a reduction in cure speeds.

Plasticizers. These are required to reduce the inherent brittleness of poly(alkyl-2-cyanoacrylates). This can be achieved by using non-copolymerizing plasticizers such as esters or higher alkyl cyanoacrylates which copolymerize with the basic adhesive monomer. Toughness properties can be improved by the inclusion of rubber toughening materials such as ABS (acrylonitrile–butadiene–styrene) or MBS (methacrylate–butadiene–styrene) copolymers. Whichever approach is adopted, toughness is only achieved at the expense of reduced cure speed.

Environmental performance

The durability (see **Durability – fundamentals**) of cyanoacrylate adhesive bonds is reasonably good on rubbers and some polymer substrates. However, on glass and metals, both thermal and moisture durability are low.

Heat resistance This can be improved by including additives in the formulation which gives rise to a more thermally stable polymer. Examples of this approach are the use of biscyanoacrylate and bismaleimide cross-linking agents. Loss in performance may also be attributed to a temperature-induced loss in adhesion. Phthalic anhydride is believed to act as a high-temperature adhesion promoter.

Moisture resistance This may be increased on metal and glass substrates by including cross-linking agents which may yield a more hydrolytically stable polymer or by using hydrophobic monomers such as fluorinated cyanoacrylates. Silane **Coupling agents** also improve moisture durability. There is also evidence to suggest that inclusion of some of the anhydrides described above has a beneficial effect.

Miscellaneous

Other modifications that can be made to cyanoacrylate adhesives include increasing viscosity by the addition of thickeners such as polymethyl methacrylate, cellulose esters or hydrophobic silicas. Colour can be imparted to the product by using selected dyes and pigments.

Applications

Cyanoacrylate adhesives will bond a wide variety of substrates with the exception of polyolefins (unless pretreated), Teflon and highly acidic surfaces. Porous substrates such as wood, paper and leather require the use of products containing accelerators. Formulations are now appearing which when used in conjunction with a so-called primer can give high bond strength on polyethylene and polypropylene. See **Industrial applications of adhesives**.

As with any adhesive, surface preparation is important. **Pretreatment of metals** is most easily achieved by solvent degreasing and grit-blasting. **Pretreatment of polymers** usually involves cleaning in a non-solvent and optional surface abrading. Glass and ceramics require surface cleaning and drying.

Advantages of cyanoacrylate adhesives include rapid curing, solvent free, high bond strength, versatility, solvent resistant and their comprising only one part.

Disadvantages of cyanoacrylate adhesives include poor thermal and moisture resistance on metals and glass, brittleness, sensitive to surface preparation, poor cure 'through gap'. In addition there can be difficulties in handling because of the danger of bonding to the skin (see **Health and safety**).

Cyanoacrylates and acrylates are compared in the article on **Acrylate adhesives**.

Select references

H W Coover, J M McIntire in *Handbook of Adhesives*, 2nd edn, ed. I Skeist, Van Nostrand Reinhold, New York, 1977, p. 569.
G H Millet in *Structural Adhesives*, ed. S R Hartshorn, Plenum Press, New York, 1986, Ch. 6.
D C Pepper, *J. Polym. Sci., Polym. Symp.*, **62**, 65 (1978).

D

Degreasing

J F WATTS

Most metal surfaces arrive from the stockholder or subcontractor with a thick layer of grease on the surface. This is often the result of mechanical forming processes which invariably rely on some form of lubricant which, although very specific at the time of processing, can subsequently be described by the generic term 'grease'. It may also have been applied as temporary protection against corrosion (see **Engineering surfaces of metals**).

Degreasing can also be important in procedures such as maintenance painting where plant which has become soiled in use must be degreased. Prior to pretreatment or painting such grease residues must be removed, or at least reduced to an acceptable level. In general, degreasing processes can be classified according to the size of the article to be cleaned (i.e. will it readily fit in a tank for immersion cleaning?), and thus whether a dip or a spray process is more applicable, and the nature of the cleaning solution employed, which may be an organic solvent, an alkaline solution, or an emulsion of organic and aqueous solutions. All cleaning methods are improved by additional agitation which may take the form of scrubbing a large article or using stirring or ultrasonic agitation in the case of a bath.

An indication of the wide variation in degreasing performance as a function of degreasing medium and the level of agitation is provided by the data of Table 12.[1]

Solvent cleaners

Organic solvents are effective degreasing agents in both the vapour and liquid phase. This has led to the design of degreasing tanks in which the solvent is heated to provide a vapour blanket above the liquid; particles to be cleaned are held for a short time in the vapour phase prior to

Table 12. Efficiency of degreasing as a function of process

Degreasing method	Cleaning efficiency (%)
1. Pressure washing with detergent solution	14
2. Mechanical agitation in petroleum solvent	30
3. Vapour degreasing in trichloroethylene	35
4. Wire brushing in detergent solution	92
5. Ultrasonic agitation in detergent solution	100

immersion in the liquid. The choice of solvent has been revised dramatically in recent years; a decade ago chlorinated hydrocarbons such as trichloroethane, trichloroethylene and perchloroethylene were widely recommended as degreasing solvents for metals. For both environmental and **Health and safety** reasons this practice has been largely superseded and a range of proprietary solvents have been successfully introduced into the market-place with improved environmental and safety properties. In situations where chlorinated solvents are still being employed they are operated in closed systems rather than with fumes being vented directly to atmosphere.

The effect of ultrasonic agitation on the cleaning process is impressive and the cleaning efficiency can be improved from 10% in a still solution at ambient temperature to 85%.[2] In the cleaning of larger articles it is necessary to resort to wiping or scrubbing. In these cases the problem is that of carry-over of grease from one article to another, and this can only be avoided by frequent replacement of pad or brush.

Alkaline cleaners

Alkaline cleaners present an attractive alternative to solvent cleaners, particularly in the case of very heavy grease deposits, without the problems of toxic vapours. They are used hot, the principle of an alkaline cleaner being the saponification of the grease layer by the alkali to produce carboxylate salts. Once this has occurred the reaction products are removed from the surface by emulsification, peptization (to prevent redeposition of grease on the fresh surface) and subsequent dissolution.[3] In addition to the alkali itself other chemicals are included in the formulation to aid wetting and emulsifying (trisodium phosphate for example) and deflocculating (alkali silicates). Alkaline cleaning can be carried out by either a spray or a dip process, but once again agitation plays an important role. The cleaning efficiency of a 5% solution of alkaline cleaner is very poor at 18% in a still solution for 24 min, but can be increased to 64% by stirring, and increased dramatically to 93% in 5 s when ultrasonic agitation is employed.[4]

Alkaline cleaners, particularly in the metal finishing industry, are sometimes used in conjunction with an electric potential; this process is referred to as electrolytic degreasing. If the workpiece is made the anode oxygen is evolved which helps loosen grease, and may dissolve the metal itself slightly, depending on the potential involved. The cathodic version of the process is more efficient and can be adjusted to provide electrochemical reduction of the surface oxide, but the cathodic production of hydrogen can cause problems of embrittlement in some steels. A satisfactory alternative is to reverse the polarity during processing; 60 s cathodic, followed by 30 s anodic, then 20–50 s cathodic.

Emulsion cleaners

The term 'emulsion cleaners' generally refers to a commercial formulation, applied directly to the soiled article, which consists of a mixture of organic and aqueous phases, they are then rinsed off with water as an emulsion of grease, water and solvent. Although they can be used in either a spray or bath process they find their widest use in maintenance cleaning of plant. Water itself is also used for this process in the guise of the ubiquitous steam-cleaning system used for removing gross amounts of soil of all types from engineering plant. Detergent solutions may also be used, but are less effective on greasy surfaces than alkaline or emulsion cleaners.

Efficacy of degreasing

In the assessment of a degreasing process gravimetric or optical methods are generally employed to determine the extent to which oils and greases have been been removed. However, such methods are very insensitive when it comes to defining a few monolayers of hydrocarbon that may remain on the surface. The question of the exact composition of metals on an atomic scale following degreasing is addressed in **Engineering surfaces of metals**.

References

1. T J Bulat, *Metal Progress*, **68**, 94–5 (1955).
2. R Walker, C L Carter, *Product Finishing*, **37** (2), 24–8 (1984).
3. R Walker, N S Holt, *Product Finishing*, **33** (4), 14–20 (1980).
4. N S Holt, R Walker, *Product Finishing*, **33** (7), 12–16 (1980).

Diffusion theory of adhesion

K W ALLEN

The diffusion theory of adhesion is basically a very simple concept, the original use of which is due to Voyutskii.[1] He was concerned with the self-adhesion (**Autohesion**) of unvulcanized (not cross-linked) rubber (see also **Rubber adhesion**). He suggested that if two such polymer surfaces are in sufficiently close contact, parts of the long-chain molecules will diffuse across the interface. They will interpenetrate and eventually the interface will disappear and the two parts will have become one. It is clear that if this is to happen the molecules must be relatively mobile. This means that there must not be any considerable degree of cross-linking and that the polymer must be above its **Glass transition temperature**.

This original idea was developed into a more quantitative form by Vasenin,[2] who applied the general theories of the diffusion of liquids to this situation. He began by considering the case where one component was oligomeric and the other was liquid, and extended this first to an oligomer and then to a polymer diffusing into itself. Eventually he continued to the case of two different polymers interdiffusing.

The mathematics involved in all these considerations is particularly complicated and is not exact, involving several empirical factors. However, Vasenin produced expressions for the force necessary to separate two polymer surfaces.

For the autohesion of a polymer to itself this is

$$F = 11v \left\{ \left(\frac{d(2+p)}{M} \right)^{2/3} K^{1/2} \right\} rt^{1/4} \qquad [1]$$

where F is the peeling force necessary to separate the two surfaces, v the vibrational frequency of a —CH— group, d the density of the polymer, p the number of chain branches in the molecule, M the molecular weight, r the rate of separation of the two surfaces, t the time of contact before testing and K a constant, characteristic of the diffusion molecule. Thus it emerges that this force is *directly* proportional to the rate of separation, and to the quarter power of the time for which the surfaces have been in contact; and *inversely* proportional to the two-thirds power of the molecular weight of the diffusing molecule.

Support for Eqn 1 is given by the agreement between theory and experiment shown in Figs 27 and 28 where the points represent experimental results, the lines theoretical predictions.[3] In Figs 27 and 28(b) the $F \propto t^{1/4}$ relationship of Eqn 1 has been fitted to the points as closely as possible. The line in Fig. 28(a) was obtained from that of Fig. 28(b) by scaling it according to the molecular weight difference using the $M^{-2/3}$ relationship.

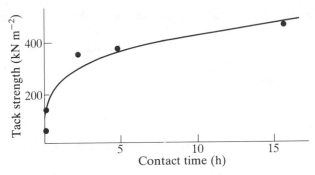

Fig. 27. Dependence of tack strength (kN m^{-2}) upon contact time (h) between nitrile rubber and natural rubber. The dots represent experimental measurements; the line is derived from Eqn 1

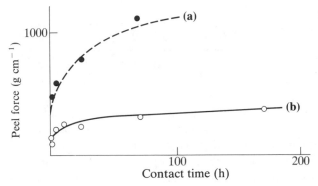

Fig. 28. Dependence of peel force (g cm^{-1}) upon contact time (h) for polyisobutylenes of molecular weight (a) 100 000 g mol^{-1} and (b) 150 000 g mol^{-1} peeled from cellophane. Dots are experimental points, lines are derived from Eqn 1

These concepts and their development have all been criticized quite sharply, but much of this criticism has been directed to claims which were never made by the originators of the theory. Later protagonists have attempted to extend the method into wider applications without adequate justification. The whole concept is concerned only with adhesion involving polymer materials and should never be extended beyond that limitation.

Much of the original work on this theory was published in Russian which renders it less than easily accessible, but a particularly helpful review is given by Wake.[3]

More recently with the development of ideas about the interpenetration of adhesives into the microstructure of metallic surfaces there has been suggestion that the mechanism involved could be regarded as one of diffusion; but that is remote from the ideas considered here.

A comparison of different theories is given under **Theories of adhesion**. Further results supporting interdiffusion are discussed under **Autohesion**, see also **Solvent welding**.

References

1. S S Voyutskii, *Autohesion and Adhesion of High Polymers*, trans. S Kaganoff, ed. V Vakula, Interscience, New York, 1963.
2. R M Vasenin, *RAPRA Translations 1005, 1006, 1010, 1075*, by R Moseley, RAPRA, Shrewsbury, 1960–63; *Adhesives Age*, **8**, 21, 30 (1965).
3. W C Wake (ed.) in *Adhesion and the Formulation of Adhesives*, 2nd edn, Applied Science Publishers, London, 1982, Ch. 5.

Dispensing of adhesives

C WATSON

Introduction

In many industrial situations, the use of adhesives has advantages over older methods of joining (see **Engineering practice, Engineering design with adhesives, Industrial applications of adhesives**), but there is sometimes a reluctance to use them because of supposed difficulties in application. They are perceived to be messy and incapable of being automatically dispensed on high-volume production lines. Modern technology has undermined the validity of both these perceptions. Effective dispensing methods are available for most adhesives, despite their diversity, which may be of chemical type (see **Adhesive classification**), of physical form (liquid, paste or solid) or a number of components. Cure times can also vary from seconds (**Cyanoacrylates**) to hours (some **Epoxide adhesives**).

A major requirement in industry is consistent quality of a manufactured article over long production runs. Accurate dispensing of the adhesive on to components to be bonded is therefore essential.

Basic dispensing principles

Dispensing involves a combination of several functions depending upon the dispensing principle used. The functions relate to the control of the 'bulk' product to supply energy to move it through flow pipes to a valve or dispenser nozzle. Control must be exerted over the valve to vary opening time, therefore quantity dispensed and over the speed of movement of the dispenser nozzle relative to the component surface. All adhesives and sealants will vary in viscosity, mainly due to the variation of ambient temperature at the point of use. In order to satisfy the need of the manufacturing process, the dispenser should be so designed that variations in viscosity will not affect the accuracy or consistency of the quantity dispensed.

Two of the most difficult types of adhesives to dispense are **Cyano-acrylate** and **Anaerobic adhesives**. Cyanoacrylates cure by interaction with the surface moisture on a component part. Anaerobics remain liquid in the presence of oxygen, but automatically cure once enclosed in a joint.

The major methods of dispensing anaerobic adhesives use pressure–time controlled valves, a cartridge and syringe system and screen printing.

Other types of adhesive may be used with some of the combinations of the above systems; the major points for consideration to be taken into account are the curing characteristics of the product, the rheology and the speed of flow required. Figures 29–33 show some typical production applications.

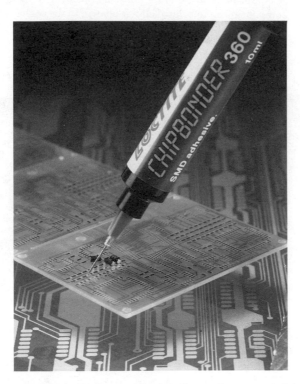

Fig. 29. A syringed adhesive used for bonding surface-mounted devices on to printed circuit boards

Fig. 30. A typical installation of an automatic machine for placement of components on to printed circuit boards. In the centre can be seen the mounted syringe for automatic application of the adhesive

Fig. 31. The bench use of a dispenser for the application of cyanoacrylate to an external light switch to bond the cover to the switch base

Fig. 32. A rotospray installation for the automatic application of an anaerobic to three bores of an engine block for the retention of cylinder liners

Fig. 33. The hand screen printer and a component on to which an anaerobic gasket has been printed for Jaguar

Pressure–time controlled system

Pressure is applied to bulk liquid and timer-controlled valves meter the
quantity dispensed. Figure 34 shows a typical pressure–time system for
adhesives where the viscosity is up to approximately 100 000 mPa.
Pressurization of the product tank pushes the adhesive through a feed
line to a dispensing valve. A timer controls the opening time of the
dispenser valve to obtain a metered quantity. Regulated pressure behind
the product and the valve opening time control the quantity dispensed.
The adhesive can be applied in dots or in bead form.

   There are a number of valves that can be used with the product tank
and control console. These are given under the headings below.

Static pinch valve The feed tube from the product tank passes underneath
the pneumatic pinch cylinder. When the valve is closed the pneumatic
cylinder closes the tube. When the cylinder opens, the product flows
through the tube to the dispenser nozzle.

   A static pinch valve can be mounted on a bench station, or alternatively,
on an automatic line. A typical bench application would be bonding the

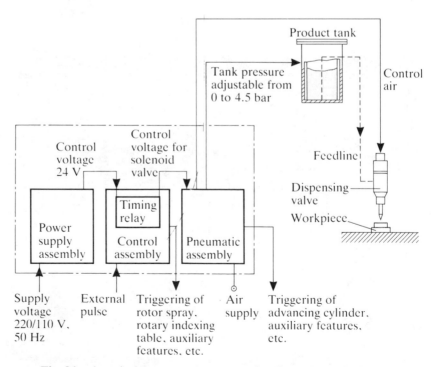

Fig. 34. A typical pressure–time system for dispensing adhesive

ends of a rubber extrusion to make up car door seals. In the case of an automatic line, a free-fall drop could be dispensed on to a component part located beneath the static pinch valve or, alternatively, a bead applied to a component part as it traverses beneath the pinch valve nozzle.

Pressure time using advancing valve The advancing pinch valve is utilized in operator-paced or fully automatic lines where the component part itself cannot be moved to the valve nozzle. The advancing pinch valve, which can be between a stroke of 25 and 150 mm (1 and 6 in), advances the application nozzle to the component part. The dispenser valve then opens allowing a metered quantity of product to flow through the valve nozzle on to the component part. The valve then closes and the advancing cylinder retracts from the nozzle. In this way dots or continuous beads of product may be applied.

Pressure time using a Rotospray unit This unit is used to apply anaerobic products to machined bores. The unit consists of an electric motor with an integral dispenser valve. At the end of the motor shaft there is a plastic cup around the periphery of which are a number of holes.

The principle of operation is that the unit is advanced into the bore to be coated, the motor turns, the valve opens and dispenses a metered quantity of product into the Rotospray cup. The product is then thrown out centrifugally through the holes in the periphery of the cup on to the bore.

Typical applications for this dispensing system would be to coat core plug holes in an engine block, electric motor end caps for the retention of bearings and the engine block for the retention of cylinder liners.

Cartridge/syringe dispensers

There are two basic methods of dispensing from a cartridge/syringe. The first is to apply an air pulse on the back of the cartridge piston, thus, dispensing a certain volume of product through the dispenser nozzle. The quantity of product dispensed using pulsed air can vary with variation of air pressure. The more complex air-pulsed systems incorporate compensating devices continually to monitor and adjust the pressure.

The second principle is to have a mechanically driven piston so that an equal volume is dispensed with each cycle. Typical applications are for dispensing surface-mounted device adhesives on to printed circuit boards where up to 20 000 dots per hour are dispensed, the dispensing of anaerobic gasketing products using a robotic head, the dispensing of gel cyanoacrylates for bonding rubber gaskets to mouldings and castings.

Screen printing of adhesives

In the **Automotive industry** there is a trend towards complete automation of production lines. In the application of adhesives/sealants for gasketing, the principles of dispensing have normally been to use a pressure–time system, or alternatively a cartridge system with a robotic head. However, a faster system is to screen print an anaerobic on to the surface of a component.

Here the basic principle is to take a polyester screen and reproduce the pattern of the gasket required from the screen. Anaerobic adhesive is dispensed in metered quantities on to the screen. The component part is located below the screen, the screen advances on to the component part or the component is raised up to the screen. A squeegee traverses across the screen, automatically pushing the anaerobic product through the pattern on to the component part. Thus completely automatic systems are being utilized to gasket component parts replacing conventional preformed manually placed gaskets.

It is important that the potential end user discuss his specific need with the adhesive supplier, who should be able to advise on the use of their own tailor-made or commercially available equipment to suit manufacturing process needs.

Select references

G Defrayne, *High Performance Adhesive Bonding*, Society of Manufacturing Engineers, Dearborn, Michigan, 1983.
C Watson, *Des. Prod. and Appl. J.*, **11**(9), 83, 1989.
M M Sadek (ed.), *Industrial Applications of Adhesive Bonding*, Elsevier Applied Science Publishers, London, New York, Basle, Ch. 4.
G Havilland, *Machinery Adhesives for Locking, Retaining and Sealing*, Marcel Dekker, New York, Basle, 1986.

Dispersion forces

K W ALLEN

The origins of forces between neutral, symmetrical molecules, such as the hydrogen H_2 or the inert gases, is not obvious. Because of the symmetry of the electron configuration there cannot be any permanent dipole; so there can be neither dipole–dipole interaction (Keesom orientation interactions) nor dipole–molecule interaction (Debye induction interactions) (see **Polar forces**). Further, there can be no Coulombic electrostatic interaction since they are electronically neutral overall. Yet there must be forces between these molecules as the existence of liquid and solid hydrogen and argon demonstrate.

This situation was first discussed and the beginning made in explaining it by F. London in 1930. While the theory is essentially based on quantum mechanics, a plausible and helpful explanation can be given in classical terms as follows.

While a neutral molecule has an entirely symmetrical electron distribution when considered as an average over a period of time, and hence no dipole moment, when it is considered at one instant the state is different. At a single instant the electrons will have a definite configuration which is most unlikely to be symmetrical so that the molecule will have an instantaneous dipole moment. This instantaneous dipole moment will induce a corresponding dipole moment in another, adjacent molecule. There will now be an interaction between these two molecules which will result in a force of attraction between them. The observed force is the result of this instantaneous force averaged over all the instantaneous configurations of the first molecule.

For isolated molecules of polarizabilities α_1 and α_2, and ionization energies I_1 and I_2, separated by a distance r, the energy of interaction is given by

$$U = -\frac{3}{2}\frac{\alpha_1\alpha_2}{r^6}\cdot\frac{I_1 I_2}{I_1 + I_2} \qquad [1]$$

The negative sign is simply a conventional way of indicating that the potential energy U is attractive. The equation is often written as

$$U = -A/r^6 \qquad [2]$$

where A is called the van der Waals dispersion attraction constant.

From these equations it can readily be seen that the attraction constant A_{12} between different types of molecule 1 and 2, is the geometric mean of A_{11} and A_{22}, the constants for interaction between molecules of the same type providing the arithmetic and geometric means of the ionization energies can be assumed to be the same:

$$A_{12} = (A_{11} A_{22})^{1/2} \qquad [3]$$

If the molecule becomes very close their electron orbitals will start to become compressed, leading to a repulsion energy which rises very rapidly as the separation decreases further. When this is taken into account the complete potential energy curve is given by the Lennard-Jones potential:

$$U = -A/r^6 + B/r^{12} \qquad [4]$$

the second term represents the repulsion and B is a constant. (An analogous expression applies for **Polar forces**.)

Table 13. Contribution of different types of van der Waals force to the interactive energies of some molecules

Substance	Dipole moment (D)*	Interaction energies (kJ mol^{-1})		
		Dipole/dipole interaction	Dipole induced interaction	Dispersion interaction
Ar	0	0	0	−1.1
CO	0.11	-4×10^5	-8×10^{-4}	−1.3
HCl	1.07	−0.2	−0.07	−1.8
NH$_3$	1.47	−6.2	−0.9	−12.9
H$_2$O	1.86	−16.1	−0.9	−5.3

* 1 D − 1 Debye unit = 3.336×10^{-30} C m.

Clearly this situation will apply to all molecules irrespective of their other characteristics, so that this force of attraction is universal: it acts irrespective of whatever other more specific interactions may be involved.

Incidentally, it has been shown that the force arising from this induced dipole is far greater than that due to the two separate instantaneous dipoles because the induced effects will be synchronized but the separate dipoles will not (see Table 13).

The title 'dispersion' arises because in the more theoretical development of this concept an important constant, characteristic of the molecule, was invoked. This constant was proportional to the electrostatic polarizability, which in turn is itself proportional to the optical dispersion of the refractive index with frequency, hence the title. However, since this word 'dispersion' can be misleading, the name of London (who was the original exponent of this theory) is frequently included in the title to give 'London dispersion forces'.

Casimir and Polder in 1948 showed that London's theory must be modified if the molecules were some distance (in molecular terms) apart. This is because the electrostatic forces are not propagated instantaneously but take a finite, if small, time to pass between the two molecules. The net result of such 'retarded' dispersion forces is that the energy of interaction between two particles at distances apart greater than ~ 200 Å is proportional to r^{-7} instead of to r^{-6}.

The relative significance of some of these interactions are shown in Table 13. These dispersion interactions are of considerable importance in adhesion and are frequently the major factor in the strengths of bonds, as is discussed under **Adsorption theory** of adhesion. They also play a significant role in surface phenomena generally, such as **Contact angles and interfacial tension**.

Select references

M Rigby et al. in *Forces Between Molecules*, ed. M Rigby, E B Smith, W A
  Wakeham, G C Maitland, Clarendon Press, 1986, Ch. 1.
T B Grimley in *Aspects of Adhesion*, ed. D J Alner, K W Allen, Transcripta Books,
  London, 1973, pp. 11–29.

Displacement mechanisms

D M BREWIS

Solids will normally be covered by a thin layer of organic materials of low cohesive strength. Freshly formed high-energy surfaces, for example of metals, will rapidly adsorb organic compounds from the atmosphere (see **Surface energy**). On other occasions metals will be covered deliberately by a fairly thick layer of protective oil or grease. Polymers will normally be covered by a layer of an additive which has migrated to the surface, or by a mould release agent, or by contamination subsequent to fabrication (see **Compatibility, Weak boundary layers**).

Sometimes these organic layers will be removed, or at least reduced, by a **Pretreatment**, but on many occasions the substrates will be used 'as received'. With suitable selection of adhesives, paints and printing inks, good adhesion can often be achieved with these 'as-received' surfaces. For example, oily steel can be bonded satisfactorily with some **Acrylic adhesives**, hot curing **Epoxide adhesives** and **Solvent-based adhesives**, see **Selection of adhesives**.

The effect of deliberately contaminating substrates has been studied. Brewis[1] spread a layer of petroleum (*c.* 2×10^3 nm thick) over HDPE which had been etched with chromic acid. The effect on joint strength is shown in Table 14. The fact that most of the beneficial effect of the pretreatment is retained indicates that most, if not all, the petroleum jelly has been absorbed by the epoxide adhesive.

Table 14. The effect of a layer of petroleum jelly
($c.$ 2×10^3 nm thick) on the lap shear strength of
HDPE bonded with epoxide adhesive[1]

Surface	Joint strength (MPa)
(a) Untreated HDPE	1.8
(b) HDPE treated with chromic acid	17.6
(c) As (b) plus contaminant	12.7

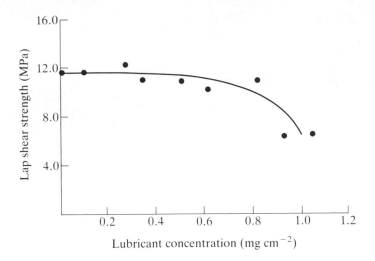

Fig. 35. The effect of lubricant concentration on the lap shear strength of
aluminium joints bonded with a one-part epoxide at 177 °C for 2 h
(Redrawn from ref. 2 with permission)

Minford[2] studied the effect of contaminating aluminium with a
commercial lubricant. The results are summarized in Fig. 35. It was found
that the lubricant had little effect on the adhesion level until its thickness
exceeded about 10^4 nm. The adhesion level then fell sharply and the mode
of failure (see **Locus of failure**) changed from cohesive within the adhesive
to apparent interfacial failure. Thus the epoxide, which required a fairly
long cure at a high temperature, was able to absorb quite large quantities
of the lubricant.

Whether contaminants are absorbed by the mobile phase will depend
on a number of factors, including the following:

1. Quantity of contaminants;
2. Chemical nature of contaminant and the mobile phase, i.e. their
    compatibility;
3. Temperature;
4. Time mobile phase is in its low-viscosity state.

References

1. D M Brewis, *J. Mater. Sci.*, **3**, 262 (1968).
2. J D Minford, *Aluminium* (Dusseldorf), **57** (10), 657 (1981).

Durability — fundamentals

A J KINLOCH

Introduction

One of the most important requirements of an adhesive joint is the ability to retain a significant proportion of its load-bearing periods of time under the wide variety of environmental conditions that are likely to be encountered during its service life. The problem of environmental attack has been discussed in depth by Brewis,[1] Kinloch[2,3] and Venables,[4] and these authors show that one of the most hostile environments for joints is water (see **Weathering**).

Locus of joint failure

When initially prepared the locus of failure of adhesive joints is usually by cohesive fracture in the adhesive layer, or possibly in the substrate materials. However, a classic symptom of environmental attack is that after such attack the joints exhibit some degree of apparently interfacial failure between the adhesive (or primer) and the substrate. However, a visual assessment of the fracture surfaces of a joint which has been subjected to environmental attack does not indicate exactly the **Locus of failure**, and techniques such as **Auger and X-ray photoelectron spectroscopy** are needed to identify precisely the locus of failure (see **Surface analysis**).

General aspects

The lack of durability of an adhesive joint is usually connected with the ingress of an attacking medium, such as water vapour or liquid. Therefore, as expected, the concentration of the medium plays an important role in determining the rate and extent of attack. Indeed, there often appears to be a minimum concentration below which no durability problems are observed. Increasing the temperature of the environment generally increases the rate of strength loss. The chemical type of adhesive employed can affect both the mechanism and rate of environmental failure. Turning to the substrate employed, then joints to high-energy substrates such as metals and glasses usually present the main problem. The substrate surface pretreatment used is a most important factor and to produce durable joints the choice of surface **Pretreatment** is often crucial. The effect of an applied stress on the durability of the joint is a complex subject. In some instances the presence of an applied stress together with water may cause a sharp crack to become blunt, and so help to prolong the lifetime of the joint. In other instances an applied stress may render interfacial molecular bonds more susceptible to attack and so reduce the durability of the bonded join. It does appear that **Fatigue** loads are particularly damaging, possibly since they cause the crack to resharpen continually as it grows under the repeated fatigue loads. Finally, as will be appreciated from the above comments, tests to ascertain how long a joint may be expected to perform satisfactorily in a given hostile environment are not well established and, currently, long-term testing is usually needed to give the designer confidence regarding the durability of bonded structures (see **Accelerated ageing**).

Mechanisms of attack

The interfacial regions have been identified as the region where an attacking medium may damage the adhesive joint and various mechanisms have been postulated to explain the experimental observations. These include (1) a loss of interfacial bonding due to a thermodynamic displacement process, (2) a loss of stability of the oxide layer to which the adhesive is bonded, for example due to hydration of the oxide, and (3) corrosion of the substrate.

Stability of the interface

The concept of work of adhesion, W^A (see **Contact angles and interfacial tension**) has been extended[2,3] to give a means of predicting the stability of an interface in the presence of liquid L. The term W_{AL} is the work of adhesion in the presence of liquid L and is essentially the difference in free energy per unit surface area between materials 1 and 2 in contact at a common interface (i.e. 'bonded') and the same materials separated along the interface, but in an environment of L. Thus W_{AL} is defined as

$$W_{AL} = \gamma_{1L} + \gamma_{2L} - \gamma_{12}$$

where γ_{12} is the interfacial energy between 1 and 2, γ_{1L} and γ_{2L} are the surface energies of 1 and 2 respectively in the presence of L (see **Surface energy**). The article, **Contact angles and interfacial tension**, gives expressions for calculating interfacial energies (such as γ_{1L} and γ_{2L}) from various contact angle measurements, so, in principle W_{AL} can be evaluated. If it is positive the interface between 1 and 2 should be stable, or if negative, unstable.

Prevention of environmental attack

As noted above, the key factor in ensuring adequately durable joints has been the development and employment of specific surface pretreatments for the substrate materials prior to bonding; examples include silane primers for glasses and metals and acid-etching and acid-anodizing methods for metals (see **Coupling agents, Pretreatment of aluminium, FPL etch, Pretreatment of titanium**). Other design considerations include (1) the reduction of stress concentrations and of fatigue loads, (2) ensuring any ingressing medium has to follow a long path length in order to gain access to the interface and (3) the use of sealants to further delay the onset of attack.

References

1. D M Brewis (ed.), *Surface Analysis and Pretreatment of Plastics and Metals*, Applied Science Publishers, London, 1982.
2. A J Kinloch (ed.), *Durability of Structural Adhesives*, Applied Science Publishers, London, 1983.
3. A J Kinloch, *Adhesion and Adhesives, Science and Technology*, Chapman and Hall, London, 1987.
4. J D Venables, *J. Mater. Sci.*, **19**, 2431 (1984).

Durability of coatings in water

P WALKER

If a surface coating is to protect or decorate, it must adhere to the substrate; adhesion then, is a particularly important property, but it must be stressed that it is also a variable property. The measured adhesion of a coating is governed initially by the resin binder, degree of cure, the nature of the substrate, the surface preparation, its age and the method of measurement. In service it is the environment that has the most effect on the measured adhesion and of the many environmental factors; water in the form of vapour or liquid is the most important. Exposure to water causes a rapid and dramatic loss of adhesion which may or may not be recoverable (see **Durability – fundamentals** and **Paint service properties and adhesion**).

This loss of adhesion may manifest itself in service as local blistering, excessive damage under wet conditions or, in extreme cases, complete stripping of the paint system. In the laboratory it has been demonstrated that coatings of the oxidative cure type (alkyd and epoxide ester) solvent evaporative (chlorinated rubber and methacrylate) and chemically cured (epoxide and polyurethane) all lose adhesion with increasing relative humidity. In the case of alkyd and epoxide ester coatings the measured adhesion values started to fall at 80% relative humidity (RH), levelling off at $\sim 20\%$ of their original value after 300–500 h at 100% RH. The

Table 15. Adhesion values of coatings after exposure to high humidity (direct pull-off)
(stainless steel substrate)

Coating	Initial adhesive (MPa)	At 90% RH (MPa)	At 100% RH (MPa)
Long oil alkyd	17.4*	12.1	2.1
Epoxide ester	33.1*	13.8	6.2
Chlorinated rubber	25.5*	24.1	6.6
Polybutyl methacrylate	29.5*	22.4	6.9
Thermosetting acrylic	39.3†	37.9	23.8
Polyurethane	24.8	7.6	4.2
Polyamide/epoxide	31.3	17.9	9.0

* Cohesive failure.
† Minimum value.

solvent evaporative coatings showed a 70% loss after 600–700 h exposure at 100% RH, while the epoxide and polyurethane coatings showed an appreciable loss of adhesion at 65% RH, falling to 20% of the original value at 100% RH. Equilibrium was not reached even after 600 h exposure.[1] Typical adhesion values after exposure for 600 h are shown in Table 15.

Water immersion results in a similar loss of adhesion although at a much greater rate, 90% of the adhesion loss occurring within the first 16 h of immersion (see **Accelerated ageing**). All the coatings tested showed considerable loss of adhesion.

It may be argued that total immersion of a single-coat paint system without primer is unrealistic and not in any way representative of a practical situation, but it has been demonstrated in a long-term experiment that loss of adhesion also occurs on full systems under natural exposure conditions.[2] In this experiment a range of multicoat systems representing the major generic classes of coating systems applied to mild steel and wood substrates were exposed to natural weathering for a period of nine years. The adhesion was measured by the torque shear test at monthly intervals *in situ* irrespective of the climatic conditions at the time of test. The data showed that the measured adhesion varied widely and wildly and also showed a direct correlation with the prevailing humidity or precipitation, high humidity, low adhesion. Intercoat adhesion failures were infrequent.

Factors known to be related to loss of adhesion are shown in Table 16. Thus the coatings showing the slowest rate of adhesion loss are characterized by low water sorption and permeability.

Loss of adhesion under water-soaked conditions occurs by diffusion of water through the bulk polymer or pores and capillaries in the coating.

Table 16. Rate of loss of adhesion compared with water sorption and permeability

Type of paint	Rate of loss of adhesion	Water sorption % in 5 days	Permeability $\times 10^8$
Thermosetting acrylic	Slow	1.1	1.3
Styrene butadiene	Slow	1.6	0.9
Chlorinated rubber	Slow	1.0	1.7
Polybutyl methacrylate	Slow	1.3	1.2
Polyamide/epoxide	Medium	3.5	3.7
Short oil epoxide ester	Medium	3.9	5.4
Long oil alkyd	Fast	4.3	40
Polyurethane	Very fast	5.7	33

The driving force for water diffusion may be temperature difference, osmotic pressure, chemisorption or physisorption. Water accumulation at the coating–substrate interface is made possible by the presence of non-bonded areas of sufficient dimensions to permit local formation of an aqueous phase.[3] It is suggested that this local aqueous phase expands laterally until either local areas of blistering occur or complete disbondment, i.e. loss of adhesion results. However, it must be stated that little work has been carried out on the actual sites of failure under water-soaked conditions and it is a distinct possibility that reported adhesion failures are more precisely cohesive failures within the coating. The mechanism of failure in this event is solvation of a boundary layer at the coating–substrate interface.[1] Whatever the precise scientific mechanism of failure, the practical reality is that water has a profound effect on the adhesion of coatings leaving them far more prone to damage and failure in service.

Articles on **Fracture mechanics test pieces** and **Wedge test** describe test specimens.

References

1. P Walker, *Official Digest*, **37**, 491, 1593 (1965).
2. P Walker, *J. Oil and Col. Chem. Assoc.*, **57**, 241 (1974).
3. H Leidheiser, W Funke, *J Oil and Col. Chem. Assoc.*, **70**, 121 (1987).

E

Electrical adhesion

K KENDALL

In 1877, Edison[1] patented a device based on the observation that a smooth metal electrode held at a high electrical potential adhered very strongly to a semiconducting surface. This electrical adhesion was directly related to the applied voltage, disappearing when the potential was removed. Unfortunately, the adhesion also failed when dust or scratches appeared on the semiconductor surface. As a result of these problems, there has as yet been no commercial exploitation of this invention.

However, the fact that certain materials adhere with a considerable force in an electric field is of great interest. It is also of practical effect in electrostatic precipitation of dust, in photocopying machines, in electrical control of powders flowing from hoppers and in electro-gelation of fluids.

Electrostatic precipitation is a widely used process for removing polluting particles from gas streams, for example in power station flues or in cement kilns. The effluent dust grains become charged as the gas flows between two electrodes, then deposit in a cake on the electrode surface. This cake sticks to the electrode with a force which rises with the electric field, but not simply proportional to the square of field strength as expected from the basic electrostatic equation for the cohesive stress σ in a dielectric:

$$\sigma = \varepsilon_0 \varepsilon_a E_a^2 / 2 \qquad [1]$$

where ε_a and E_a are the average permittivity and field strength in the cake and ε_0 the permittivity of free space. This equation cannot be correct because it predicts an adhesive pressure which is orders of magnitude too low. Typically, McLean[2] found an adhesive stress of around 100 Pa for an applied field of 1 kV mm^{-1} with an $E^{2/3}$ dependence.

The cause of this electrical adhesion has not been explained completely, but appears to be related to the electrical conductivity of the powder.

Only for particles with resistivities between 10^4 and 10^9 ohm-metre is the adhesion apparent. Below this the particles discharge on contact with the electrode, acquiring a charge of opposite sign, and jumping off the electrode, the so-called pith-ball effect. For larger resistivities, localized electric breakdown occurs and no adhesion develops.

McLean[2] came to the conclusion that it was the concentration of electric field at the small contact spots between the semiconducting grains which caused the large adhesion observed. It has been shown[3] that this field concentration gives the correct magnitude of the adhesion stress, and also the two-thirds power dependence on applied voltage. Figure 36 shows some results for the tensile stress required to break a contact between a smooth disc of semiconducting magnesium titanate and a polished steel plate as the voltage across the contact was varied. At zero potential there was no adhesion, but as the voltage was raised, the adhesion stress σ became substantial, giving a reasonable fit to the theoretical equation

$$\sigma = (\varepsilon_0\varepsilon_a/8\pi)^{1/3}[VEd/\pi D^2(1-v^2)]^{2/3} \qquad [2]$$

where V is the applied voltage, E the elastic modulus of the material, v its Poisson ratio, d the contact diameter and D the disc diameter. A familiar example of electrical adhesion is the photocopier developed by Xerox. In this device small particles of semiconductor adhere to a plate which has been selectively discharged by exposing it to light.

Perhaps the most novel application of this electrical adhesion of solids is the control of flowing particles. For example, the phenomenon has been used to stop the flow of powder out of a silo by applying a strong field

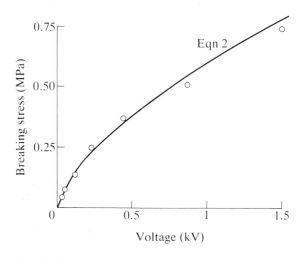

Fig. 36.   Results for electrical adhesion of a ceramic disc to a metal plate

across electrodes at the open base. Alternatively, the flow was metered by pulsing the electric field in a controlled fashion.

A similar device made use of a dispersion of semiconducting particles in an insulating fluid. Winslow[4] in 1947 first patented such a fluid made by mixing fine silica particles in oil. This fluid flowed easily under ordinary conditions, but gelled when a strong electric field, about 2 kV mm^{-1}, was applied across it. Winslow showed by microscopic study that strings of particles were formed between the electrodes when the voltage was imposed, these strings effectively solidifying the fluid. Such electro-rheological fluids have been used experimentally in clutches, brakes, valves, active damper systems and robots.[5]

References

1. T A Edison, British patent 2909 (1877).
2. K J McLean, *J. Air Pollution Control Assoc.*, **27**, 1100 (1977).
3. K Kendall, *J. Phys. D: Appl. Phys.* **23**, 1329 (1990).
4. W M Winslow, US Patent 2 417 850 (1947).
5. H Block, J P Kelly, *J. Phys. D.: Appl. Phys.*, **21**, 1661 (1988).

Electron microscopy

D E PACKHAM

Whereas in **Optical microscopy** visible light is used to interact with a specimen by reflection from its surface or by transmission through its thickness, an electron microscope uses electrons in an analogous way. A heated wire cathode provides a source of electrons which are focused by electromagnetic lenses. The column of the microscope has to be evacuated as electrons are rapidly absorbed in air at atmospheric pressure.

The electrons may be scanned over the specimen and the secondary electrons collected to form an image: this occurs in **Scanning electron microscope**, discussed elsewhere.

The interaction of electrons with matter generates X-rays. In **Electron probe microanalysis** (EPMA) these are analysed to yield information about the elemental composition of the specimen.

This article is concerned with the transmission electron microscope (TEM) in which the electrons pass through the specimen and may be focused to form an image. Because of the much smaller wavelength of electrons compared with visible light, the electron microscope has much higher resolution than an optical microscope. The resolution depends on the particular instrument, specimen and operating conditions, but a figure of 5 Å may be taken as indicative of what is often obtained.[1]

Some transmission microscopes have X-ray detectors and so can give elemental information in a manner analogous to EPMA. In scanning transmission electron microscopes (STEM) the electron beam can be scanned. When used in conjunction with X-ray detectors, elemental information can be obtained from a very small area of the specimen. The STEM can generally be operated in a SEM mode when it has enhanced resolution – 1.5 nm compared with 5 nm for conventional SEM.[2]

Because of the low penetrating power of electrons, specimens for examination in transmission have to be very thin, depending on the circumstances tens or hundreds of nanometres in thickness. Surface features on thick specimens can be examined by making a thin replica. A standard technique is to evaporate a thin film of carbon on to the surface of interest, to 'shadow' it by depositing an even thinner layer of platinum at a different angle, and then to remove the deposited replica, for example, by dissolution of the original specimen.

Applications to problems in adhesion

The TEM in its various forms is widely used in materials science, and some of its applications are studies of adhesion. It has helped to elucidate failure and bonding mechanisms in composites (see **Fibre matrix adhesion–carbon fibres**). An important use is in the examination of the structure of surface oxides. Several techniques have been employed.

The replica method has been used for examining the pore structure of films formed in the **Anodizing** of aluminium. Figure 37 represents a piece of anodized aluminium which had been bent so that part of the anodic film has spalled off. The deposition and shadowing directions are indicated. Replicas made in this way show the structure of both the top surface of the anodized layer and the fracture surface: Fig. 38 shows features of the pores described in the article **Anodizing**.

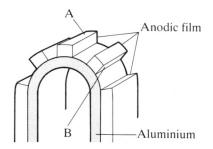

Fig. 37. Preparation of sections through an anodic film for replication for TEM examination: carbon deposition direction A, shadowing direction B

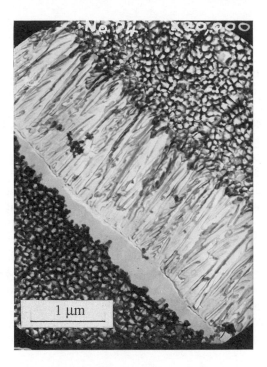

1 μm

Fig. 38. Electron micrograph of an anodic layer formed on aluminium in phosphoric acid. Top right and bottom left show pore openings in the top surface, the central band shows a section through the pores

Surface oxides may also be examined directly by removing them from the base metal. Iodine in dry methanol will dissolve most metals, leaving the oxide unaffected. Electron diffraction, especially if combined with elemental analysis provides a powerful method of studying the structure and composition of a surface layer on a metal.

Venables and colleagues have used the STEM in its SEM mode to study films formed on aluminium and titanium.[2] This high-resolution technique has been applied to layers produced by **Anodizing** and the **FPL etch** treatment. By producing micrographs in stereo pairs fine details can be seen and isometric drawings produced. This has enabled Venables to argue for the importance of mechanical interlocking (see **Mechanical theory**) in adhesion to porous anodized surfaces.

References

1. P J Goodhew and F J Humphries, *Electron Microscopy and Analysis*, 2nd edn, Taylor & Francis, London, 1988.
2. J D Venables, *J. Mater. Sci.*, **19**, 2431 (1984).

Electron probe microanalysis

B C COPE

One result of the bombardment of a material by a high-energy electron beam is the emission of X-rays. The energy of these X-rays is characteristic of the nucleus of origin. Hence if the energy of the emission is assessed the elemental composition of the target material may be determined. This is the principle of electron probe microanalysis (EPMA).

Early electron probe microanalysers were usually stand-alone instruments, but the instrumentation currently produced normally takes the form of an add-on option to a scanning electron microscope (SEM) (see **Scanning electron microscopy**), the same beam being used to create the visible image and to generate X-rays, thus an area of interest may be subjected to both topographical and compositional investigation.

The specimen for analysis must be a solid that is stable in a vacuum and which is not subject to beam damage or charging. Many polymers are therefore inherently unsuitable for investigation, although the more conductive materials may be able to be examined at relatively low accelerating voltages. Metals are ideally suited to the technique, especially when polished or sectioned, but glasses, ceramics, semiconductors may all yield reliable results.

Since the analysis is usually carried out in an SEM the same limitations on specimen size apply. Most instruments are limited to samples of few centimetres in diameter, but particular demands sometimes lead to the construction of specialized instruments. For example, tribologists interested in automotive engine wear may commission SEM/EPMA combinations with specimen stages large enough to accommodate a piston or other substantial component.

The X-rays generated may be examined by either energy-dispersive or wavelength-dispersive spectrometer. Energy-dispersive X-ray analysis (EDX) employs relatively robust equipment capable of detecting elements from sodium upwards. Wavelength-dispersive X-ray analysis (WDX) is less common, more expensive and employs more demanding equipment, but it has the ability to detect elements from boron to uranium.

Instrument parameters and the nature of the sample combine to set the volume selectivity of a particular analysis. The volume sampled can range from the order of 1 μm^3 up to several cubic micrometres. Sensitivity varies greatly with the element in question and the matrix in which it is found, but may be around 1000 ppm for EDX and about one order of magnitude better for WDX. Although qualitative investigations are relatively straightforward, quantitative work requires calibration against standards held in matrices similar to that under investigation.

The scope of EPMA in investigating problems associated with adhesion is limited by its relatively deep sampling depth (*c.* 1 μm) and moderate spatial resolution (also *c.* 1 μm). However, it can be of use in studying layers on a substrate surface, such as **Primers** and some resulting from **Pretreatment**. It can help determine the **Locus of failure** and show the distribution of adhesive on substrate surface. Identification of adhesive is eased, if it incorporates a filler or additive containing an element not present in other components.

Figure 39 shows part of the resin side of a failure surface for rubber-toughened epoxy resin (see **Toughened adhesives**) bonded to oxidized steel. The right-hand part of the figure is the electron image (SEM picture), the left-hand side an image constructed from the iron X-ray signal. The resin in the lower region of the specimen is clear in the electron image. The upper region appears different and shows a strong iron signal, indicating an area of cohesive failure within the iron oxide.

The resin side of a broken epoxy–zinc is given in Fig. 40. The electron image only reveals the resin, but the zinc X-ray image shows local concentrations of zinc in the resin. It is likely that some dendrites on the metal surface have broken and remain embedded in the resin, giving an unusual example of a mixed failure mode (see **Microfibrous surfaces**).

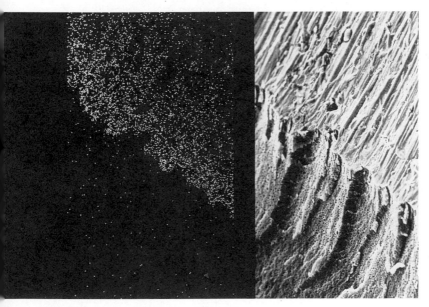

Fig. 39. The resin side of a broken oxidized steel–rubber reinforced epoxide bond: right, electron image; left, iron X-ray image. The top part of the specimen shows cohesive failure within the oxide (P J Hine, S El Muddarris, D E Packham, unpublished micrograph, see *J. Adhesion Sci. Tech.* **1**, 69 (1987))

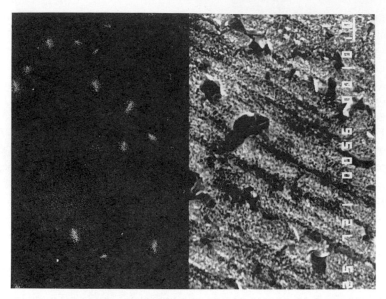

Fig. 40.   The resin side of a broken bond between zinc and epoxide resin:
left, electron image; right, iron X-ray image. Note the local
concentrations of zinc within the resin. (P J Hine, S El Muddarris,
D E Packham, unpublished micrograph, see *J. Adhesion*, **17**, 207
(1984))

A particular attribute of the SEM/EPMA instrument is the ability to
present data in a striking form. A topographical micrograph may be
obtained and compared with a composition map where the concentration
of a chosen element or elements may be displayed in grey tones or in
(computer-generated) colour. Thus, for example, segregation in metals,
the inclusion of foreign bodies or selective etching may be readily
illustrated.

In the linescan technique a line may be scanned across the sample and
data presented as a histogram of concentration of a particular element
against position. It may sometimes be useful to simply plot count frequency
against energy level to give a wide-range spectrum whose peaks define
the elements present in the area under analysis.

X-ray analysis facilities can also be used in transmission **Electron
microscopy**.

Select references

P J Goodhew, F J Humphreys, *Electron Microscopy and Analysis*, 2nd edn,
Taylor & Francis, London, 1988.
V D Scott, G Love, *Quantitative Electron Probe Microanalysis*, Ellis Horwood,
Chichester, 1983.

Electrostatic theory of adhesion

K W ALLEN

The concept of regarding two surfaces which are being separated as the two plates of an electrostatic condenser is due to Deryaguin et al.[1] He considered particularly the peeling of a pressure-sensitive tape from a rigid substrate and the subsidiary phenomena of noise and emission of light (visible if the peeling is carried out in the dark) which accompany the separation.

In a quantitative development of this, the energy of a charged condenser in terms of the potential and the separation of the plates is identified with the work of adhesion involved in peeling a tape from a substrate. This is compared with conditions for an electrostatic discharge in terms of the potential difference and the separation of two electrodes with various pressures of gas between them (using a double logarithmic plot). Deryaguin and Krotova[1] reported excellent agreement between values predicted in this way and experimental values of peel strength when PVC tape was peeled from a glass surface in an atmosphere of argon.

However, this whole argument has been sharply criticized on the grounds that it is essentially cyclic. It involves equating the work of adhesion with the energy of the system regarded as a condenser and, at the end of the calculation, an argument demonstrating that they are equal. Additionally, it is tacitly assumed that the idealized relationship for electric discharge (derived in quite different circumstances) is applicable, although careful examination suggests that this may not be true. Also, the whole discussion fails to take any account of the energy which is dissipated in a peeling situation in deforming the flexible component. Finally there are arguments about the whole origin and nature of the electric double layer which is involved.

However, the crucial point is that this theory does give some explanation for the well-established fact that the work of adhesion varies with the nature and pressure of the gaseous atmosphere in which it is measured, as well as for the noise and light emitted.

Certainly there is support for the principle, even if not for the detailed development, from Weaver's work[2] on the adhesion of thin metal films to insulating surfaces (see **Scratch test**).

In addition Deryaguin[3] has emphasized the point that the force of attraction between the plates of a condenser is independent of their separation, in contrast to the distance–force relationships for other types of attraction. Thus while at small distances electrostatic forces may be negligible by comparison with others, as distances become greater the relative significance will change and the electrostatic forces become important.

Quite recently Possart,[4] with some elegant work using potential contrast **Scanning electron microscopy**, has conclusively demonstrated the existence of an electric double layer at the interface between polymer and estimated its magnitude. Using these results and a theoretical treatment, he has provided support for the significance of this electrostatic component to adhesion while recognizing that it is only a part of the whole phenomenon. See also **Rubber adhesion**.

A comparison of different theories is given under **Theories of adhesion**.

References

1. B V Deryaguin, N A Krotova et al., *Proc. 2nd Int. Congr. Surface Activity*, Vol. III, Butterworths, London, 1957, p. 417; B V Deryaguin, V P Smilga, in Min. of Technology, *Adhesion: Fundamentals and Practice*, Elsevier, Amsterdam, 1969, p. 152.
2. C Weaver in *Aspects of Adhesion 5*, ed. D J Alner, University of London Press, London, 1969, p. 202.
3. B V Deryaguin, Yu P Toporov in *Physicochemical Aspects of Polymer Surfaces*, ed. K I Mittal, Plenum Press, New York, 1983, p. 605.
4. W Possart, *Int. J. Adhesion and Adhesives*, **8**, 77 (1988).

Ellipsometry

R GREEF

Ellipsometry is a branch of oblique-incidence reflection spectroscopy, in which the change of polarization of the light is analysed, rather than just its change of intensity. It derives its name from the fact that the most general state of polarization of light is elliptical, of which the more generally known states, linearly and circularly polarized light, are particular cases. The polarization state of obliquely reflected light is controlled by the optical constants of the reflecting surface, and is also very sensitive to films of foreign materials. Because ellipticities can be measured with high resolution and accuracy, submonolayer thicknesses can be detected, and measurements of film thickness can be made quickly and reliably.

The technique, therefore, has found application in many diverse fields of thin-film measurements, and has the potential to be even more widely applied with the advancing capability, and declining cost, of the available instrumentation.

Theory

The theory and practice of ellipsometry has been comprehensively treated by Azzam and Bashara.[1] The following brief synopsis can only serve to highlight the main points of the technique.

The basic idea of ellipsometry is depicted in Fig. 41. Collimated monochromatic light passes first through polarization optics which define its polarization state. It is then incident on the test surface, which may be immersed in a medium transparent at the wavelength of the probe light beam. The light reflected at the specular angle is then analysed by further polarization optics.

The change of state of polarization caused by the reflection is described by two quantities, which are both angles: Δ, a phase change term, and ψ, the tangent of which describes the amplitude change. For a clean, bare surface, these quantities are related to the optical constants n and k of the substrate in a one-to-one mapping. Ellipsometry is therefore one of the principal ways of evaluating the optical constants of materials.

When the surface is covered by one or more films, Fresnel analysis applied to each interface in turn followed by vector summation of the resulting multiply reflected beams leads to the Drude equations, which predict the Δ- and ψ-values for a stack of films for which the thickness and refractive indices are known.

This classical analysis technique applies to the ideal case of isotropic films with smooth interfaces. Graded-index films can be treated by simple extension, replacing the film in the calculation by a 'pile of plates', each of uniform properties, with the individual refractive indices varying from plate to plate. Rough interfaces can also be treated in some circumstances by replacing the interface with a film of index intermediate between those of the component materials – the so-called effective medium theory. This theory is dependent on the rough surface being approximated by an intimate mixture of the grains of the component materials of a size smaller

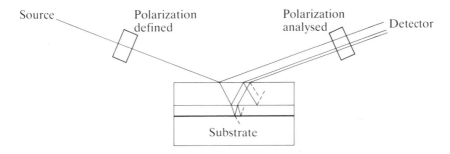

Fig. 41. Basic optical set-up for ellipsometry

than the wavelength of the measurement light. More general theories of reflection from rough surfaces are under development but not presently available. None the less, empirical correlations can be found which are useful in making qualitative predictions of the microscopic structure of rough surfaces on the basis of ellipsometric observations.

Applications

In the last 20 years there have been significant improvements in the instrumentation that is available commercially, and this process has continued to accelerate with the advent of the desk-top computer. The inversion of the Drude equations, that is the estimation of unknown thicknesses or optical constants from ellipsometric measurements, relies upon the application of computer-intensive search and optimization methods which are, except for the most complex multilayer systems, now well within the capabilities of personal computers. This has resulted in the fast-widening scope of ellipsometry as reflected in the number of publications in which the technique is dominant.

Applications that are relevant to the topic of adhesion include the determination of the growth kinetics and densification of polymer films at surfaces, quality control of anodization and other surface-modification pretreatments, measurement of adsorption and determination of molecular orientation and compaction of adsorbed layers and Langmuir–Blodgett films, detection of surface damage in plasma-etching processes, and measurements of thinning of lubricant films. Metallization from the solution or vapour phase can be studied up to the stage at which the metal becomes opaque (around 40 nm thick for most metals).

Reviews of these applications are available,[2,3] but because of the rapidly developing nature of the subject they are soon outdated. The best way of locating literature references on ellipsometry at present is by computerized database search. Using the keyword fragment 'ellipsomet' in conjunction with keywords for the specific application excludes almost all spurious references.

References

1. R M A Azzam, N M Bashara, *Ellipsometry and Polarized Light*, North-Holland, Amsterdam, 1977.
2. W E J Neal, *Appl. Surf. Sci.*, **2**, 445 (1979).
3. R Greef in *Comprehensive Chemical Kinetics*, Vol. 29, ed. R G Compton, Elsevier, Amsterdam, 1989, Ch. 10.

Emulsion and dispersion adhesives

G C PARKER

Introduction

Following a period of sustained growth after their introduction many years ago, the use of emulsion and dispersion adhesives is now widespread in many industrial and commercial operations. This situation is considered very likely to continue owing to the many advantages offered by these products when compared to other types of adhesive, not least in terms of environmental considerations.

Types of emulsion adhesives

Probably the most widely used industrial emulsion or dispersion adhesives are those based on polyvinyl acetates, commonly referred to as PVAs. These products are normally manufactured by a process referred to as emulsion polymerization whereby, basically, vinyl acetate monomer is emulsified in water and, with the use of catalysts, is polymerized. The presence of surfactants (emulsifiers) and water-soluble protective colloids facilitates this process resulting in a stable dispersion of discrete polymer particles in the aqueous phase.

These PVAs are not normally suitable for use as an adhesive in their basic unmodified state as they tend to form brittle films and have limited adhesion capabilities. They are, however, compatible with a wide range of compounding ingredients which are capable of modifying their properties very considerably. Plasticizers and thickeners are two of the materials most commonly used in the formulation of PVA-based adhesives, but many other modifiers may be incorporated as required, resulting in a very wide range of PVA-based adhesives being available to meet the specific demands of many different applications.

It must be stressed, however, that while a substantial proportion of emulsion adhesives are based on PVA, which is referred to as a homopolymer as if it is the polymer of a single material, vinyl acetate, copolymer-based emulsions are becoming of increasing significance and importance in the market-place.

Copolymers, as the name implies, are produced from the polymerization of two different materials. Probably one of the most widely used copolymer emulsion adhesives are those based on vinyl acetate ethylenes, commonly referred to as VAEs (see **Ethylene vinylacetate copolymers**). These are produced by the copolymerization of vinyl acetate and 10–20% ethylene, the resulting polymer base possessing some superior properties over the PVA-based emulsions referred to above. These superior properties relate principally to the increased inherent flexibility of the dry VAE film due to the internal plasticization effect of the ethylene component in the polymer which enhances adhesion to many 'difficult' surfaces. There are, however, a number of other polymers and copolymers which are used as the formulating basis for alternative specialized emulsion adhesive systems.

Setting mechanism

Having dealt with the basic chemistry of emulsion and dispersion adhesives, the method by which they function should now be considered.

If a thin 'wet' film of emulsion adhesive is applied between two porous surfaces, and these three components subjected to an adequate degree of compression, the initial state of the adhesive is one of a water phase containing discrete and stable polymer particles. **Wetting** of the substrate surfaces then takes place followed by the commencement of penetration of the water phase into the substrates. As this occurs the solids content of the adhesive film rises and it becomes more viscous and tacky. Capillary forces in the substrate then continue to draw water (and, dependent on substrate porosity, possibly some polymer solids) from the adhesive until its solids content rises to the point where it coalesces into a continuous film.

Maximum bond strength will not be attained, however, until all the water present in the adhesive film and the substrates have been removed and dispersed into the atmosphere, the rate of which will be dependent upon many factors. The situation detailed above deals with two porous substrate surfaces. While this is not an essential requirement for the successful use of emulsion adhesives, it must be stressed that when these products are to be utilized in the conventional manner (i.e. bonds made using 'wet' adhesive), at least one of the substrate surfaces must be porous to permit the escape of the water phase from the adhesive.

Methods of application

The objective of any application system is to deposit a controlled quantity of adhesive in the correct position on a substrate at the right time. A number of types of applicator may be used to achieve this including wheel, extruder, spray, roller, dauber or printing techniques.

These different applicating systems all place differing demands on the adhesive in terms of its viscosity and rheology characteristics, and also subject the adhesive to varying degrees of shear. It is important, therefore, to ensure that the properties of an adhesive are compatible with the equipment with which it is to be applied and, if high shear forces are likely to be exerted on the adhesive, that it possesses a sufficient degree of shear stability.

Advantages and limitations

Some of the main advantages of emulsion adhesives include the relatively fast bond development (dependent upon specific substrate and bonding conditions) compared to that obtained with solution adhesives. The carrier phase (water) is cheap and environmentally safe (see **Health and safety**). Copolymer systems in particular can be formulated to give very good adhesion to many 'difficult' substrate surfaces.

Among their limitations is a slow setting speed, which is faster than that of water-based solution adhesives but is significantly slower than that of **Hot melt adhesives**. Further, when used in the conventional 'wet' application mode, at least one substrate surface must be porous.

The water resistance of the dry adhesive film obtained from emulsion adhesives, while fair, tends to be adversely affected by the presence of small residual quantities of protective colloids which are water soluble. They are used as **Adhesives in the textile industry** and for bookbinding and woodworking (**Wood adhesives – basic principles, Wood adhesives – edge banding**).

References

1. A E Corey, P M Draghetti, J Fantl in *Handbook of Adhesives*, 2nd edn, ed. I Skeist, Van Nostrand Reinhold, New York, 1977, Ch. 28.
2. C E Blades in *Handbook of Adhesives*, 2nd edn, ed. I. Skeist, Van Nostrand Reinhold, New York, 1977, Ch. 29.
3. M S El-Aasser, J W Vanderhoff (eds), *Emulsion Polymerisation of Vinyl Acetate*, Applied Science Publishers, London, 1981.

Engineering advantages

R D ADAMS

Adhesive bonding allows the designer new freedoms in joining methods. In welding, similar materials have to be used, and the (metal) substrates generally have to be heated to a temperature above their melting-point, whether this is by flame, electric arc, friction, or whatever. Mechanical methods of fastening use screws, bolts, rivets, etc. and need pre-drilled or prepared holes. In addition, mechanical fasteners cause localized stress concentrations, which encourage the formation of fatigue cracks.

Adhesive bonding allows the loads to be transferred over a large area. Even so, there will still be stress peaks at the ends of the joint owing to unavoidable discontinuities.

But one of the major advantages of adhesive bonding is that it enables dissimilar materials to be joined, thus allowing plastics panels to be attached to a metal frame, or carbon-fibre reinforced plastics to be joined to titanium.

In many applications, welds and mechanical fasteners cause surface intrusions which can be unacceptable in the surface finish of a motor car (see **Automotive applications**). The bonnet (hood) and boot (trunk) is almost invariably nowadays made by bonding a skin to an inner frame to provide a smooth but rigid structure.

In aerospace applications, careful surface preparation is required to ensure maximum joint strength, while the automobile industry insists on their adhesives being suitable for bonding oily surfaces. In general, appropriate surface preparation should always be used: this may be no more than cleaning and/or abrading, but it may require a chemical etch or the electrodeposition of surface layers. The adhesive supplier should always be consulted on the correct surface preparation to be used for any given application.

Select reference

W A Lees, *Adhesives in Engineering Design*, Design Council, 1984.

Engineering design with adhesives

C WATSON

In this article are discussed ways in which adhesive technology can influence the approach of the design engineer.

Industry is, ideally, a partnership of people from quite different disciplines working together to turn a design brief into a production reality. Strong links have to be forged internally between those involved in design and manufacturing and, externally, with the suppliers of materials. Some of these are summarized in Fig. 42. According to Roger Bishop, writing in *Eureka*: 'Product excellence, ease of manufacture and profit are the motivating factors in ensuring a proper account is taken of material technologies in product design.' Adhesive technology is able to contribute to this process.

Benefits of using adhesives for joining

The list below gives well-established benefits which can result from the use of adhesives in preference to such joining methods as riveting, bolting, spot-welding, ultrasonic welding, welding or mechanical clipping:

1.   Reduction in weight;
2.   Ability to join thinner materials;
3.   Better stress distribution over joint area;
4.   Improvement in joint strength;
5.   Ability to join dissimilar materials;
6.   Scope for sealing as well as joining;
7.   In some cases repairs are more readily effected;
8.   Increased flexibility in choice of component material;
9.   Elimination of extra components such as keys, circlips, etc.;
10. Elimination in some cases of machining operations;
11. Facilitation of manufacturing process;
12. Aesthetic improvement.

A number of case histories are now described illustrating the advantages available to the designer from a skilful use of adhesive technology.

Floppy disc cover This first example is a simple case of the manufacture of a paper floppy disc cover and illustrates the importance of the link

INVOLVEMENT OF EXPERTISE AT
DESIGN BRIEF STAGE (KNOWLEDGE)

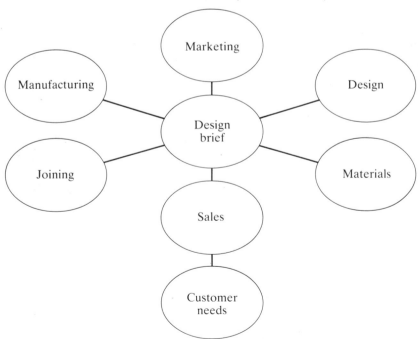

Fig. 42.   Some factors affecting engineering design

between design and manufacturing. The designer originally chose a solvent-based adhesive (see **Solvent-based adhesives**) to bond the flaps on the three sides. The process involved applying the adhesive to both faces, and waiting until the solvent flashed off – in this case 10 min. During the flash-off time the covers had to be stored and the solvent extracted.

To achieve ease of manufacture with total continuous automation, a single component cyanoacrylate (see **Cyanoacrylate adhesives**) is now used. It is applied automatically to one of the joint faces and the faces mated immediately. The adhesive develops initial handling strength in 4 s.

Loudspeaker The loudspeaker as shown in Fig. 43 has eight or so adhesive joints. In order to meet production requirements of one every 10–15 s, it is necessary to use automatically applied fast-curing adhesives. The alternative previously employed was to use solvent-based and/or two-component mixed products, but both were slow in terms of total process time.

Before changing to the fast-curing adhesives, one large European manufacturer was considering building a factory extension to provide extra space for 'work in progress' following an increase in production demand. It can be seen from the above that 'work in progress', becomes a major influence in the choice of adhesives, where performance criteria are similar.

Coaxial assemblies Often the application of adhesives technology to the manufacture of such components allows the elimination of components

SPEAKER ASSEMBLY

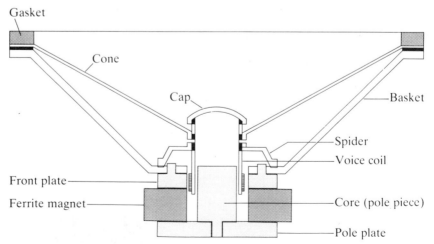

Fig. 43. Loudspeaker assembled by adhesive bonding

and the machining operations leading to a reduction in weight. Figure 44 shows a variety of cylindrical components – pulley to shaft, gear to shaft, trunnion end into roller and sprocket on to shaft.

Considering the first case, pulley to shaft, it can be seen that in order to prevent the pulley moving around the shaft a key is used. Keyways have to be cut in the pulley and the shaft, weakening both. To prevent the pulley moving along the shaft, a set screw is used necessitating the use of additional material on the side of the pulley. Using adhesives, the design will be simpler. The key and machining of keyways could be eliminated. (This could enable a reduction in the shaft size due to elimination of stress concentration in the case of gear to shaft application.) The set screw can be eliminated and with that the integral ring. The other components could be redesigned with analogous advantages.

Ink distributors A world leader in office machinery has recently adopted the use of an **Anaerobic adhesive** in the manufacture of ink distributors. After an extensive study it was found that this adhesive could save £15 000 per year in manufacturing cost by replacing solder on four joints to join shaft ends to a distributor tube. In two further joints a slow-setting sealant has also been replaced by the fast-curing anaerobic material.

The cost of implementing the change was less than £1500. The result is a reduction in cost and aesthetically better product manufactured more easily.

Fig. 44.   Some coaxial components with potential for redesign involving adhesive bonding

Flexibility in choice of component material

Adhesives manufacturers are continually trying to develop adhesives to meet the needs of industry. One group of plastics that have been difficult to bond are polyolefins and related low-energy substrates (see **Surface energy**). They could not be bonded without elaborate surface preparation such as **Flame treatment** or **Plasma pretreatment, Corona discharge treatment** or oxidative chemical methods.

Why then are these materials so attractive to the designer? Providing the performance criteria can be met by polyolefins, they are less costly than other plastics. According to an ICI source, polyethylene and poly-propylene, cost 0.06 and 0.07 pence per cubic centimetre respectively compared with, for example, 0.40 pence per cubic centimetre for nylon.

A new **Primer** has been developed which enables a cyanoacrylate to be used to bond polyolefins.[1] The primer is applied to one surface, adhesive to the other – the bond is virtually instantaneous. This will allow the designers flexibility in choice of plastics, enabling less costly polyolefins to be selected.

These are but a few examples illustrating ways in which adhesive technology can influence the approach of the design engineer in the choice of materials and manufacturing processes.

Reference

1. Loctite Technical Information, *New Primer for Polyolefins – Loctite 757.*

Select references

M M Sadek (ed.), *Industrial Applications of Adhesive Bonding*, Elsevier Applied Science Publications, London and New York, 1987.

G Defrayne, *High Performance Adhesive Bonding*, Society of Manufacturing Engineers, Dearborn, Michigan, 1983.

G Haviland, *Machinery for Locking, Retaining and Sealing*, Marcel Dekker, New York and Basle, 1986.

Engineering surfaces of metals

J F WATTS

The surfaces of metals used in practical adhesion situations are very far removed from the concept of a clean metal surface; the substrate to which adhesives and coatings manufacturers exhort us to apply their product.

The surfaces of metals, unless very noble (gold or platinum) will invariably have a layer which reflects the interaction of the metal with the atmosphere. This may take the form of a simple oxide, but is more likely to be of an oxyhydroxide form, and its composition may be mixed cations or the major or minor cations of the alloy. In addition material may be absorbed from the atmosphere to form a loosely bound surface layer (a potential **Weak boundary layer**) or there may be extraneous material at the surface which is merely left over from a previous processing step—lubricant from mechanical processing and temporary protectives for corrosion protection. However thorough the cleaning by **Degreasing** or **Abrasion treatment** some of this material will invariably remain attached to the metal surface. Added to this gamut of chemistry at the metal surface we have various morphological and metallurgical features such as surface roughness, grain boundaries, dislocations and intermetallic precipitates it becomes clear that the engineering surfaces of metals are very complex indeed.

As far as adhesion of an organic phase is concerned it is the presence of an unsuspected layer at the surface of a metal that is often responsible for poor adhesion performance. The metallurgical aspects may have important secondary roles, for instance the presence of a precipitate may interfere with the development of a uniform anodic layer on aluminium (see **Anodizing**) and grain boundaries may be the route by which a minor alloying element is enriched at the surface, but they are rarely responsible for poor adhesion *per se*.

In order to develop a picture of the chemistry of metal surfaces let us first consider the interaction of the metal with atmospheric oxygen and water vapour. In a pure oxygen most metals will develop a thin film which may, or may not confer protective properties upon the alloy depending on its composition. On exposure to the atmosphere the situation is more complex, indeed early descriptions of the surface of metals following atmospheric exposure indicated the presence of much adventitious material such as salts, sulphides, polar and non-polar organics, and adsorbed water in addition to metal oxides and hydroxides resulting from a chemical reaction between substrate and environment.[1,2] More recent work making use of the surface analysis methods of **X-ray photoelectron spectroscopy** and **Auger electron spectroscopy** has indicated that such a model, although essentially correct, is unnecessarily complicated. The oxide layer is, on some alloys (e.g. mild steel), an oxyhydroxide phase, on others (stainless steel for example) a mixed oxide is covered with an extremely thin layer of a hydroxide. These inorganic layers are covered with a more labile mixture of adsorbed water and organic contamination which has been deposited during a cleaning process or during subsequent air exposure.[3]

The method of surface cleaning employed will have a marked effect on the levels of such adventitious material.[4] In general dry **Abrasion treatments** carried out with clean abrasive provide the lowest levels of contamination, but invariably produce a very rough surface profile (see **Roughness of surfaces**). In contrast **Degreasing** procedures with either organic solvents or aqueous reagents tend to produce surfaces high in carbon and its associated chemisorbed water. The direct correlation between these two contaminants indicates that they are associated together at the surface perhaps by **Hydrogen bonding**.

In terms of surface cleaning or degreasing practices the requirements are that gross levels of extraneous material are removed. This includes friable oxides, mill scale, grease, residues of corrosion inhibitors and protective wraps. This is readily achieved by the use of the appropriate surface treatment.

The remaining layers described above are more tenacious and are not easily removed. The question of whether such material remains as a discontinuity at the metal oxide–polymer interface has yet to be unequivocally resolved, but it is clear that in some cases, such as coatings stoved at elevated temperatures, the presence of water promotes the growth of the inorganic layer (oxide, oxyhydroxide, or hydroxide) while the carbonaceous phase is absorbed by the polymer.

As long as a well-defined **Weak boundary layer** is not present it seems that most organic coatings and adhesives will at least tolerate the presence of some 'contaminant' material on the metal surface.

References

1. F R Eirich in *Interface Conversion for Polymer Coatings*, eds P Weiss, G D Cheever, American Elsevier, New York, 1968, pp. 350–74.
2. J C Bolger, A S Michaels in *Interface Conversion for Polymer Coatings*, eds P Weiss, G D Cheever, American Elsevier, New York, 1968, pp. 3–52.
3. J E Castle in *Corrosion Control by Coatings*, ed. H Leidheiser, Science Press, Princeton, 1979, pp. 435–54.
4. J E Castle, J F Watts in *Corrosion Control by Organic Coatings*, NACE, Houston, 1981, pp. 78–86.

Epoxide adhesives

J COMYN

Epoxide adhesives have this name because one of their components, the resin, contains epoxide rings: that is, a three-membered ring with two carbon atoms singly bonded to an oxygen atom. The second component is a hardener or curing agent, and hardening is by chemical reaction which produces a cross-linked polymer. Cured epoxides are hard and rigid.

Epoxide adhesives adhere to a large number of materials including metals and glasses, and can also be used as the matrix-resin in **Composite materials**. They generally show high strength, but, unless modified, rather limited toughness (see **Toughened adhesives**). They are widely used in **Aerospace, Automotive**, building, electrical and woodworking applications (see **Wood adhesives**). Advantages over other **Reaction-setting adhesives** are that no volatile products are released on cure, and shrinkage is low. Their **Glass transition temperatures** represent an upper service temperature, and epoxides which can be cured at room temperature generally have lower glass transition temperatures than those which require high temperatures for cure. Upper service temperatures for unmodified epoxide resins are around $95-100\,^{\circ}\mathrm{C}$.

Like other rigid adhesives they have good resistance to shear forces but are weak in peel. Rubber-modified epoxide adhesives are now available which have improved peel strengths (see **Toughened adhesives**). Epoxides are compared with acrylics in **Acrylic adhesives** and with polyurethanes in **Structural adhesives**.

The most commonly used epoxide resin is the diglycidylether of bisphenol-A (DGEBA). This is made from epichlorohydrin and the disodium salt of bisphenol-A in the following reaction:

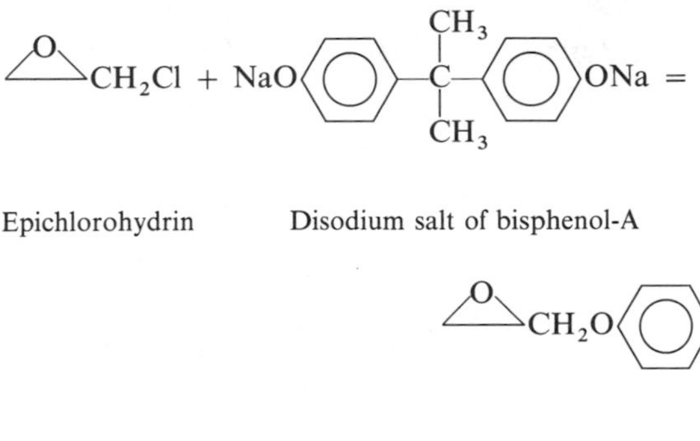

Epichlorohydrin   Disodium salt of bisphenol-A

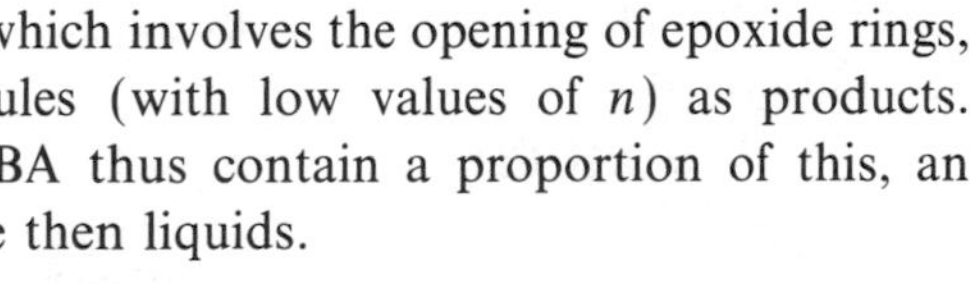

DGEBA

There is a further reaction which involves the opening of epoxide rings, to give the following molecules (with low values of n) as products. Commercial grades of DGEBA thus contain a proportion of this, an advantage being that they are then liquids.

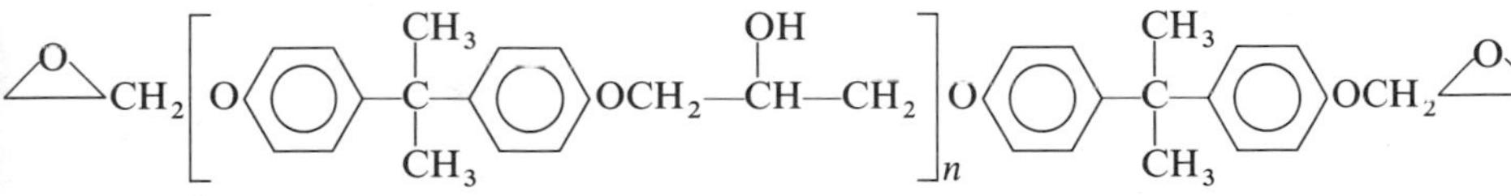

The structures of two other epoxide resins of commercial significance, but which are of lesser importance than DGEBA are shown below.

(3,4-epoxycyclohexyl)methyl-3,4-epoxycyclohexane carboxylate   Ciba–Geigy araldite CY 175

While there are only a few epoxide resins of commercial importance, the number of substances which have been used as hardeners is large: aliphatic amines, aromatic amines and acid anhydrides are the groups most commonly represented.

Amine hardeners

Amine hardeners are mixed with resins in such quantities that there is approximately one amine-hydrogen for each epoxide ring; a primary amine and epoxide will react as follows in a **Condensation polymerization**.

$$-NH_2 + CH_2\!-\!CH\!-\!CH_2\!-\!O- \quad = \quad -NH\!-\!CH_2\!-\!CH\!-\!CH_2\!-\!O-$$

Reaction of first amine-hydrogen

$$-NH\!-\!CH_2\!-\!CH\!-\!CH_2\!-\!O- \quad + \quad CH_2\!-\!CH\!-\!CH_2\!-\!O-$$

Reaction of second amine-hydrogen

$$= -N \begin{cases} CH_2\!-\!CH\!-\!CH_2\!-\!O- \\ CH_2\!-\!CH\!-\!CH_2\!-\!O- \end{cases}$$

Some typical aliphatic amine curing agents are triethylenetetramine which is 6-functional and bis(aminopropyl)tetraoxaspiroundecane which is 4-functional:

$$NH_2CH_2CH_2NHCH_2CH_2NHCH_2CH_2NH_2$$

Triethylenetetramine

$$NH_2CH_2CH_2CH_2\!\!\left\langle\!\!\begin{array}{c} O \; O \\ X \\ O \; O \end{array}\!\!\right\rangle\!\!CH_2CH_2CH_2NH_2$$

Bis(aminopropyl)tetraoxaspiroundecane

Mixtures of epoxide resins with aliphatic amines may be cured at room temperature, the process usually requiring many hours to complete. Cure with aromatic amines requires an elevated temperature, typical conditions being 2 h at 150 °C. Some typical examples are shown below.

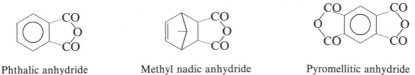

1,3-diaminobenzene   4,4'-diaminodiphenylsulphone

4,4'-diaminodiphenylmethane

Acid–anhydride hardeners

Cure with acid anhydrides is by condensation polymerization and it requires an elevated temperature. Examples of some acid anhydride hardeners are

Phthalic anhydride   Methyl nadic anhydride   Pyromellitic anhydride

Other hardeners

Thiol hardeners can be used for rapid cure. Dicyandiamide

$$H_2N—C(=NH)—NH—C\equiv N$$

is only slightly soluble in DGEBA at room temperature, but becomes much more soluble at cure temperatures which are about 150 °C. It can thus be incorporated into one-part adhesives which can be stored for periods of about a year before use.

All the hardeners dealt with so far react by condensation polymerization in which there are roughly stoichiometric quantities of epoxide and hardening groups. Tertiary amines and complexes of boron trifluoride can be used in catalytic quantities, and initiate a ring-opening **Addition polymerization**.

Radiation-cured adhesives include epoxides polymerized via a cationic mechanism.

Select references

C D Wright, J M Muggee in *Structural Adhesives, Chemistry and Technology*, ed. S R Hartshorn, Plenum Press, New York, Ch. 3.
C A May (ed.), *Epoxy Resins, Chemistry and Technology*, 2nd edn, Marcel Dekker, New York, 1988.

Etch primers

J L PROSSER

An etch primer is one having a specific chemical reaction with the metal surface to promote adhesion and control corrosion. It comprises a polyvinyl butyral resin in a predominantly alcoholic solution, phosphoric acid and zinc tetroxychromate; it may be formulated to be a one-pack or two-pack material. The two-pack variety is the original one; the second constituent is an aqueous solution of phosphoric acid; the resin may be admixed with a phenolic resole or an epoxy resin. Many variants in resin and pigment are employed.

The mode of action is said to be the following: the chromate pigment dissolves in the aqueous/alcoholic phosphoric acid to yield chromate and zinc ions; oxidation of the alcohol by chromate yields trivalent chromium, and this ion complexes with hydrogen phosphate ions, water molecules and hydroxyl groups of the resin to give an insoluble cross-linked resin complex which additionally reacts with the metal surface.

The primers are extensively used on aluminium, but are applied also on steel. Reference should also be made to **FPL etch** where sulphochromating solutions for pretreating aluminium are discussed.

Select references

W A Riese, *Metall-Lacke*, Verlag Colomb, 1965.
P Walker in *The Performance of One-pack Etch Primers*, FATIPEC X, 1970, p. 425.

Ethylene–vinyl acetate copolymers

D E PACKHAM

Ethylene–vinyl acetate copolymers (EVA) are used in many adhesive applications.

Ethylene and vinyl acetate are copolymerized using a radical initiator (see **Addition polymerization**). Figure 45 shows the chemical structure of EVA. The reactivity ratios of both monomers are close to unity, so copolymers have a random structure, and can be prepared with composition ranging from almost 0 to almost 100% vinyl acetate (VA). (The compositions in this article are expressed as percentages by weight.) Bulk polymerization is widely used for polymers of 40–50% VA, and an emulsion process for those over 60% VA. In the intermediate range (15–60%), solution polymerization may be used.[1] Both low (10–40%) and high (60–95%) VA materials have adhesive applications, the former as **Hot melt adhesives**, the latter as emulsion adhesives (see **Emulsion and**

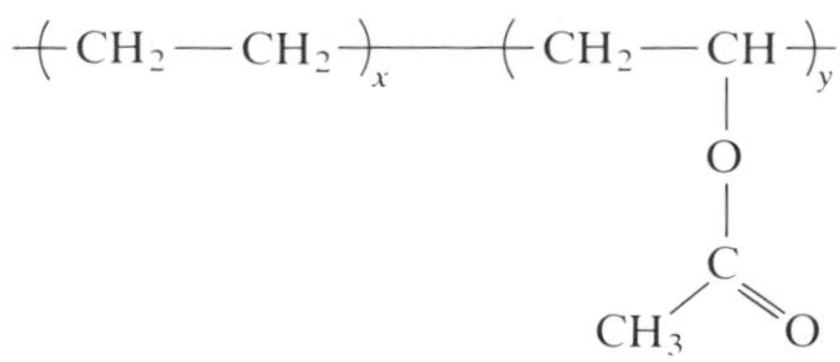

$$+CH_2-CH_2\overset{}{\underset{x}{)}} \qquad +CH_2-CH\overset{}{\underset{y}{)}}$$

Fig. 45. Chemical structure of the ethylene and vinyl acetate units in EVA copolymers

dispersion adhesives). The term 'vinyl acetate–ethylene' is sometimes used to describe the copolymers rich in vinyl acetate.

Properties

As with homopolymers, the properties of EVAs are affected by the molecular weight and degree of short- and long-chain branching. Here particular emphasis is placed on properties which depend upon vinyl acetate content.

Polyethylene is of course crystalline, and copolymers of low VA content are semi-crystalline thermoplastics, which can be thought of as modified polyethylenes. The effect of varying vinyl acetate content is to vary the crystallinity and the polarity. The crystallinity depends partly on the extent of chain branching, but the vinyl acetate content exerts a strong effect. As bulky VA groups are introduced into the chain the crystallinity falls, becoming zero at around 50% VA (Figure 46). As the crystallinity decreases, the elastic modulus and yield strength fall and the elongation at break increases (Fig. 47).

The mechanical properties of EVAs were found by Hatzinicolaou and Packham to exert a strong influence on their adhesion to metals.[2] Figure 48 shows a close correlation between the strain energy density of four EVAs and the peel loads when applied as hot melts to copper with a **Microfibrous surface**. Although such a straightforward connection was not found under all circumstances, the peel energies found could be rationalized using an analysis based on an extension of Eqn 7 in the article **Peel tests**.

Polyvinyl acetate is a glassy, amorphous polymer with **Glass transition temperature** about 28 °C. The effect of increasing the proportion of ethylene is to reduce the strong forces between the chains which, in the homopolymer, are associated with a high concentration of polar acetate groups. Consequently the glass transition temperature falls and is as low as 25 °C for a VA content of 65%.[3,4] The properties at room temperature change as a result: modulus and tensile strength fall and extension to break increases (Fig. 49).

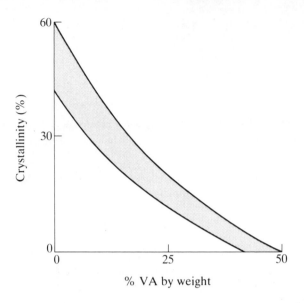

Fig. 46. Envelope showing the relationship between crystallinity and vinyl acetate content (after ref. 1)

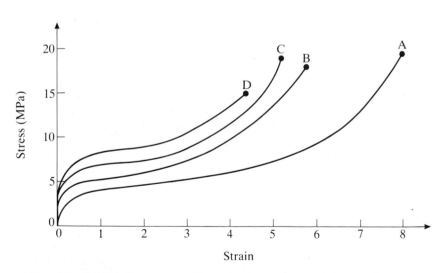

Fig. 47. Schematic representation of stress–strain behaviour of some EVAs, VA content weight per cent A 28, B 18, C 12.5, D 9 (after ref. 2)

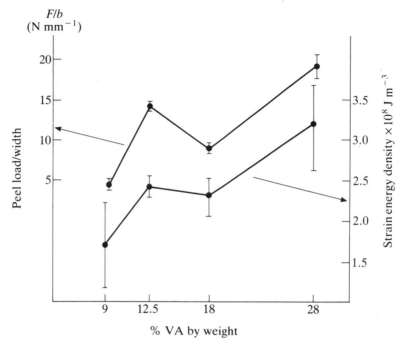

Fig. 48. Comparison of strain energy density at failure for four EVAs with the peel loads measured for backed strips of the polymers peeled from chlorite-oxidized copper[2]

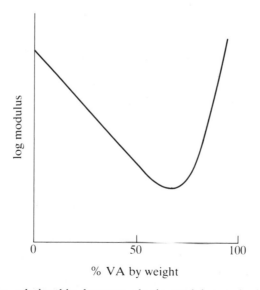

Fig. 49. The relationship between elastic modulus and vinyl acetate content (after ref. 4)

Adhesive applications

Low vinyl-acetate content EVAs are used as **Hot melt adhesives**. A typical copolymer composition is 28% VA, but materials in the range 15–50% VA are used. They may occasionally be employed on their own,[2] but are most commonly blended with hydrocarbon waxes and tackifying resins such as esterified wood resins. Their uses include packaging, woodworking, (see **Wood adhesives – basic principles**), bookbinding and carpet-backing. Domine and Schaufelberger[5] give some typical formulations.

   High VA copolymer emulsion adhesives are often modified to suit the specific application intended, for example with addition of plasticizers or cross-linking agents.[3] Applications include laminating PVC and other films, packaging and in furniture and joinery (see **Wood adhesives – joints for furniture**) and textile bonding: **Adhesives for textile fibre bonding** and **Adhesives in textile coating**.

References

1. G W Gilbey in *Developments in Rubber Technology – 3*, eds A Whelan, K S Lee, Applied Science Publishers, London, 1982, p. 101.
2. T A Hatzinicolaou, D E Packham, *Adhesion*, **9**, 33 (1984).
3. C E Blades in *Handbook of Adhesives*, 2nd edn, ed. I Skeist, Van Nostrand Reinhold, New York, 1977, p. 484.
4. N L Zutty, J A Faucher, S Bonotto in *Encyclopedia of Polymer Science and Technology*, Vol. 6, eds H F Mark et al., John Wiley, New York, 1967, p. 396 *passim*.
5. J D Domine, R H Schaufelberger in *Handbook of Adhesives*, 2nd edn, ed. I Skeist, Van Nostrand Reinhold, New York, 1977, p. 495.

Extrusion coating

R J ASHLEY

This versatile technique is generally used where a ply of polyethylene or copolymer thereof is required in a structure. Other polymers may be used in specialized areas but the handling can become more difficult. The process is widely used within the **Packaging industry** for the coating of paper, board, foils, cellulose film and thermoplastic films. The most common coating resin used is low-density polyethylene, but this now extends to copolymers such as **Ethylene–vinyl acetate**, ethylene–acrylic acid, polypropylene, high-density polyethylene and ionomers (e.g.

Surlyn). The acrylic acid-based materials and ionomers are used in areas where enhanced adhesive strength is required such as resistance to difficult environments.

The coatings are usually applied to a substrate to provide a heat-sealing medium, but they also serve to reduce the permeability to water vapour and to improve the mechanical properties of certain base materials.

The substrate to be coated is unwound from a braked or driven turret off-wind and passes into a priming unit (see **Primers for adhesive bonding**). The priming unit consists of a gravure applicator and a drying oven and applies chemical adhesion promoters from a solvent base. The solvent is removed from the primer coat in the oven leaving a dry coat weight in the range $0.2–1.0$ g m^{-2}. An alternative to priming is to treat the substrate by an electronic discharge (see **Corona discharge treatment**) and this is most frequently used where plastic films are used or **Flame treatment** (see **Pretreatment of polyolefins**).

After priming or treating the substrate is drawn into the primary nip of the coater which comprises a steel cooling drum about 1 m in diameter (the chill roll) which is cooled with a high-volume flow of water, and a hard rubber back-up roller that forces the substrate against the chill roll.

Molten polymer is then extruded through a straight slot die, essentially similar to that used for polymer film casting and is drawn vertically down into the primary nip. The distance between the die and the nip (air gap) is critical since it influences the degree of oxidation and subsequently adhesion of the coating. This distance is generally between 150 and 300 mm. Recent modifications of the technique concern the use of an ozone shower or specific active species introduced to the polymer melt curtain to improve adhesion.

Since the polymer is molten when drawn into the nip, the surface of the coating conforms to that of the chill roll and both matt and polished rolls are used depending on the properties required of the final coating. The coated web passes around the drum and is peeled away by a secondary nip roller, edge trimmed and electronically treated if required for further processes.

Extrusion coating is relatively low cost for long production runs and has the advantage over adhesive laminating of polymer films in that very thin layers can be applied (especially of polyethylene). Typical coating thicknesses would be in the range $5–30$ μm.

There are three major problems with the process that limit its application: (1) good gauge control is necessary, (2) adequate oxidation of the melt must be achieved for good adhesion and (3) excessive oxidation or too high an extrusion temperature gives rise to odour problems in the coating. The use of primers contributes to achieving good adhesion levels while operating at minimum melt temperatures.

An associated technique is to use the extrusion coating in melt form as an adhesive interply between two substrates to be laminated. This is particularly useful for the lamination of paper to foil. A combination of extrusion lamination and coating can be used for the preparation of more complex structures with single machine pass.

See also **Coextrusion, Tie-layers, Packaging industry**.

Select reference

R J Ashley et al., *Industrial Adhesion Problems*, eds D M Brewis, D Briggs, Orbital Press, Oxford, 1985, Ch. 8.

F

Fatigue

R D ADAMS

The process of fatigue in such well-characterized materials as metals is one which produces a large scatter of results: in polymers, the situation is worse. Adhesive joints consist of polymers sandwiched between two adherends. Not only is the adhesive material one which is subjected to uncertain geometric effects (bond line thickness, sharp or blunt corners, surface condition), but the stress distribution is highly non-linear in even perfectly characterized geometries.

The fatigue process in joints is one which is uncertain both in crack initiation and in propagation. It is therefore very difficult to estimate the fatigue life of a bonded joint. Instead, if the loading situation is likely to be one involving fluctuating loads, good practice should be followed. First, an adhesive should be chosen which is toughened: brittle adhesives should be avoided as these allow for easy crack propagation. Woven carriers, as sometimes used in film adhesives, should be avoided as these tend to provide localized stress raisers and a ready crack-propagation path. Whereas long joints are little stronger than short joints in monotonic loading owing to the stress concentrations at the overlap ends, in fatigue loading a longer joint provides a longer path for the crack to have to traverse before final failure. Thus, longer joint lengths are beneficial in fatigue, despite the opposite advice for static loading.

Some test methods are listed in Appendix 2.

Fibre composites – introduction

B C COPE

Any material consisting of more than a single phase can justifiably be described as a **Composite material**. In plastics technology the term is usually reserved for two-phase systems consisting of a reinforcing fibrous material, which is usually inorganic but which may be polymeric, dispersed in a continuous polymeric matrix. Such materials offer exceptionally high levels of strength, stiffness and impact strength combined with a density lower than those of structural metals and alloys. Strength and stiffness arise from the properties of the reinforcing fibres which are very small in diameter and consequently substantially free from the flaws that normally reduce the strength of materials from high theoretical values to the low practical values familiar in bulk samples.[1,2] The fibres are typically 50 times stronger and stiffer than the matrix with strength sometimes reaching 4 GPa and tensile modulus exceeding 400 GPa.

The composite, then, is seen to consist of very strong fibres embedded in a matrix which, because of the presence of stress-concentrating cracks, has a much lower strength than the fibres. The matrix serves to maintain fibre position and orientation, transmit shear forces, protect the fibre surface and transfer loads to the reinforcement. For this reason fibre–matrix adhesion is essential.

The relationship between strength and modulus of the composite and the corresponding properties of the fibre and matrix depends on the fibre length and orientation.[1] For continuous unidirectional fibres, the modulus of the composite in the fibre direction may easily be shown to be equal to the modulus of the fibres times their volume fraction plus the modulus times the volume fraction of the matrix. This assumes that fibre–matrix adhesion is adequate to transmit the stresses involved. The strength in the fibre direction is essentially the fibre strength times volume fraction.[1]

The greatest volume fraction of fibre that it is possible to incorporate in most composites is not much more than 0.5 so the maximum possible strength of a composite is about 50% of the theoretical strength of the reinforcement. In practice the attainable figure is rather less than this because of fibre imperfections, fibre damage, incomplete wetting-out and other defects. The matrix material thus contributes little directly to the strength of the composite but is essential to hold the fibres in place and to transfer shear force.

It is obvious that in order to obtain a high volume fraction of fibre the fibres must be unidirectional in their orientation, and are thus able to contribute nothing to the strength of the composite when it is stressed so that there is no resultant of that stress in the fibre direction. These unidirectional composites are immensely strong in tension, and consequently in bending, but easily split lengthwise.

For most applications resistance to stresses in more than one direction is essential and so a unidirectional composite is not acceptable. Strength and stiffness in both dimensions of a plane, as in a car body panel, or isotropically in all three dimensions is normally required. Consequently the orientation of the fibres must be modified to provide resultants in the required directions. This may be achieved in two directions by the use of woven or knitted cloths or with non-woven, random felts or mats. In three dimensions random orientation of the reinforcement is usually the only possibility. However, the maximum strength attainable drops sharply in multi-directional composites, not only because of the smaller fraction of fibres contributing to resisting the stress but because the maximum packing density of the fabric decreases and hence the volume fraction of reinforcement decreases. Despite this fall-off the strength of the fibres is such that a very substantial reinforcement still results.

Toughness in composites arises from the bonding between the matrix and the fibre. Under impact loads cracks start in the matrix and propagate until the stress concentration ahead of the crack reaches a matrix–fibre interface. There, if the strength of interphase bonding is of the right order, delamination occurs, with the dissipation of large amounts of energy, and, by blunting the crack, removes the source of stress concentration.

The fibre–matrix interface

The optimum level of bonding between the phases of a composite is usually difficult to determine and to achieve (see **Fibre–matrix adhesion – assessment techniques**).

High matrix and interfacial shear strength are desirable to keep the ineffective fibre length (that part of the fibre not contributing to the strength and stiffness of the composite) as low as possible. Against this, when a composite is loaded in plane tension only, the interface should be as weak as possible to maximize strength. The situation is unusual but may be met in filament-wound pressure vessels.

When a crack propagates in a composite it is desirable that the interface should fail in transverse tension or shear ahead of the crack, thus blunting it and halting its progress. The maximum efficiency in this operation depends on a moderate level of interphase bonding, but in practice it is difficult to achieve bonding good enough to seriously degrade impact strength.

Small fibres are inherently unstable in compression unless well bonded to the matrix, so for good shear and compressive strength the interface should be as strong as possible. In transverse tension fibres act as stress concentrates and hence a weak interface is particularly deleterious.

Composites made with large-diameter isotropic fibres such as boron usually have a much higher compressive strength than tensile strength. Those with small-diameter isotropic fibres such as glass are usually stronger in compression than in tension, whereas those with small-diameter anisotropic fibres, such as aramids, are usually stronger in tension, with a low compressive strength that is particularly sensitive to poor interfacial bonding.

In general a weak interfacial bond reduces tensile strength and very seriously reduces compressive and shear strengths, but toughness increases with decreasing bond strength.

Overall it is usually desirable to maximize interfacial bonding.[3] The technique adopted to achieve this depends on the fibre in question. With glass the application of a size containing a coupling agent consisting of a methacrylic chromic chloride complex or an organosilane is the usual practice. The **Coupling agents** are designed to bond to hydroxy groups on the glass surface and to be reactive towards the appropriate thermosetting matrix resin (see **Fibre–matrix adhesion** – **glass fibres**).

Carbon fibres are frequently subjected to an oxidative etching process to improve adhesion, but a size is sometimes used (see **Fibre–matrix adhesion – carbon fibres**). Sizing is generally ineffective with aramid fibres because of lack of susceptible sites, but plasma deposition of amine groups is said to improve adhesion.

Coupling agents currently in use are capable of, in some circumstances, eliminating interfacial failure and their use may more than double the flexural strength of a laminate and greatly improve its water resistance; see **Durability**. Their use does, however, demand control as too thick a coating or incorrect cure can lead to failure to improve laminate performance. The titanates, sometimes called coupling agents, are better described as surfactants and although they may improve dry strength they do little for wet durability.

Some of the main constituents used in these materials are discussed under **Fibre composites – fibres and matrices**, and practical aspects are considered under **Fibre composites – joining** and **Fibre composites – processing techniques**.

References

1. B Harris, *Engineering Composite Materials*, Institute of Metals, London, 1986.
2. J G Morley, *High-performance Fibre Composites*, Academic Press, London, 1987.
3. E P Plueddemann, *Composite Materials*, Vol. 6, *Interfaces in Polymer Matrix Composites*, Academic Press, New York, 1974.

Fibre composites—joining

A J KINLOCH

Introduction

There are two main classes of fibre-composite polymeric materials: those based upon thermosetting matrices, such as **Epoxide adhesives, Phenolic adhesives** and unsaturated polyester matrices, and those based upon thermoplastic matrices, such as poly(ether-ether ketone), polyphenylene sulphide and polyamides (see **Fibre composites − introduction, Fibre composites − matrices and fibres**).

Thermosetting-based polymer fibre composites

These materials generally possess relatively high **Surface energy** and their surfaces are relatively polar in nature (see **Wetting and spreading**). Further, they do not, of course, form oxides or corrode in moist environments. Therefore, as might be expected, this combination of properties results in surface pretreatments being required prior to adhesive bonding which simply remove contaminants such as mould lubricants or general dirt.

The simplest method is obviously to rely upon using a clean moulding surface, and then to bond directly to this as-moulded surface. However, it is difficult to ensure that no significant concentrations of mould **Release** agents, which are typically based upon fluorine- or silicone-containing materials, remain on the fibre-composite surface. As shown in the review by Kinloch,[1] if a relatively high concentration is left on the fibre-composite surface then it is not readily removed by any subsequent treatment and poor adhesive joints result (see **Weak boundary layers**).

Many fibre-composite surfaces are bonded where the as-moulded surface is the bonding surface, due attention having been paid to the choice and level of release agent used in the processing of the fibre composite. However, in applications where the strength of the fibre composite is critical then a pretreatment step is usually employed to ensure the cleanliness of the as-moulded surface. There are two main techniques used to achieve this: (1) the peel-ply method and (2) **Abrasion treatment** and **Degreasing**, often conducted after a peel-ply surface has been exposed. In the former method one ply of fabric, such as a polyester film, is installed at the bonding surfaces during manufacture of the fibre-composite component and just prior to bonding the peel ply is removed. Again, it is important to choose a peel ply which does not leave a significant concentration of release agent behind on the bonding surface.

Thermoplastic-based polymer fibre composites

More recently fibre composites have been commercially introduced which are based upon carbon fibres in a thermoplastic matrix. Such composites may be formed using thermoplastic processing techniques and have superior impact properties to those based upon current thermosetting resins. However, recent work by Kodokian and Kinloch,[2] has revealed that bonding thermoplastic fibre composites using abrasion and solvent-cleaning pretreatment techniques, with epoxy or acrylic adhesives, leads to significantly lower joint strengths, compared to conventional composites. The reasons for this appear to be lack of sufficient wetting (see **Wetting and spreading**) and intrinsic adhesion of these adhesives to the thermoplastic fibre-composite substrate. A most effective method for overcoming this problem is to pretreat the composite surface using **Corona discharge**. This treatment introduces polar groups into the surface of the composite and greatly increases the surface free energy of the materials.

On the other hand, if it is possible to use a very high-temperature hot-melt adhesive, then a film of the appropriate thermoplastic may be successfully used, with no corona pretreatment being necessary.

Absorbed water and repair

Another problem that may arise upon bonding fibre composites is that absorbed moisture in the composite may be evolved during the adhesive bonding cycle and lead to air voids in the adhesive layer. The obvious method of overcoming this problem is to dry the composite prior to bonding. This is the usual route followed, although it can be difficult in practice when undertaking repair work. Also, when undertaking adhesive repairs, it may be necessary to remove absorbed hydraulic oils, etc. from the composite before satisfactory adhesion can be achieved (see **Repair methods**).

Design of fibre-composite joints

The use of overlap joints loaded in tension is a very common joint design, but all these types of joints lead to the introduction of transverse (out-of-plane) tensile, or cleavage, stresses (see **Joint design**). Such stresses are most important when designing joints with fibre-composite substrates since these substrates are highly anisotropic and possess relatively low interlaminar tensile strengths. Thus, with such bonded joints it is important to try and minimize the transverse (out-of-plane) tensile stresses which are acting when the joint is loaded. This is discussed in detail by Kinloch[1] and Adams and Wake[3] and often involves ensuring that the eccentricity in the loading path is minimal and paying attention to the design of the ends of the joints, for example tapering the substrates and including an adhesive fillet in the joint design.

References

1. A J Kinloch, *Adhesion and Adhesives, Science and Technology*, Chapman and Hall, London, 1987.
2. A J Kinloch, G K A Kodokian, J F Watts, *Phil. Trans. Roy. Soc. Lond.*, **A 338**, 83 (1992).
3. R D Adams, W C Wake, *Structural Adhesive Joints in Engineering*, Elsevier Applied Science, London, 1984.

Fibre composites — matrices and fibres

B C COPE

The attractions of these particular **Composite materials** and the critical influence of interfacial adhesion are considered under **Fibre composites — introduction**. The range of materials which may be used as fibres and matrices serves to extend the properties which may be obtained.

Fibres

It may be shown that fibres used for reinforcement should be as small in diameter and as high in aspect ratio as possible; very short fibres simply act as stress concentrators in all modes of stressing except compression. The upper limit to fibre length is purely a matter of practicalities; long fibres do not flow well when the matrix is moulded, but continuous fibres can be wound into radially symmetrical structures, such as pipes, with great success. The materials listed under the headings below are commonly used.

Glass Cheap and versatile, glass is by far the commonest reinforcement. Ordinary sodium silicate glass, 'A-glass', is rarely used now as the greater durability in damp conditions of the less alkaline E-glass more than compensates for its extra cost. Other, more exotic formations such as C- and S-glass are used in advanced composites.

Glass is available in the form of rovings (untwisted bundles of strands), yarn (twisted bundles of strands), and fabrics made from each. Yarn-based fabrics are the most expensive but yield the stronger composites. The cheapest fabric is the chemically bound, non-woven, two-dimensionally random, chapped strand mat, but where better mechanical properties are needed tapes, scrims, square-weaves, twills, satins and various knitted constructions are preferred.

Surface treatments used are discussed elsewhere: **Coupling agents, Fibre-matrix adhesion – glass fibres**.

Carbon fibre This is sometimes referred to as 'graphite', and is much more expensive than glass, but is lighter, stiffer and, to a degree, electrically conducting. Being black and opaque carbon does not give the translucent composites that glass is capable of providing. Carbon-based composites have a high fatigue resistance.

As with glass a range of cloth constructions is available. Further details of the fibres and their surface treatment are discussed under **Fibre–matrix adhesion – carbon fibre**.

Aramids These are aromatic polyamides, a class of materials now being marketed as engineering thermoplastics. The material most familiar as a reinforcement is Du Pont's Kevlar. Like carbon it is lighter than glass and gives very stiff composites, having good fatigue and creep resistance. It has the advantage of a light colour. Kevlar is difficult to cut, as it has a marked tendency to fibrillate axially. This is a disadvantage in the fabrication of composites, but imports a degree of resistance to ballistic impact.

Boron Elemental boron may be grown into long crystals that may be used as reinforcing material. It is extremely stiff, but its very high cost rules it out for most applications outside military aerospace projects.

Hybrid fabrics containing two or more fibres are available to permit great subtlety in reinforcement placement and orientation.

Matrix polymers

Unsaturated polyesters The cheapest matrix resins, the polyesters, are versatile and undemanding. Choice of suitable hardeners enables them to be cross-linked at temperatures from outdoor winter ambient to 150 °C. Cure times range from a few minutes to several hours depending on conditions. Polyesters are forgiving enough to be used by amateurs and are available in chemical-resistant and flame-retardant grades.

Vinyl ester resins These are epoxide resins with unsaturated carboxylic acids reacted in to provide sites for free radical cross-linking. They provide a combination of ease of processing with good chemical resistance.

Epoxide resins These may be formulated to cross-link in the cold, but for optimum properties must be post-cured or cured hot. They offer greater strength, rigidity, toughness and durability than polyesters, but at a higher price. Their chemical resistance is usually superior to that of polyesters. Epoxides are commonly used with the higher-priced forms of reinforcement.

Polyimides These expensive resins are hot cured and used primarily in aerospace and military applications where their superior thermo-chemical properties justify their cost premium.

Phenolic resins These were long considered to be unsuitable for composite production as their condensation curing mechanism produces water that vaporized at normal curing temperatures of over 100 °C. The pressure needed to prevent this water introducing porosity into resulting laminates was sufficient to seriously degrade the reinforcing fibres. Recently acid catalysed curing systems have made possible the product of phenolic composites by low-pressure techniques. Phenolic composites are sold primarily on their fire resistance.

Thermoplastics Many engineering thermoplastics, and even such materials as polypropylene, are produced in glass-reinforced, and some cases, carbon-fibre-reinforced, grades. These are filled with short fibres (longer ones readily break up on passage through compounding machines) randomly oriented, but typically still show at least twice the stiffness of the unfilled equivalents. Recently, ICI have started to sell a range of materials ('Verton') with somewhat longer fibres and consequently better properties, although some deterioration still occurs as the fibres suffer damage in the moulding machine. Cloth reinforcement is not usually possible, but **Polyether ether ketone** (PEEK) based and other advanced prepregs are available.

   Fibre composites – joining and **Fibre composites – processing techniques** give details of practical aspects of these materials.

Select references

K Partridge (ed.), *Advanced Composites*, Elsevier, London, 1989.
F P Gerstle in *Encyclopedia of Polymer Science Engineering*, Vol. 3, eds H F Mark et al., Wiley–Interscience, New York, 1985, p. 776.

Fibre composites – processing techniques

B C COPE

Before the advantage of these **Composite materials** (see **Fibre composites – introduction**) can be realized, they have to be fabricated. Depending on the particular materials involved (see **Fibre composites – matrices and fibres**) specially developed processing methods may be necessary.

Processing techniques

Hot curing materials with randomly arranged reinforcement are available in grades suitable for processing by compression, transfer or injection moulding. Thermoplastics are usually injection moulded. Processes that are peculiar to composites are necessary when dealing with continuous or cloth reinforcement. Some are outlined under the headings below.

Pultrusion In this analogue of extrusion a bundle of fibres is impregnated with resin and then pulled through a heated die.

Contact moulding This group of processes applies no pressure, other than atmospheric, as the matrix resin cures. At its simplest contact moulding involves nothing more sophisticated than an individual with a brush and a bucket of resin. Reinforcing fabric is laid on to a simple mould, made from timber, plaster or other cheap material, and then wetted out manually with liquid resin. Developments employ resin sprays or vacuum impregnation, to reduce labour content and improve consistency. A large mechanical laminating machine is employed by Vosper-Thorneycroft to lay up the hulls of the Royal Navy's ocean-going mine-hunters that are currently believed to be the largest one-piece composite, or indeed plastics structure of any kind, in production.

Filament winding, the application of a pre-impregnated continuous length of fibre around the outside of a mould to produce pipes, pressure vessels and similar articles is also a form of contact moulding.

Cold press moulding This is used chiefly with polyesters and features a closed mould that can be locked after a fast-setting resin has been poured onto a fibrous preform.

Prepregs These consist of reinforcement pre-impregnated, usually with hot curing resin fully compounded, if necessary, with hardener. Some, such as the polyester-based sheet moulding compound and bulk moulding compound are processed by conventional compression moulding using very low pressures. However, when woven or knitted reinforcement is used compression moulding may not be appropriate and is replaced by a technique in which cut-out layers of prepreg are stacked in a metal mould that is then closed and heated in an oven or autoclave. Very large mouldings for aerospace applications are made by this technique.

Some other practical aspects of the use of these materials are discussed under **Fibre composites – joining**.

Select references

A B Strong, A P Loskoka, *Fundamentals of Composites: Manufacturing, Materials and Applications*, SME, Dearborn, Michigan, 1989.

W B Goldsworthy in *Encyclopedia of Polymer Science and Engineering*, Vol. 4, eds H F Mark et al., Wiley–Interscience, New York, 1986, p. 1.

P K Mallich, S Newman (eds), *Composite Materials Technology, Processes and Properties*, Hauser, Munich, 1990.

Fibre—matrix adhesion – assessment techniques

F R JONES

The performance of all **Composite materials** is dependent on the adhesion between the different phases. The significance of this adhesion is discussed in **Fibre composites – introduction**: in this article some ways of measuring it are described.

The quantification of the interfacial bond strength between fibres and matrix has been recently reviewed in Jones.[1] A number of methods have been employed, but four tests are currently more widely used.[2]

Indirect measurements

The most common indirect test involves the determination of interlaminar shear strength (ILSS) by a short beam shear test in three-point bending. Full specification of the test has been given by Curtis.[3] For typical results see Fig. 54 of **Fibre matrix adhesion – carbon fibres**. The disadvantage of this test is that once the interfacial shear strength exceeds the shear strength of the matrix, failure occurs predominantly in the resin and the ILSS reaches a plateau. This has been confirmed by scanning **Secondary ion mass spectrometry**.[4] The values of ILSS consequently vary between systems and are also subject to variabilities in the interphase structure and fibre surface chemistry.[4]

Direct measurements

Multiple fragmentation test[5] In this test a single filament is embedded in a matrix resin and tested under tension until fragmentation into successive shorter lengths is complete and the fibre length has reached equilibrium.

The interfacial shear strength τ_i is given by a consideration of the stress transfer between fibre and matrix

$$\tau_i = \frac{d_f \sigma_{fu(l_c)}}{2l_c} \qquad [1]$$

where d_f is the fibre diameter, l_c the equilibrium critical length determined in the test and $\sigma_{fu(l_c)}$ the fibre strength at the critical length l_c for the system under consideration.

This test has a number of disadvantages as follows:

1. A tendency to select the strongest single filament from the bundle.
2. Tedious measurements of the Weibull statistics of single-filament strengths are necessary for the determination of an accurate value of the fibre strength at the critical length $\sigma_{fu(l_c)}$.
3. The test is limited to matrices whose failure strain is significantly greater than the fibre failure strain. This can present difficulties in comparing performance in differing matrices and environmental factors which cause matrix failure strain to decrease. The stress-transfer mechanism at the interface is further complicated by the possibility of matrix yield.
4. The stress state exhibited by a single fibre is an unrealistic description of that exhibited in a composite. This latter has been examined by Wagner.[6]

Pull-out test In this test the force to pull a single filament out of a resin block or drop is determined. The average shear strength τ_i can be related to the embedded length l_e in a similar manner to Eqn 2:

$$\tau_i = \frac{P_f}{\pi d_f l_e} \qquad [2]$$

where P_f is the maximum pull-out load.

Pitkethly and Doble[2] have discussed the problems associated with this test, which mainly arise from the mechanism of stress transfer and the build-up of stress in the fibre from its ends. They used a theoretical analysis to provide an accurate extrapolation of P_f/l_e data to determine τ_{max}. In this way inaccuracies associated with the measurement of l_e can be mitigated.

Table 17. Typical values of interfacial shear strength (τ_i) as measured by the various techniques

Fibre	Matrix	Treatment	τ_i*	τ_i† (MPa)	τ_i‡	τ_{max}¶	ILSS (MPa)
Aramid	Epoxy	None	—	8	—	—	55
E-glass	Epoxy	Silane	79	—	56	—	—
E-glass	Polyester	Silane	28	—	23	—	—
E-glass	Vinyl ester	Silane	18	—	—	—	51
AR-glass	Vinyl ester	Silane	14.5–16	—	—	—	41–51
AR-glass	Vinyl ester	None	14.5	—	—	—	38
Carbon HS	Epoxy	None	12	28	28	124	24–70
Carbon HS	Epoxy	100%	37–44	65	—	151	80–100

NB. These figures have been obtained from various sources.
* Fragmentation test.
† Pull-out test.
‡ Microindentation test – debonding strength.
¶ Pull-out test.

Microindentation or push-in test In this experiment, a micro-indentor is used to push the fibre into the matrix. Its major advantage is that it can be applied to fibres in a polished section of a fibre composite without the need for special specimen fabrication. The interfacial shear strength is related to the displacement of the fibre into the matrix l_d and the appropriate load P:

$$l_d = \frac{2P^2}{\pi^2 d_f^3 \tau_i E_f} - \frac{2\gamma}{\tau_i} \qquad [3]$$

where E_f is the fibre modulus and γ the fracture surface energy.

Mandell and McGarry have applied a finite-element analysis to the model. This approach has been adopted by the Dow Company for the development of a commercial instrument, which records the force at which debonding occurs; see Caldwell et al.[2]

Comparison of test results

The values of τ_i obtained by these different tests are given in Table 17. These figures are given for illustrative purposes and demonstrate the need for further understanding of mechanisms involved in each test. The apparently low results for silane treated glass fibre composites reinforces the complexity of the interface (see **Fibre matrix adhesion – glass fibres**). Similar variabilities in the performance of differing grades of carbon fibre are also discussed in the literature.

Aramid fibres

As is illustrated in Table 17 the interfacial shear strength of Aramid fibre composites is generally low. Several attempts to improve the interfacial bond have been attempted. These involve grafting appropriate functional groups on to the surface of the fibres. However, the improvement is limited because the crystal structure of Aramid fibres ensures that the transverse strength is low. Consequently, improvements in interfacial shear strength simply shifts the locus of the failure from adhesive to cohesive within the fibre. In which case, mechanical performance is generally not improved. However, there is some benefit to the moisture sensitivity of the fibres and composites.

References

1. P J Herrera-Franco and L T Przal, *Composites*, **23**, 2 (1992).
2. R Jones (ed.), *Interfacial Phenomena in Composite Materials*, Butterworths, London, 1989.
3. P T Curtis, *CRAG Test Methods for Engineering Properties of FRP*, RAE Technical Report 88012, 1988.
4. P Denison, F R Jones, A Brown, P Humphry, J Harvey, *J. Mater. Sci.*, **23**, 2153 (1988).
5. L T Drzal, M J Rich, P F Lloyd, *J. Adhesion*, **16**, 1 (1982).
6. H D Wagner, L W Steenbakkers, *J. Mater. Sci.*, **24**, 3956 (1989).

Fibre—matrix adhesion – carbon fibres

F R JONES

The performance of all **Composite materials** is dependent on the adhesion between the different phases. In this article, aspects relevant to carbon-fibre reinforced plastics are discussed. There is a parallel article on **Fibre matrix adhesion – glass fibres**. A discussion of the fibre—matrix interface can be found in **Fibre composites – introduction**.

Carbon fibre structure

Carbon fibres, from acrylic or polyacrylonitrile precursors, are defined by the heat-treatment temperature used for carbonization. Thus Type I (high-modulus) fibres have experienced 2400 °C and Type II (high strength) 1500 °C. These designations arose because, for the original fibres, maximum fibre strength was achieved at the latter temperature. With the identification of the strength limiting flaws, Type A fibres, which have been carbonized over the temperature range 1000–1200 °C were developed.

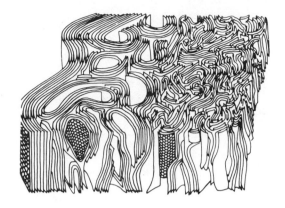

Fig. 50.   Schematic PAN carbon fibre structure. (After S C Bennett, D J Johnson in *Proc. 5th London Carbon and Graphite Conf.*, vol. 1, SCI, London, 1978, p. 377, reproduced with permission)

In the latter fibres the graphitic structure is much less developed so that the surface chemistry of the fibres is strongly dependent on the previous thermal history. For example, the Type I fibres are known to exhibit a core-sheath structure, with a high degree of order to the circumferential graphitic planes at the fibre surface. The microstructure in the core of the fibre is predetermined by the polyacrylonitrile precursor, which forms an imperfect ladder polymer as a result of the intramolecular cyclization of the pendant nitrile groups which occurs during the oxidation stage. Since the precursor polymer is largely atactic, these cyclized sequences occur over a short range so that on carbonization, a turbostratic structure results, as shown in Fig. 50.[1]

Surface treatment

A number of surface modifications have been examined but anodic oxidation in an aqueous electrolyte (e.g. ammonium bicarbonate) is preferred. This is largely a result of the control which can be achieved with electrochemistry as much with the convenience of adding an electrochemical bath to the production line. A typical unit is schematically given in Fig. 51. The degree of oxidation for a particular fibre type can be quantified by reference to the current density (typically $100\,C\,m^{-2}$). However, much of the research is reported in terms of a fraction of the commercial surface treatment per cent degree of fibre treatment (DFT), which is set at 100%. Variation in surface treatment can be normalized accordingly by adjusting the current density proportionally.

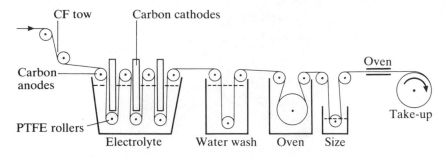

Fig. 51. Typical arrangement for electrochemical surface oxidation of carbon fibres

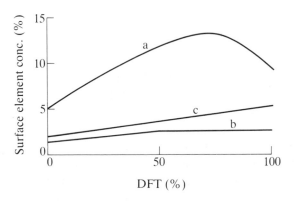

Fig. 52. The surface concentration of (a) oxygen atoms, (b) oxygen atoms combined as carboxylic acid groups, (c) adsorptive sites, as a fraction (in %) of 40 000 carbon atoms contained in a graphitic segment $\simeq 3.5$ nm thick.[2] DFT – degree of fibre treatment

Surface microstructure, microchemistry

Surface characterization of oxidized carbon fibres has been undertaken using various techniques. It is well established that the surface oxygen concentration increases, as shown in Fig. 52. The chemical state of the oxygen has been the subject of many papers, but as shown in Fig. 52 a significant proportion exists as carboxylic acid groups. The remaining oxygen is in the form of adsorbed water, phenolic hydroxyl and/or ketonic groups. It is generally accepted that these functional groups are located at the edges of the basal planes in the graphite lattice. A correlation between surface chemistry and interfacial bond strength has not been proven. Furthermore, a significant increase in surface area during surface treatment could also not be detected. Denison and Jones[2] approached this problem by establishing a model segment for quantification of the surface derived groups (see **X-ray photoelectron spectroscopy**). In this way it was possible to show that the surface of treated Type II fibres was saturated with acid groups and their average separation was smaller than theoretically possible. As a result they postulated that the functional groups were located within the micropores in the surface. Detailed study of Type I fibres confirmed that the concentration of carboxylic acid groups rose to a maximum with surface treatment, while the dimensions of the micropores increased. Following the demonstration that thermal desorption of the oxygen-containing groups *in vacuo* at 1400 °C, did not always lead to a reduction in interlaminar shear strength (see **Fibre matrix adhesion – assessment techniques**), and that the measured acidity was not affected (i.e. the acid groups reformed by reaction with water during analysis), it was concluded that active carbons, devoid of functional groups, could also be involved in adhesive bond formation.[3] It has also been observed that molecular-thin layers of epoxy resins can be chemically bound to the surface.[4] A schematic model of the surface of Type I carbon fibres before and after surface treatment is given in Fig. 53. The spacing of these micropores is considered to be determined by the crystallite size distribution. The micropore dimension of ≈ 0.7 nm demonstrates that they occur at the twist boundaries between these crystallites. Transmission **Electron microscopy** of embedded fibres has also demonstrated that the fibre–matrix adhesion occurs predominantly at the edges of the exposed basal planes. The different surface microstructures for Type A and Type I fibres have led to different interpretations; however, it is clear that the adhesion mechanism involves contributions from functional group chemistry, stereochemistry of the resin molecules and the presence of micropores whose dimensions must be large enough to allow access to the edges of the basal plane where the reactivity exists. For Type A fibres the presence of the graphitic skin is uncertain and consequently these fibres may have varying degrees of perpendicularly organized basal planes, exposed by surface treatment. The schematic in Fig. 53 should be varied accordingly for the different grades.

After surface oxidation the fibres are coated with a polymeric size, which is usually an epoxy resin. Whether this resin coating can be dissolved efficiently into the matrix is unclear. This aspect of the fibre–matrix interaction in carbon fibre composites is under-researched, but the possibility of a matrix–size interface also needs to be considered.

The effect of surface treatment on the mechanical properties of carbon fibre composites Brittle fibres in a brittle matrix can have appreciable toughness because the cracks can get diverted along the fibre–matrix interface. If

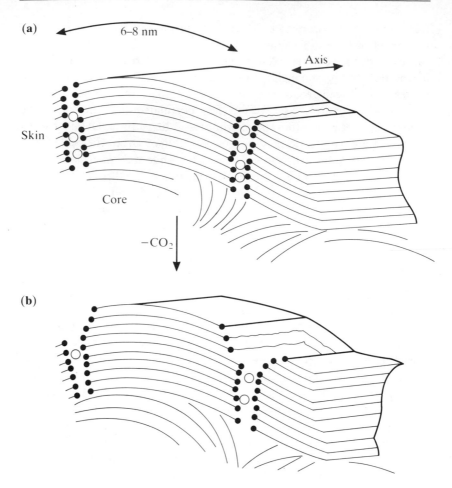

Fig. 53. The schematic model of Type I (high-modulus) carbon fibre surface before (a) and after (b) electrochemical oxidation. (●) Adsorptive sites which may or may not have functional groups attached; (○) adsorbate molecules (e.g. H_2O)

the bond is weak the composite will not support loads in shear or compression, but when the bond is too strong, the material will be brittle. These aspects are illustrated in Fig. 54 where it is seen that the interlaminar shear strength reaches a plateau but the notched tensile strength decreases monotonically with surface treatment. Optimization and careful control of the interfacial bond strength is therefore essential for each fibre—resin system. It is also apparent from Fig. 54 that test methods other than fracture tests have to be used accurately to assess the interfacial bond strength (see **Fibre matrix adhesion— assessment techniques**).

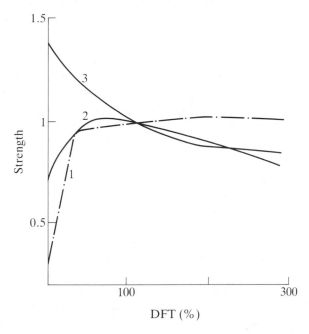

Fig. 54.   Effect of surface treatment (degree of fibre treatment, DFT) on
the mechanical performance of Type II carbon fibre epoxy
composites. (1) Interlaminar shear strength of a unidirectional
composite, (2) impact strength, (3) notched tensile strength, of a
$(0°/+45°/0°)_s$ laminate, normalized to commercial treatment at
100% DFT

References

1. J-B Donnet, R C Bansal, *Carbon Fibres*, 2nd edn, Marcel Dekker, 1989.
2. P Denison, F R Jones, J F Watts, *J. Mater. Sci.*, **20**, 4647 (1985); *Surf. and Interface Anal.*, **9**, 431 (1986); **12**, 455 (1988).
3. P Denison, F R Jones, A J Paul, J F Watts in *Interfacial Phenomenon in Composite Materials*, ed. F R Jones, Butterworths, London, 1989, pp. 105–10.
4. P Denison, F R Jones, J F Watts, in *Interfaces in Polymer, Ceramic and Metal Matrix Composites*, ed. H Ishida, Elsevier, New York, 1988, p. 77–85.

Fibre–matrix adhesion – glass fibres

F R JONES

The properties of glass-fibre reinforced plastics, like those of all **Composite
materials**, are dependent on the adhesion between the different phases;
see **Fibre composites – introduction**.

Glass-fibre forms

While several glass-forming compositions (see **Fibre composites – matrices and fibres**) are used to manufacture reinforcements for composites, E-glass is the most industrially important. The fibres are drawn from a low-sodium, aluminoborosilicate glass melt at 1250 °C. The drawn fibres of ≈ 15 μm diameter are immediately cooled with a water spray and coated with an aqueous 'size' in contact with a rubber roller. The 'size' or 'finish' is crucial to the handleability of the fibres and their **Compatibility** with the resin matrix. Each bushing produces approximately 200 individual filaments which are assembled into a strand or tow. In conventional manufacturing processes, a number of these are combined in a separate process to give rovings of appropriate tex (typically 600, 1200 or 2400 tex, 1 tex = 1000 grams per m). When carried out in a separate stage the rovings take on a slight twist. Consequently for aligned fibre composites such as pultruded sections and prepreg manufacture, direct rovings formed at the bushing are used in these applications.

To cater for the variety of composite fabrication techniques, fibres may be formed into woven rovings, knitted preforms or continuous random mats (CRM) or chopped into lengths of > 20 mm to < 5 mm for use in moulding compounds or assembling into chopped strand mat (CSM). For these textile processes the polymeric component of the 'size' is required to maintain strand integrity and filament strength. The random continuous and chopped strand mats require additional polymeric binders.

The glass-fibre 'finish'

The 'finish' which is applied as an aqueous emulsion consists of the following: (1) adhesion promoter or **Coupling agent**; (2) protective polymeric size; (3) additional polymeric binder (emulsion or powder); (4) lubricant. Items (2)–(4) impart good handleability and controlled wet-out kinetics with matrix resins and are therefore chosen for compatibility with the fabrication process. For specialist applications, such as environmental resistance, the chemical nature of the 'size' and binder are crucial and selected accordingly (see **Durability**). Good economical design can be achieved by combining fibres with differing 'finish' in different laminae. Typical coatings are given in Table 18.

Surface treatment for adhesion

The manufacturing process described above involves bringing hot glass into contact with water immediately prior to surface treatment. This results in a unique surface chemistry. A comparison of the bulk and surface

Table 18. Typical glass fibre finishes

Glass type	Polymeric size	Polymeric binder	Application
E	PVAc	—	General-purpose roving
E	PVAc	PVAc emulsion	General-purpose CSM
E	Polyester	—	Environmentally resistant GRP
ECR	Polyester	—	Environmentally resistant GRP
E, ECR, S, R	Epoxy	—	High-performance composites
E	Epoxy/polyester copolymer	—	High-performance composites with wide range of compatibility
E, ECR	Polyester	Various – powder	CSM – environmental resistant GRP CSM – processing with controlled wet-out CRM
E	Polyurethane	—	Roving for thermoplastics – short-fibre moulding compounds (e.g. nylon)
C	—	Polyacrylate Polystyrene	Reinforcing veils for gel coats, chemically resistant barrier layers

CSM = chopped strand mat, CRM = continuous random mat, GRP = glass fibre reinforced plastic, PVAc = polyvinylacetate.

Table 19. Typical bulk and surface (XPS) elemental composition of E-glass

Element	Atomic compositions (%)		
	Bulk	Fibre surface	Plate surface
Si	18.6	29.6	29.9
Al	6.1	5.0	8.0
Mg	2.2	1.8	—*
Ca	6.3	3.4	7.9
B	4.1	0.0	—*
F	0.4	0.4	—*
O	61.8	59.9	54.2
Na/K	≈ 0.01	0.0	—*

* Below the detection limit in X-ray photo electron spectroscopy (XPS). Data from differing sources – slight variations possible.

chemistry of E-glass (Table 19) shows that the surface is silica rich. In addition, it is reported that a multimolecular layer (*c.* 20 monolayers) of water is adsorbed by **Hydrogen bonding** through surface hydroxyl groups. The adhesion promoter therefore has the following function: (1) to displace adsorbed water; (2) to create a surface which can be fully wetted with resin; (3) to develop strong interfacial bonds between the fibre and matrix, which minimize the interfacial free energy and maximize the work of adhesion (see Young's equation in **Wetting and spreading**). Thus the adhesion promoter has a primary role to interact with both the fibre surface and the matrix and is consequently called a **Coupling agent**. The above requirements are best achieved for most matrices, with organo-silanes. However, chrome complexes and titanate coupling agents are also available.

For measurement of adhesion see **Fibre–matrix adhesion – assessment techniques**.

The structure of the silane coating

The application of organosilane coupling agents has been described by Plueddemann.[1] More recently, the adhesion mechanisms have been reviewed by Jones.[2]

The silane coupling agents have a simplified structure

$$
\begin{array}{c}
R' \\
\backslash \\
O \\
\backslash \\
\quad Si\!-\!R \\
O \diagup \quad | \\
R' \diagup \quad O \\
\qquad | \\
\qquad R'
\end{array}
$$

where R' is C_2H_5 or CH_3 and R is a reactive or resin-compatible functional group such as a vinyl (for unsaturated polyester, epoxy or amino for epoxy resins). In the aqueous size, the $>Si\!-\!OR'$ group hydrolyses. This is often acid catalysed by adjusting the pH to *c.* 3.5 with acetic acid. As shown in reaction [i] further condensation of the silanol groups can occur leading to a complex deposit.

$$
R\!-\!\underset{\displaystyle OR}{\overset{\displaystyle OR'}{Si}}\!-\!OR' \xrightarrow[-ROH]{+H_2O} R\!-\!\underset{\displaystyle OH}{\overset{\displaystyle OH}{Si}}\!-\!OH \xrightarrow{+H_2O} R\!-\!\underset{\displaystyle O}{\overset{\displaystyle OH}{Si}}\!-\!O\!-\!\underset{\displaystyle OH}{\overset{\displaystyle O}{Si}}\!-\!R \qquad [i]
$$

Since these reactions occur at the glass surface, there is a competition between surface silanol-coupling agent condensation, step-growth polymerization and cyclic oligomer formation.

The structure of the deposit has been the subject of chemical analysis over the last decade as sensitive analytical techniques became available. While individual coupling agents (varying R) polycondense to differing degrees, it is generally accepted that three distinct layers are deposited on to the glass surface as observed by differing hydrolytic stabilities: a strongly chemisorbed layer at the immediate glass surface with loosely chemisorbed and physisorbed overlayers. The latter can amount to a significant proportion of the deposit (80 out of 100 monolayers in the case of γ-methacryloxysilane (γ-MPS)[3]) and comprises cyclic and linear oligomers. The intermediate layer is a misnomer since its enhanced hydrolytical stability arises from the formation of a three-dimensional network polysiloxane. In the case of γ-MPS, the strongly chemisorbed layer as well as being bound to the glass surface through siloxane bonds may be homopolymerized through the functional group R.

Mechanism of adhesion at the glass-fibre–resin interface

Several adhesive mechanisms (cf. **Theories of adhesion**) have been proposed to account for the enhancement of interface-dominated properties (such as retained tensile strength after environmental conditioning): (1) the chemical bonding theory; (2) the deformable layer hypothesis; (3) the surface wettability hypothesis; (4) the restrained layer hypothesis; (5) the reversible hydrolytic bonding mechanism.

The simplistic chemical bonding theory is usually cited but cannot explain all the observations, not least the efficacy of γ-aminopropyltriethoxysilane (γ-APS) under conditions when chemical bonding through the amino group cannot be readily understood. The last-mentioned theory (5) allows for reversible reformation of chemical bonds on drying out of aqueous conditioned specimens.

It is now becoming clear that the adhesive bond arises from a combination of chemical bonding and the formation of an interpenetrating network between the polysiloxane and matrix resin. The former arises because the physisorbed oligomers dissolve into the resin leaving a porous structure with exposed functional groups into which the matrix resin can diffuse and copolymerize. This structure is schematically illustrated in Fig. 55. The presence of the so-called interphase has been confirmed by scanning **Secondary ion mass spectrometry** (see Thomason in ref. 2) and ^{13}C NMR.[4] The latter work also confirmed the reaction of the amino group in γ-APS with polyamide acid end-groups. A consequence is that the concentration of silane in the 'size' emulsion can significantly influence the adhesion. The structure of the interphase in the composite is further complicated by the presence of the dissolved size and/or binder in the matrix and the interpenetrating network. As a result, the heterogeneity

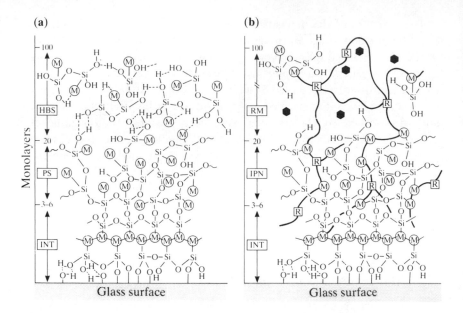

Fig. 55.   Schematic diagrams of (a) a polysiloxane deposit on glass fibre;
M is the methacryloxypropyl group in γ-methacryloxypropyl
trimethoxy silane and (b) the composite interface. Here R–M
is the interpenetrating copolymer with the resin matrix (R),
(●) is the dissolved binder and/or size; PS = polysiloxane,
HBS = hydrogen bonded oligomeric silanes, IPN = interpenetrating network, INT = interface

and thermomechanical properties of the interphase can vary with curing
temperature (and glass type) leading to a modification of the stress transfer
mechanism between fibre and matrix (see also **Theories of adhesion**).

There is a related article on **Fibre–matrix adhesion – carbon fibres**.

References

1. E P Plueddemann, *Silane Coupling Agents*, 2nd ed, Plenum Press, New York,
    1991.
2. F R Jones (ed.), *Interfacial Phenomena in Composite Materials*, Butterworths,
    London, 1989.
3. H Ishida, J L Koenig, *J. Polym. Sci. Polym. Phys. Ed.*, **18**, 1931 (1980).
4. T L Weeding, W S Veeman, L W Jenneskens, H Angad Gaur, H E C Schuurs,
    W G B Huysmans, *Macromolecules*, **22**, 706 (1989).

Filled polymers

J R G EVANS

The addition of particulate fillers to polymers may serve to reduce cost, increase elastic modulus, confer flame retardency or produce electrical conductivity. In the same way, composites may be prepared with particular dielectric, magnetic or piezoelectric properties.[1] In general, the addition of particulates reduces impact strength, elongation at failure and may reduce failure stress. Considerable effort is therefore devoted to modification of the interface of particulate composites in order to improve mechanical properties. High-modulus fibres such as glass, carbon, Kevlar, boron or silicon carbide may also be incorporated in polymers for enhancement of specific modulus and specific strength (see **Fibre composites – introduction**). The properties of the resulting composite are dependent on the shear strength of the fibre–matrix interface and the resistance to environmental deterioration is dependent on resistance to hydrolysis of the interface, notably in the case of glass-fibre reinforced plastics. In general low interfacial shear strength in fibre reinforced plastics results in low mechanical strength accompanied by improved toughness arising from the additional energy absorption associated with fibre pull-out or crack bifurcation. Improved interfacial bonding often results in high mechanical strength but lower toughness.[2] In particulate filled composites low interfacial shear strength often results in both reduced strength and toughness.

Broadly, two strategies for improving mechanical properties can be identified. In the first place matrix and filler can be selected to provide strong interfacial adhesion such as that arising from acid–base interactions (see **Acid–base interaction**). Secondly, a range of silane, zirconate, titanate and zirco-aluminate **Coupling agents** is available for enhancement of interfacial shear strength.

Coupling agents are intended to provide hydrogen bonding to hydrated oxides, and in some cases silanol groups form primary siloxane bonds to siliceous surfaces. Silane adsorption depends on the nature of the mineral surface and, in general, minerals with low isoelectric points in water (i.e. 'acidic' minerals such as silica and its glasses) present better adhesion with silane coupling agents than surfaces with high isoelectric points (such as magnesium or calcium-based minerals).[3] Coupling agents should also be selected to react chemically with the resin, but in the case of polyolefins, which constitute a high tonnage output from the plastics industry, this is not immediately possible. Chemical reaction may be initiated by peroxides, by using azide coupling agents or by the use of an interpenetrating polymer network such as that afforded by amino-functional silanes. In addition to mechanical properties, viscosity and hence processing behaviour of polymers is also influenced by filler additions and may be modified by coupling agents. Unfortunately, depending on the filler and polymer, coupling agents may either increase or reduce viscosity.

The selection of polymers to provide acid–base interactions has been explored by Fowkes and co-authors[1] and by Mansen (see **Acid–base interactions**, ref. 5). For a range of composites based on **Ethylene–vinyl acetate copolymers** and surface-treated silicas there was an overall correlation between relative tensile strength, elastic modulus, elongation at break and the extent of the acid–base interaction. Similar conclusions were drawn from a study of zinc and wollastonite fillers in **Epoxide**, polyvinylchloride, polystyrene and polymethylmethacrylate resins.

The effects of the interfacial adhesion on the flexural strength, elastic modulus and toughness of glass-bead-filled epoxy and polyester resins of varying filler loadings were analysed by a **Finite element** method and experimentally by Sahu and Broutman[5] who used a mould **Release** agent and a coupling agent to decrease and increase the adhesion respectively. Elastic modulus was unaffected by interfacial adhesion below 50 vol% filler as was the impact energy. Flexural and tensile strength, on the other hand, were significantly increased by filler matrix adhesion.

References

1. H S Katz, J V Milewski (eds), *Handbook of Fillers and Reinforcements for Plastics*, Van Nostrand Reinhold, New York, 1978.
2. D Hull, *An Introduction to Composite Materials*, Cambridge University Press, 1981.
3. E P Plueddemann in *Interfaces in Polymer Matrix Composites*, ed. E P Plueddemann; *Composite Materials*, Vol. 6, eds L J Broutman, R H Krock, Academic Press, London, 1974.
4. F M Fowkes, *Rubber Chem. and Technol.*, **57**, 328 (1984).
5. S Sahu, L J Broutman, *Polym. Eng. Sci.*, **12**, 91 (1972).

Finite element analysis

A D CROCOMBE

Principles

The finite element technique is an approximate numerical method for solving differential equations. Within the field of adhesive technology it is used almost exclusively to determine the state of stress and strain within a bonded joint. As in many other numerical methods solution is obtained by assuming an arbitrary form of trial function or guess. The trial function consists of a number of unknown parameters which are selected to make the function a best fit to the true solution. What sets the finite element method apart from the other numerical methods is that the region of interest is discretized into a number of small finite-sized elements (hence the name of the technique). The trial function is then defined in a piecewise manner over each of these elements separately with certain continuity requirements enforced between trial functions at element boundaries. In the case of stress analysis the unknown parameters which make up the trial function are structural displacements defined at points called nodes (which usually lie on the element boundaries). Having found the displacements, the strains and hence the stress can easily be reconstituted.

What follows is a very brief symbolic synopsis of a linear finite element analysis. The stresses σ and the strains ε at an arbitrary point in an element can be expressed in terms of the nodal displacements δ as

$$[\varepsilon] = [B][\delta] \qquad [\sigma] = [D][\varepsilon] = [D][B][\delta]$$

Virtual work principles are used to find expressions for internal nodal forces that are equivalent to the stress distribution and these in turn are summed and equated to the external nodal forces F to give the finite element equations which involve a stiffness matrix $[K]$:

$$[F] = [K][\delta]$$

where

$$[K] = \int_{vol} [B]^{T}[D][B]\, dv$$

The stiffness matrix is simply a function of the geometry of the structure and the material properties and can be calculated readily. The force vector will usually be known and thus the displacements can be found by solution of the resulting set of simultaneous equations. There are a number of introductory texts on the subject, but ref. 1 provides more in-depth reading.

Application to adhesive joints

The finite element method has been used to analyse a large number of adhesive joints including double, single and thick lap shear, scarf, peel, double cantilever, cracked lap shear, edge notched flexure, tubular lap, butt and many others. It would be beyond the scope of this synopsis to deal adequately with this, but a good source of more detailed reference is found in refs 2 and 3 and elsewhere in this handbook (see **Stress in joints**). Instead it is intended to address the problem of how the finite

This region shown enlarged below

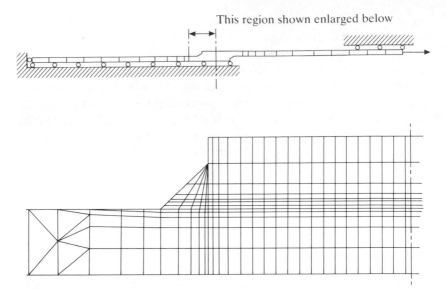

Fig. 56. Typical finite element mesh for an adhesive joint

element method should be used to analyse an adhesive joint and what results can be obtained.

The elements that are most frequently used in adhesive joint analysis are known as membrane elements. These have either three or four sides and usually model either linear or quadratic variation of displacement, depending on the number of nodes. These elements are used to model the cross-section of the joint and an example of such a mesh is shown in Fig. 56. Clearly an assumption has to be made about the out-of-plane behaviour and this is usually taken to be plane strain.

If there is some concern over the assumptions involved using membrane elements or joint loading also occurs in the out-of-plane direction, then it is possible to use full three-dimensional elements to model the joint. However, because of the computational overheads involved, such elements are rarely used.

Before carrying out a finite element analysis it is necessary to assign appropriate material properties to the elements. These are usually known, or well documented for the adherend, but may have to be determined by test for the adhesive; however, consult **Literature on adhesion**. There are a number of tests which can be used for such purposes including bulk tension and torsion, thick-lap shear, **Napkin ring test** and others (see **Tensile tests, Shear tests** and **Testing of adhesives**).

The amount of material data required depends on how the material is assumed to behave. A linear elastic analysis requires only the modulus and Poisson's ratio. This would be used to assess the general level of stress in a joint and might successfully predict the strength of a joint with a brittle adhesive, which might typically be based on maximum stress levels. An elasto-plastic analysis requires data defining yield and post-yield behaviour. This would be required if joint strength predictions for joints with more ductile adhesives are required, often based on maximum levels of strain. Finally visco-plastic analyses require data defining the relaxation or creep of the adhesive (see **Viscoelasticity**), and such a class of analysis would be required if a joint were to be exposed to sustained levels of load (see **Creep**).

Having defined the mesh and the material properties it is necessary to run the analysis. Often a commercial finite element package can be used, but these can be limited in certain situations. These include modelling both adhesive yield (which is sensitive to the level of hydrostatic stress) and some adhesive viscoelastic responses). Further details of these can be found in refs 3 and 4 respectively.

As the number of elements increases so does the accuracy of the results. However, such mesh refinement highlights a problem in modelling the end of the adherend. For simplicity this is usually taken as being perfectly sharp, but with increasing mesh refinement the stresses and strains in this area become singular. Clearly it is difficult to use maximum levels of stress or strain to predict joint strength and this is an area of current research.

In addition it is possible to use the finite element results to evaluate **Fracture mechanics** parameters such as energy release rate, *J*-integral, stress intensity factor, etc. Further reference to this aspect can be found in ref. 5.

References

1. O C Zienkiewicz, R L Taylor, *The Finite Element Method*, McGraw-Hill, New York, 1989.
2. F L Matthews, P F Kilty, E W Godwin, A review of the strength of joints in fibre-reinforced plastics. Part 2. Adhesively bonded joints, *Composites*, **13**, 29–37 (1982).
3. R D Adams, W C Wake, *Structural Adhesive Joints in Engineering*, Elsevier Applied Science Publishers, London, 1984.
4. J N Reddy, S Roy, NOVA: a non-linear visco-elastic analysis program for adhesive joints, *Adhesion Science Review, Proc. 5th Annual Progress Review*, Virginia Technical Centre for Adhesion Science, 1987.
5. G P Anderson, S J Bennett, K L De Vries, *Analysis and Testing of Adhesive Bonds*, Academic Press, New York, 1977.

Flame-sprayed coatings

H REITER

Background

Surface treatments have been used to enhance the appearance or properties of components since the early days of metalworking. In modern engineering they range from **Conversion coatings**, involving a chemical reaction with the substrate, to deposition coatings where the bond with the substrate is physico-mechanical in nature. Plastic coatings belong to the latter category.

The idea of spraying metals is attributed to a Swiss engineer, Dr M U Schoop, who worked at the beginning of the century; by the 1920s the technique had been developed into a commercial process. Since then progress has been mainly in the development of equipment and better process control, as well as the introduction of new materials particularly for wear and high temperature corrosion resistance. The high melting temperatures of carbides and refractory oxides led to the development of the plasma flame gun which allowed materials such as tungsten carbide, alumina, zirconia and many other refractory oxides to be successfully sprayed. The flame spraying of plastics, however, did not come into prominence until the early 1950s when the demand for corrosion protection of North Sea oil rigs provided a new impetus for plastic coatings.

Principles of flame spraying

Flame spraying is a deposition process involving essentially four basic features:

1. The material to be sprayed is heated (usually in powder form) in a suitable source, which can be an oxy-gas flame, an electric arc or a plasma arc. For plastics, oxy-gas flames based on hydrogen, acetylene or propane provide an adequate heat source.
2. The plastic or molten material is projected on to the substrate at high velocity (up to sonic velocity) by means of a carrier gas.
3. The hot particles on striking the surface form splats which interlock and bond together, thereby gradually building up a coating of the desired thickness.
4. The coating bonds to the substrate by mechanically interlocking with a roughened surface (see **Mechanical theory**).

The capital costs for thermal spray equipment are low. The main features of a thermal spray gun are illustrated in Fig. 57. Powder is

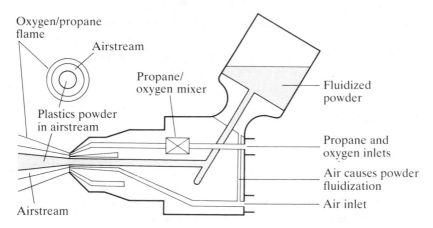

Fig. 57. Schematic diagram of a thermal spray gun

introduced into the gas stream either from a small hopper attached to the hand gun or from a main container and transported into the flame, where it is heated to a molten or semi-molten state. After heating the particles are propelled by the gas stream on to the surface to be coated, usually a distance of 25–50 mm from the gun. Early equipment for spraying plastic materials was essentially the same as that for metals, but in recent years special spray guns have been developed to overcome the problems of overheating the coating materials and to better cope with the different flow characteristics of plastics compared to metal powders. Powders used for electrostatic coatings (20–50 μm) are usually too fine for flame spraying as the particle would tend to overheat, but those used in fluidized beds (80–200 μm) are generally suitable. Fully automated equipment for in-house spraying of large structures is now available in addition to the more common hand-held equipment.

For good adhesion the substrate surface must be free from surface contamination such as scale, rust, grease or paint. This is usually achieved by grit-blasting, which has the added advantage of providing a surface suitable for mechanical keying of the coating (see **Mechanical theory**). It is essential that spraying is carried out as soon after grit-blasting as possible. This is particularly important in the case of steel substrates where the more reactive nature of the surface after grit-blasting leads to rapid oxidation (rust formation), particularly in damp atmospheres. It is also advisable to preheat the substrate in the range 150–200 °C.

Materials

The most commonly applied materials are polyamide (nylon 11) and **Ethylene–vinyl acetate copolymer** (EVA) the latter now becoming widely used for sealing concrete structures. In addition many other thermoplastic materials are used for local repair of damaged coatings. The abrasion resistance of coatings can be enhanced by incorporating mineral fillers into the polymer powder (see **Filled polymers**).

Applications

The majority of coatings are applied for corrosion protection against a wide range of chemical and environmental conditions. While the cost per square metre of plastic coatings is higher than, for example, a zinc coating and painting, plastic coatings are generally thicker, more durable and an insurance against explosion hazards, which makes them very attractive to the oil industry. Thermal spraying also has the advantage of being able to coat structures which could not normally be dipped or plated because of their size. A further advantage of the process is that it can be applied on site to structures after final assembly as well as for repair work.

Thermally sprayed coatings also find use in the food industry where they can satisfy the stringent hygiene requirements for processing plant.

Select references

W E Ballard, *Metal Spraying and Flame Deposition of Ceramics and Plastics*, 4th edn, Griffin, London, 1963.
J Bishop, Equipment for the production of flame sprayed plastic coatings, *Surf. J. Int.*, (1), 97–9 (1986).
J D Hearn, A Weld, R D White, Development of thermal spray equipment for plastic coatings, *9th International Thermal Spraying Conference*, The Hague, 1980, pp. 211–18.

Flame treatment

D BRIGGS

Flame treatment is mainly used for pretreating plastic articles of fairly thick section, such as blow-moulded bottles and thermo-formed tubs. Its early application to polyolefin film treatment was not sustained because of process control difficulties and safety problems; however, these have been overcome in recent times and flame treatment competes with **Corona discharge treatment** in this area once again.

An attractive feature is the ability to handle complex shapes, such as the bottle illustrated in Fig. 58. Each burner consists of a large number of closely spaced jets, usually in staggered rows. In the case of the bottle treatment is achieved by bringing it into position on a line, rotating it and then moving the burners into their fixed geometry for a short time (<1 s). Clearly this type of treatment is suited to robotics, and large items

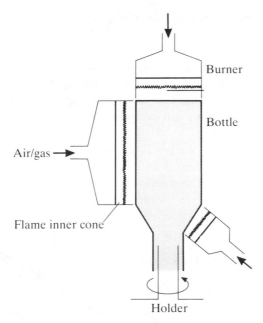

Fig. 58. Schematic diagram of set-up for flame treatment of blow-moulded bottles

such as plastic car bumpers are treated in this way (to achieve good paint adhesion).

The most important variables in the process are the air:gas ratio, the air/gas flow rate, the nature of the gas, the burner–surface separation and the exposure time. The gas used is either mains (mainly methane) or bottled (propane or butane). For optimum treatment an oxidizing flame should be used, i.e. with an excess of oxygen over that required for complete combustion. For a given air:gas ratio, treatment level in a given exposure time increases as the volume of mixture burned increases. It is clearly important that the exposure time is not sufficient for the surface polymer to melt (the flame temperature is $\sim 2000\,°C$); in the case of film treatment the film is treated as it passes over a cooled roll.

Surface analysis shows that very high levels of surface oxidation are possible, even under sub-optimal treatment conditions. The mechanism of oxidation is not fully understood. Both thermal and 'plasma' radical oxidation is possible.

Select reference

D Briggs in *Surface Analysis and Pretreatment of Plastics and Metals*, ed. D M Brewis, Applied Science Publishers, London, 1982, p. 210.

Footwear applications of adhesives

S G ABBOTT

The footwear industry makes widespread use of adhesives, appreciating their ability to provide a flexible, uniform and unobtrusive joint between dissimilar materials. Adhesives have replaced mechanical fastenings wherever practicable; the main exceptions are the sewing of uppers and nail attachment of heels.

Figure 59 summarizes the principal applications of adhesives in footwear. These can be considered under two headings, sole attaching and ancillary operations. The former is the more technically demanding as

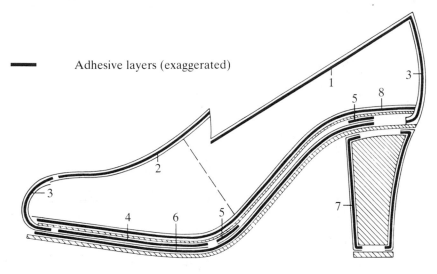

Use	Adhesive types	Typical peel strength (N mm^{-1})
1. Topline folding	Hot melt – polyamide	
2. Linings	Latex – natural rubber; hot melt – polyamide, EVA	0.2–0.5
3. Toe puff, heel stiffener	Hot melt coating – EVA	0.5–1
4. Lasting	Hot melt – polyamide, polyester	0.5–1
5. Shank	Hot melt – EVA, polyamide	–
6. Sole attaching	Solvent-based – polyurethane, polychloroprene	3–7
7. Heel covering	Solvent-based/latex – polychloroprene	0.5–1
8. Insock	Latex – natural rubber, acrylate	0.2–0.5

Fig. 59. Typical uses of adhesives in footwear

sole bond integrity is critical for satisfactory service life. With the ancillary operations, ease of use and retention of aesthetics may be as important as level of bond strength which is often limited by the surface strength of the materials to be bonded.

Adhesives for sole attaching

About 90% of UK-produced footwear has soles attached with adhesive. Soles are mostly rubber or thermoplastics such as PVC, with small proportions of other materials including polyurethane and ethylene–vinyl acetate (EVA). Leather uppers feature in half the output and predominate in men's shoes, but other types include polyurethane-coated fabrics and textiles.

Process requirements Adhesives based on solutions of rubbery polymers are used due to their ability to wet, penetrate and bond surfaces as diverse as roughed leather and smooth PVC. The adhesive polymers are not usually reactive, although an isocyanate curing agent is sometimes added to enhance heat resistance or adhesion.

In the 'stuck-on' process, adhesive is applied to the prepared surfaces of sole and upper and allowed to dry. At bonding, the film on at least one surface, usually the sole, is heat-reactivated to 75–90 °C and will coalesce under pressure with the second film to form the bond. This, after 10–15 s pressing, must have sufficient initial 'green' strength to resist any tendency for the sole to spring away. In the 'moulded-on' process, the soling compound is injection-moulded (PVC, SBS rubber) or directly vulcanized (natural, SBR or nitrile rubbers) on to the adhesive-coated upper to form the sole.

In all cases, good control of the bonding process is important.[1]

Development of adhesives and primers Sole bonding developed in the 1930s and 1940s using natural rubber or nitrocellulose adhesives. The main growth came during 1950–65 with the development of polychloroprene adhesives compounded with metal oxide–phenolic resinates to improve **Tack**, heat resistance and adhesive strength.[2]

Polyurethane adhesives were introduced in the mid-1960s as they could satisfactorily bond plasticized PVC due to inherent plasticizer resistance, and have now largely displaced polychloroprenes. The adhesives are usually based on linear polyurethanes produced from polyester diols and 4,4'-diphenylmethane-diisocyanate. Resins and chlorinated rubber may be added to improve tack and heat resistance.[2] Use on thermoplastic (SBS) and vulcanized rubbers was facilitated by the SATRA-developed halogenation process[3] which saturates the butadiene in the rubber making the surface compatible with polyurethane adhesives. Low-toxicity halogenating solutions such as Satreat are now in factory use world-wide. Primers for other materials such as EVA and nylon have been developed which are less toxic than the isocyanates used previously.

In the moulding-on of rubber soles, the adhesive is a solvent solution of the same type of vulcanizable rubber as the soling itself.

Roughing of upper materials continues to be necessary to remove both finishes and weakly attached outer layers – the grain of leather and coatings of synthetic materials.

Solvent hazards The inflammable, toxic nature of the solvents in primers and adhesives means that attention has to be given to maintenance of safe factory atmospheres. Aqueous dispersions of polyurethanes have been considered for sole attaching, but lack versatility and show lower bond strengths on some materials. Reactive polyurethane hot melts show promise but will require investment in new equipment. Methods of destroying solvents, such as catalytic combustion, rather than venting to atmosphere are the subject of current research (see also **Health and safety**).

Adhesives for ancillary operations

Natural rubber latex, thickened or compounded with resins to improve tack, has been used for many years and favoured for clean working and good initial 'grab'. Application is either to one surface with immediate wet bonding, or to both surfaces with bonding after allowing to dry. During the last 30 years, **Hot melt adhesives** have augmented and partly displaced latex, offering rapid application and bonding, often by in-machine processes. The principal types are dimer acid/diamine polyamides, terephthalate polyesters and **Ethylene–vinyl acetate** (EVA) **copolymers**.[4]

Topline folding. Hot melt adhesive is extruded along the edge of the upper as it is folded. Polyamides are used due to their quick-setting properties and flexibility.

Linings. Attachment is by latex, or by polyamide or EVA hot melt coating – either continuous or in dot form to maintain handle and moisture permeability. Bonding may be immediately after application or by heated platen press.

Toe puff and heel stiffener. These are usually thermoplastic-impregnated fabrics carrying a layer of EVA hot melt, and are bonded to the upper by heated platens or clamps.

Lasting. The upper is shaped by pulling over the last and securing to the insole. Hot melts have largely displaced metal tacks except in the heel area. Polyamides are used at the side of the shoe for flexibility and quick-setting polyesters at the forepart.

Shank. This metal or wooden stiffener is either riveted to the insole or attached by dipping each end into EVA or polyamide hot melt adhesive.

Heel covering. As an alternative to lacquering, polystyrene heels of women's shoes may be covered with leather or synthetic material to match the upper. Solvent-based polychloroprene is applied to the heel by dipping or brushing and polychloroprene latex to the cover. The bond is made cold after allowing to dry.

Insock. Latex is applied to the insock which is bonded by wet stick. Acrylic emulsions or solvent-based polyurethanes are sometimes used with difficult materials. **Pressure-sensitive adhesives** are used, but creep resistance must be high to avoid slippage in wear.

This list is not exhaustive and adhesives are used for a variety of other tasks such as temporarily fitting components before sewing and folding of strapping. Cyanoacrylate 'superglues' (see **Cyanoacrylate adhesives**) have found a role in bonding decorative trims.

Some test methods are listed in **Appendix 1**.

References

1. S G Abbott, *Int. J. Adhesion and Adhesives*, **5**, 7 (1985).
2. R S Whitehouse in *Synthetic Adhesives and Sealants*, ed. W C Wake, John Wiley, Chichester, 1987, Ch. 1.
3. D Pettit, A R Carter, *J Adhesion*, **5**, 333 (1973)
4. A Hardy in *Synthetic Adhesives and Sealants*, ed W C Wake, John Wiley, Chichester, 1987, Ch. 2.

Fourier transform infra-red spectroscopy – FTIR

B C COPE

Infra-red (IR) spectrometry has long been one of the major analytical techniques used by the organic chemist, and hence the adhesive technologist. The traditional dispersive instrument is relatively cheap and robust and does not require expert technicians for its operation. It may be used to identify unknown materials, to detect surface contamination (through use of multiple internal reflectance (MIR) techniques (see **Infra-red spectroscopy of surfaces**) and to give information on stereo-regularity, molar mass and crystallinity. However, the dispersive spectrophotometer has several drawbacks; it takes a few minutes to scan the IR wavelength range and hence produce the spectrum, a relatively large sample is needed and it is difficult to produce well-balanced spectra of strong absorbers.

The Fourier transform (FT) spectrophotometer has been in existence for some considerable time without achieving any real importance, but in the last few years the development of FTIR instruments with built-in personal computers has dramatically enlarged the scope of the technique and reduced the level of skill and training needed for its practice.

The main difference between dispersive and FT spectroscopy is that in the latter the slow scanning by monochromator is replaced by a short burst of IR radiation that passes through an inferometer before the sample. The FT thus generated is converted to a spectrum by the instrument's computer and stored in a memory. The information to produce the spectrum itself may be stored permanently on disc and used to print out the spectrum whenever it is required.

An immediate advantage of FTIR is that a very high accuracy of wavenumber is combined with an excellent signal : noise ratio. An ultimate result is that very small samples can be made to yield usable spectra; a microgram may be sufficient. The spectrum is produced in much less than a second permitting kinetic studies and the elucidation of transient phenomena. This high-speed operation is particularly useful in allowing FTIR to be used as a diagnostic detector in gas chromatography, high-performance liquid chromatography and gel permeation chromatography.

It is possible to build the IR optics of an FTIR spectrophotometer into a specially adapted microscope. Thus very small particles, inclusions and segregated species, down to around 5 μm diameter, may be targeted by the beam and analysed separately from their surroundings.

The incorporation of a photoacoustic spectroscopy system (PAS) into FTIR obviates the need for specimen preparation imposed by conventional IR and enables spectra to be recorded from irregular bulk specimens such as, in the context of adhesives, hot-melt flakes or spread adhesives. The technique depends on the transformation, by the sample, of IR energy into heat. The consequent thermal expansion is detected as a photoacoustic signal when the intensity of the incident beam is modulated in the acoustic frequency range. This signal is picked up by a microphone and amplified and processed to yield the spectrum.

However the spectrum is obtained, the dedicated computer built in to the spectrophotometer allows great flexibility in manipulation. Unknown spectra may be compared with the contents of a self-generated or commercially available library and a tentative identification displayed on the screen. The accuracy of wave number maintenance permits scale expansion of characteristic regions within spectra and highlights small variations from reference. The computer also enables the user to subtract a reference spectrum from an unknown thus permitting the resolution of mixtures and laminates.

These advantages of FTIR make it particularly valuable in addressing problems related to adhesion. Carbon black-filled rubbers had been almost impossible to examine by IR spectroscopy, but with FTIR spectra of the polymer can be obtained (see **Rubber adhesion, Rubber: fillers**). Much better quality spectra can be obtained by surface techniques such as ATR and diffuse and specular reflectance spectroscopy (see **Infra-red spectroscopy of surfaces**). This aids identification of thin surface layers whether produced by surface treatments or residues after bond failure (see several articles on **Pretreatment** and **Primers, Locus of failure**). Silane **Coupling agents** have been identified by the technique.

Select references

H Ishida, *Fourier Transform Characterisation of Polymers*, Plenum Press, New York, 1987.
H W Siesler, K Holland-Moritz, *Infra-red and Raman Spectroscopy of Polymers*, Marcel Dekker, New York, 1980.

FPL etch

D E PACKHAM

There are many references in the literature to this method of surface preparation of aluminium. The name refers to the US Forest Products Laboratory which published a specification based on etching the metal at *c.* 65 °C in a solution of sodium dichromate and sulphuric acid, sometimes loosely referred to as 'chromic acid'. In the UK a similar Ministry of Supply Aircraft Process Specification known as DTD 915 has been widely used. Over the years a number of slightly different solutions ('chromic acid etches') and procedures have been developed.[1-3] Clearfield gives recipes for six such solutions.[2]

When optimized, these are suitable preparations for both adhesive bonding and painting: their effect is to give a reproducible, oxidized surface on the aluminium. Initial adhesion is generally good, and **Durability** fair, but not as good as with **Anodizing** pretreatments (see **Pretreatment of aluminium**).

The precise nature of the layer formed, and the subsequent adhesion and durability, depend on the composition of the solution used, the time and temperature of etching, rinsing conditions and the particular alloy being treated.[1,2,4] For example, the fresh solution does not give good results, unless it is artificially 'aged' by adding some aluminium;[1,2] the presence of traces of fluoride can reduce bond durability;[1,5] close attention to temperature control is necessary;[4] rinsing procedures can exert a critical effect on bond strength.[1] Optimization of conditions, together with careful control of process variables, are therefore necessary to achieve satisfactory results.

The morphology of the surface has been studied by transmission **Electron microscopy**.[5] Under optimum conditions it consists of a cell structure with oxide whiskers protruding from the surface. Venables[5] comments that microscopic interlocking (see **Mechanical theory**) appears to be a crucial factor in the adhesion. It may be that the sensitivity to processing conditions, mentioned above, is a result of the production of a morphology with less potential for interlocking.

The use of ferric sulphate as an oxidizing agent in the place of chromates has been developed,[2] and overcomes some of the disposal problems associated with the toxicity of FPL and similar etching solutions (see **Health and safety**).

Chromic acid etching and other techniques are compared in **Pretreatment of aluminium**.

References

1. A C Moloney in *Surface Analysis and Pretreatment of Plastics and Metals*, ed. D M Brewis, Applied Science Publishers, London, 1982, p.175.
2. H M Clearfield, D K McNamara, G D Davis in *Adhesion and Adhesives*, ed. H L Brinson, *Engineered Materials Handbook*, Vol. 3, ASM (Ohio), 1990, p. 259.
3. ASTM D2651 in *Annual Book of Standards*, American Society for Testing and Materials, Philadelphia, published annually.
4. D E Packham, *Aspects of Adhesion* **6**, 127 (1971).
5. J D Venables, *J. Mater. Sci.*, **19**, 2431 (1984).

Fracture mechanics

A J KINLOCH

Introduction

Adhesive joints usually fail by the initiation and propagation of flaws and, since the basic tenet of continuum fracture mechanics is that the strength of most real solids is governed by the presence of flaws, the application of such theories to adhesive joint failure has received considerable attention. The reader is referred to reviews by Broek,[1] Kinloch and Young[2] and Williams[3] for detailed discussions of the general principles of fracture mechanics and to a review by Kinloch[4] on the application of fracture mechanics to the failure of adhesive joints.

The main aims of the various fracture mechanics theories are to analyse mathematically the loads at which the flaws propagate and describe the manner in which they grow. The source of naturally occurring flaws, sometimes referred to as 'intrinsic flaws' may be voids, cracks, dirt particles, additive particles, inhomogeneities in the adhesive, etc. which may be initially present at a critical size or develop during the fracture test. Fracture mechanics has proved to be particularly useful for such aspects as characterizing the toughness of adhesives, identifying mechanisms of failure and estimating the service life of 'damaged' structures – the damage being in the form of cracks, air-filled voids, debonds, etc. and having arisen, for example, from environmental attack (see **Durability, Weathering of adhesive joints**), **Fatigue** loading, subcritical impact loads (see **Impact resistance**). Two main, interrelated conditions for fracture have been proposed.

Energy-balance approach

The energy criterion arises from Griffith's work which supposes that fracture occurs when sufficient energy is released (from the stress field) by growth of the crack to supply the energy requirements of the new fracture surfaces. The energy released comes from stored elastic or potential energy of the loading system and can, in principle, be calculated for any type of test piece. This approach therefore yields a measure of the energy required to extend a crack over unit area and this is termed the fracture energy or critical energy release rate and is denoted by G_c. It is important to note that the value of G_c cannot usually be directly equated to that needed solely to rupture molecular bonds, since the value of G_c also encompasses the dissipative-energy losses incurred around the crack tip, i.e. viscoelastic- and plastic-energy losses in the crack-tip damage zone. (cf. **Peel test**). In the case of bonded structures exhibiting bulk linear-elastic behaviour, the fracture criterion may be written as

$$G_c = \frac{F_c^2}{2b}\frac{dC}{da}$$

where F_c is the load at the onset of crack propagation, b is the specimen thickness, a the crack or debond length and C the compliance of the structure and is given by displacement/load. This equation is the foundation of many linear-elastic fracture-mechanics (LEFM) calculations of G_c, since if C is determined as a function of a, either experimentally or theoretically by an analytical or numerical technique, then dC/da may be found. If the bonded structure does not exhibit bulk linear-elastic behaviour then the fracture energy approach is still valid, but a different form of equation to that given above is needed to describe the value of G_c in terms of the loads and displacements applied to the joint.

Stress-intensity factor approach

The second approach comes from the work of Irwin who found that the stress field around a sharp crack in a linear-elastic material could be uniquely defined by a parameter named the stress-intensity factor, K, and stated that fracture occurs when the value of K exceeds some critical value, K_c. Thus, K is a stress-field parameter independent of the material, whereas K_c, often referred to as the fracture toughness, is a measure of a material property. However, this approach becomes far more difficult to employ when the crack is either in a thin adhesive layer or is located at the adhesive–substrate interface, and the reader is referred to articles by Kinloch[4] and Wang, Mandell and McGarry[5] for further details on this very complex issue.

See also articles on **Fracture mechanics test specimens, Blister test, Peel tests, adhesion – fundamental and practical**.

References

1. D Broek, *Elementary Engineering Fracture Mechanics*, Noordoff, Leyden, 1987.
2. A J Kinloch, R J Young, *Fracture Behaviour of Polymers*, Applied Science Publishers, London, 1983.
3. J G Williams, *Fracture Mechanics of Polymers*, Ellis Horwood, Chichester, 1984.
4. A J Kinloch, *Adhesion and Adhesives, Science and Technology*, Chapman and Hall, London, 1987.
5. S S Wang, J F Mandell, F J McGarry, *Int. J. Fracture*, **14**, 39 (1978).

Fracture mechanics test specimens

A J KINLOCH

Introduction

A basic aim of **Fracture mechanics** is to provide a parameter for characterizing crack growth which is independent of test geometry. Therefore, in order to investigate this requirement experimentally a wide range of different geometries have been developed in order to ascertain values of the various fracture parameters under modes II (in-plane shear) and III (anti-plane shear), as well as the more usual mode I (cleavage, or tensile, opening) case. (Although it should be noted that when a crack is propagating along a bimaterial interface the basic meaning of mode I, mode II and mode III is rather different; see ref. 1 for a discussion on this aspect.) The reader is referred to several papers[1,2] for further details, and to a separate article on the **Blister test** specimen.

Flexible joints

The energy-balance approach is generally the most applicable to flexible joints, since away from the crack tip the adhesive or substrates may not exhibit linear-elastic behaviour and so the stress intensity factor approach is invalid. The most common test methods are shown in Fig. 60.

Peel test (Fig. 60(a)). Assuming the strain in the tab is negligible (e.g. if (1) the peel forces are very low or (2) a fabric- or plastic-backed rubbery adhesive or a relatively thick metallic substrate is the peeling member) and plastic bending of the tab does not occur, then adhesive fracture energy (critical strain energy release rate) is given by

$$G_c = P(1 - \cos \alpha)$$

where *P* is the peel force per unit width, commonly termed the peel strength or peel energy and α is the peel angle. This is further discussed in **Peel tests**.

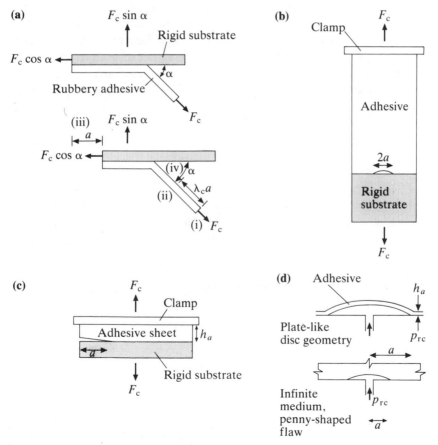

Fig. 60. Common fracture mechanics test methods

Simple-extension test piece (Fig. 60(b)). Adhesive fracture energy is given by

$$G_c = (\pi/\lambda_c^{1/2})aU_{dc}$$

where λ_c is the extension ratio in the rubber sheet, a the crack length and U_{dc} the strain energy density in the rubber sheet at the onset of crack growth.

Pure-shear test specimen (Fig. 60(c)). G_c may be expressed as

$$G_c = h_a U_{dc}$$

Rigid joints

A large number of test specimens have been designed to determine the crack growth behaviour of structural adhesive joints and some of the most common designs are shown in Fig. 61. Expressions for the adhesive fracture energy G_c are given below, but it should be noted that the value of dC/da can always be determined experimentally as a function of crack length and, for linear-elastic joints, such values may then be used together with the equation stated in the article on **Fracture mechanics** in order to ascertain the value of G_c. If the stress-intensity factor at fracture is required, then either the geometry factor must be ascertained by analytical or numerical methods or it must be deduced from the value of G_c. In the latter case it is important to use the correct modulus value. For a crack propagating in the adhesive layer then the value of the adhesive's modulus should be used. However, when the crack propagates at, or very close, to the interface the problem becomes complex and the reader is referred to ref. 1 for further details.

Double cantilever beam (Fig. 61(a)). This gives a mode I failure when the crack is in the adhesive layer. The value of dC/da is not constant, and under a constant load the value of G_I increases as the crack gets longer. It is used in ASTM D 3433 specification. For linear-elastic behaviour and thin adhesive layers, the value of dC/da may be expressed analytically by

$$\frac{dC}{da} = \frac{8}{E_s b} \left| \frac{3a^2}{d^3} + \frac{1}{d} \right|$$

where E_s is the modulus of the substrate material.

Tapered double cantilever beam (Fig. 61(b)). Again mode I, but the value of dC/da can be made constant by tapering of the arms of the specimen. Thus G_I is independent of crack length and this design is well suited to tests where the value of a is difficult to measure, e.g. environmental tests (see **Durability**) and dynamic fatigue tests. Also used

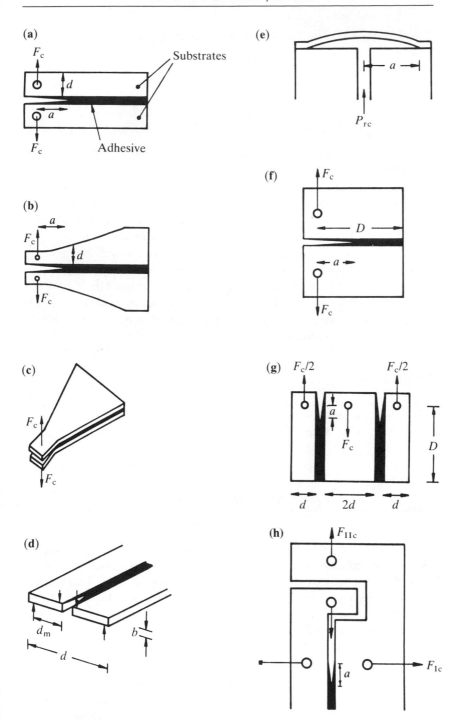

Fig. 61. Test specimen designs for determining crack growth behaviour

in ASTM D 3433. The value of G_{Ic} is given by

$$G_{\text{Ic}} = (4F_c^2 m)/E_s b^2$$

where

$$m = \frac{3a^2}{d^3} + \frac{1}{d}$$

where m is the geometry factor and E_s is the modulus of the substrate.

Width-tapered beam (*Fig. 61(c)*). The value of dC/da is now made constant due to a variation in width with crack length, rather than the beam height – this is very useful when using fibre-composite substrates.

Double-torsion (*Fig. 61(d)*). The value of dC/da is again constant and the force may be applied by a simple compression loading jig.

Blister test specimen (*Fig. 61(e)*). See the **Blister test** article.

Compact tension (*Fig. 61(f)*) *and Compact shear* (*Fig. 62(g)*). The former is a mode I specimen while the latter gives mode II.

Independently loaded mixed-mode (*Fig. 61(h)*). Cleavage and shear loads can be applied independently to produce either mode I or mode II conditions, or a combination of both.

References

1. A J Kinloch, *Adhesion and Adhesives, Science and Technology*, Chapman and Hall, London, 1987.
2. E H Andrews, A J Kinloch, *Proc. Roy. Soc.*, **A332**, 385 (1973).

Friction – adhesion aspects

A D ROBERTS

The two basic laws of friction have been known for a long time: the frictional resistance is proportional to the applied load and it is independent of the area of the sliding surfaces. Coulomb (1781) considered the possibility that friction might be due to molecular adhesion between the surfaces, but rejected the idea because if this were so the friction should be proportional to the area of the sliding bodies, whereas it was found to be independent of it. Nowadays, and especially after the investigations of Bowden and Tabor,[1] we accept that adhesion plays a role in friction, along with other factors such as ploughing hysteresis. Thus, in general, the frictional force involves one component associated with overcoming the adhesion between the surfaces, and another related to bulk deformation such as ploughing and hysteresis. The balance between the two depends on the materials involved and the conditions of the experiment. Had Coulomb known about rubber friction he might have taken the adhesion idea more seriously because its friction is manifestly related to the area of contact.

This article describes the adhesion component of friction by reference to rubber, an essential point being what is the real area of contact between solids when placed together. Experiments show that surfaces contain hills and valleys which are large compared with molecular dimensions (see **Roughness of surfaces**). The majority of solids are supported on the summits of the highest hills, so that the area of intimate contact is very small. Rubber is exceptional in that being relatively soft it can drape over the hilltops and, with enough applied load, begin to fill the valleys so that the real area of contact begins to approach the geometric area of the sliding bodies. Using optically smooth surfaces minimizes the micro-irregularities and makes possible investigations of how adhesion relates to friction.

Schallamach waves

If an optically smooth-surfaced rubber sphere is made to slide over dry plate glass, wrinkles in the rubber surface can be seen rapidly moving across the contact zone. These have been described by Schallamach as 'waves of detachment'. The rubber is in intimate contact between the waves and for this particular situation it is possible to relate the sliding friction to the interface adhesion through an expression involving a rate-dependent peel energy (see articles on **Rubber adhesion** and **Peel tests**). As a wave progresses across the contact region, energy Γ_p is required to peel rubber from the glass. Neglecting subsurface energy dissipation, then in steady-state sliding at a speed V, the frictional stress is given[2] by

$$F = \Gamma_p \omega / \lambda V$$

where the waves move with a speed ω and are spaced apart by a distance λ. Experiments suggest this expression is accurate to about 10% over the speed range for which the waves propagate.

Waves of detachment have also been found for wet contact. In water there is much less adhesion. The waves transport water within them through the contact zone rather than air as in dry contact. About half the wet friction can be accounted for using the above expression. The waves also occur when rubber slides on sufficiently cold ice. These observations suggest the waves may arise for a variety of conditions and be relevant to the performance of rubber articles possessing a smooth surface.

Significance of area of contact

The friction coefficient of a reasonably smooth spherical rubber slider on a flat track can be related via the contact area to operating parameters because the geometric and real contact area are similar. If the contact area is given as a function of the normal load W by the classical theory of Hertz,[3] and if the frictional force is proportional to this area of contact, then the friction coefficient can be written as

$$\mu = \pi \tau (9R/16E)^{2/3} W^{-1/3}$$

where R is the slider radius, τ the interfacial shear strength and E Young's modulus of the rubber. Knowledge of the interfacial shear strength is needed, and this will depend upon such factors as whether surfaces are clean and dry (i.e. high adhesion), contaminated by dust or lubricated (low adhesion). Model experiments suggest[2] that friction is proportional to the Hertzian stress parameter $R^{2/3} W^{-1/3}$ so that the above expression can be used to estimate the friction coefficient to a reasonable approximation.

With increasing surface roughness the friction coefficient becomes less sensitive to load, and the absolute level of friction declines. No satisfactory analysis exists for the 'adhesive' friction of such surfaces. The rougher a surface, the more overt the viscoelastic response of the rubber (see **Viscoelasticity**). In the absence of adhesion the friction coefficient can depend upon load raised to a positive index due to ploughing hysteresis.

Friction plays a part in **Powder adhesion**.

References

1. F P Bowden, D Tabor, *The Friction and Lubrication of Solids*, Oxford University Press, Oxford – Part I, 1950; Part II, 1964.
2. A D Roberts (ed.), *Natural Rubber Science and Technology*, Oxford University Press, Oxford, 1988.
3. S P Timoshenko, J N Goodier, *Theory of Elasticity*, 3rd edn, McGraw-Hill, New York, 1970.

Fusion welding by mechanical movement

M A GIRARDI

The joining of thermoplastics by softening may be achieved by external heating (see **Fusion welding – external heating**) or by processes involving mechanical movement. This article is concerned with the latter which include ultrasonic welding, spin welding and vibration welding.

Ultrasonic welding

Ultrasonic welding is a bonding process in which high-frequency mechanical vibrations, i.e. ultrasonic vibrations, are used for joining thermoplastics materials. The parts to be joined are held together under pressure and subjected to ultrasonic vibrations, usually of 20 or 40 kHz frequency. The consequent alternating stresses generate heat in the plastic at the joint interface by a combination of surface and intermolecular friction. The ability to weld a component successfully is governed by the design of the equipment, the mechanical properties of the material to be welded and the design of the components.

Two types of joint are made by this process (Fig. 62): (1) projection, (2) shear. The projection joint is the most commonly used. This is provided by a wedge-shaped protrusion, known as an energy director. This gives a high concentration of vibratory energy at its apex, resulting in a rapid heat build-up. Molten material from the energy director flows across the joint interface and fuses on cooling. In shear joints, such an energy director is not used. Heat operation is assumed to occur mainly by frictional (shear) forces at the interface.

Ultrasonic joints are also classified as 'near' or 'far' field. In near field the sonotrode (welding horn), is close to the joint interface (<6 mm) while in far field the sonotrode is >6 mm away from the joint interface. The greater the distance from the weld interface the greater the attenuation of the ultrasonic motion caused by damping, resulting in a reduction of the energy arriving at the interface. This is also critical, depending on the thermoplastic material. If it is semi-crystalline the further away the sonotrode is from the interface, the greater the damping of the ultrasound, hence the weldability will be lower. Semi-crystalline thermoplastics are regarded as being more difficult to weld than amorphous ones. Hence it is usually recommended that semi-crystalline materials are welded in near field.

Friction (spin) welding

Spin welding is a process which produces the heat required to melt the joint faces by friction between the two halves of the part to be welded; one half is rotated relative to the other under a compressive load. The process is restricted to parts having a circular joint, with no requirement for an angular alignment between the two halves. The main process parameters are frictional speed, weld pressure and weld times. The effect these have on the quality of the weld has not been studied systematically, but it is possible to estimate the effects of process conditions.

The main advantages of friction welding is the speed, simplicity and reproducibility of the process. Very little preparation is necessary

(a)

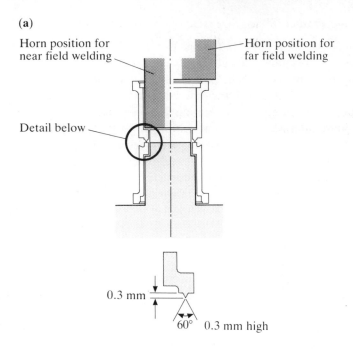

Horn position for near field welding

Horn position for far field welding

Detail below

0.3 mm

60° 0.3 mm high

(b)

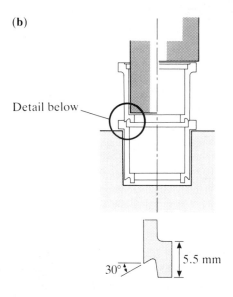

Detail below

5.5 mm

30°

Fig. 62. (a) Ultrasonic projection joints; (b) 60° shear joints

beforehand, although moisture can reduce joint strength, and therefore some materials, for example ABS nylon, need to be dried prior to welding. The major disadvantage is that the process is restricted to welding of circular cross-sections.

Vibration welding

Vibration welding is also known as linear friction welding. As with conventional friction (spin) welding, frictional heat is generated by relative movement between the two parts to be welded which are held under pressure. The difference between the process is that the movement applied is linear. Once molten material has been generated at the joint interface, vibration is stopped and the parts are aligned to allow the weld to consolidate on cooling. The main process parameters are weld frequency, amplitude and weld time. Most industrial machines operate at a fixed weld frequency of 120 Hz, although 240 Hz machines are also available, with variable frequency. The amplitude of vibration is less than 5 mm, with a weld time of 1–10 s. A typical vibration welding machine is illustrated in Fig. 63. Recently, vibration welders have been developed with an in-built dual force cycle. The addition of this has allowed a shorter weld cycle, with a maximum weld strength.

Vibration welding has the advantages of being fast, and requiring only simple tooling. As it is insensitive to surface preparation all thermoplastic materials may be welded. Complex linear joints may be formed, but the process is expensive, and is only suitable for high-volume production.

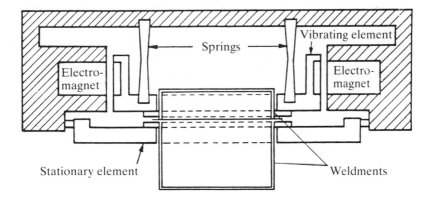

Fig. 63. Principle of construction of electromagnetic vibration welding machine

Select references

H Potente in *Ultrasonic Welding – Principles and Theory*, ed. M N Watson, TWI. Abington, 1988, pp. 66–72.
G Menges, H Potente, Studies on the weldability of thermoplastic materials by ultrasound, *Welding in the World*, **9**, 47–55 (1971).
S B Dunkerton in *Friction Welding of Plastics – Principles and Application*, ed. M N Watson, TWI, Abington, 1988, pp. 73–82.
V K Stokes, *Vibration Welding of Thermoplastics* – Part 1: Phenomenology and Analysis of the Welding Process, General Electric Technical Information Series No. 86, CRD, 223, Nov. 1986.
M G Dodin, Welding mechanisms of plastics: a review, *J. Adhesion*, **12**, 99 (1981).

Fusion welding – external heating

M A GIRARDI

Thermoplastic polymers soften on heating and therefore may in principle be brought together to form a bond. This is the basis of heat sealing and fusion welding. The heat may be generated by relative movement of the parts to be joined (see **Fusion welding by mechanical movement**) or by external heating which is the subject of this article.

Hot-plate welding

Hot-plate welding is perhaps the simplest welding technique used with plastics. It is widely used both in mass production and for large structures such as pipelines. The principles of the process are simple (Fig. 64). A heated plate is clamped between the surfaces to be joined until they soften. The plate is then withdrawn and the surfaces are brought together again under controlled pressure for a specific period. The fusion surfaces are allowed to cool, forming a joint which normally has at least 90% of the strength of the parent material.

Hot-plate welding is capable of reliable high-quality joints, but it is a relatively slow process if required for a high-volume mass production. However, its simplicity and ability to weld all thermoplastics makes it attractive where time is not a restriction, and is widely used in applications where high-integrity joints are required.

Hot-gas welding

Hot-gas welding is similar to welding of metals, the parts to be joined are heated by a stream of hot gas, the rod is pushed into the joint and heated until it softens sufficiently to fuse with the workpiece. The process is illustrated in Fig. 65. A good weld can only be achieved by using a filler rod of the same composition as the parent material.

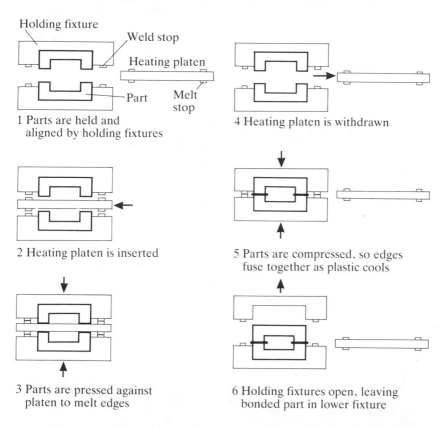

Holding fixture

Weld stop

Heating platen

Part

Melt stop

1 Parts are held and aligned by holding fixtures

2 Heating platen is inserted

3 Parts are pressed against platen to melt edges

4 Heating platen is withdrawn

5 Parts are compressed, so edges fuse together as plastic cools

6 Holding fixtures open, leaving bonded part in lower fixture

Fig. 64. Sequence of heated tool (hot plate) welding technique

The most important applications are in the fabrication of sheet or plastic form, such as air ducting. It is also used for welding plasticized (soft) films, for example lining of reservoirs. Another important application is the repair of damaged areas in thermoplastics. In the automotive industry there is an increasing use of thermoplastics for body panels (see **Automobile applications**). When damaged, welding can often effect a repair of high standard in terms both of strength and appearance. This is generally cheaper than replacement with a new component.

The principal advantage of hot-gas welding is that large, complex fabrications can be constructed. However, it is slow and weld quality is entirely dependent on the manual skill of the welder.

High-frequency welding

Radio-frequency (RF), high-frequency or dielectric welding is a widely used technique for sealing thin films. Dielectric welding works on

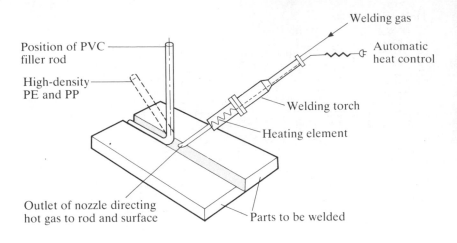

Position of PVC filler rod

High-density PE and PP

Welding gas

Automatic heat control

Welding torch

Heating element

Outlet of nozzle directing hot gas to rod and surface

Parts to be welded

Fig. 65. Principle of hot gas welding of plastics

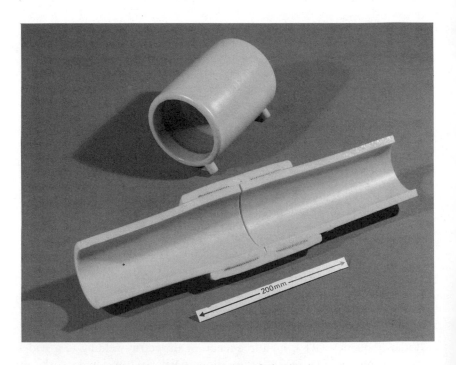

Fig. 66. Implant welding of plastic pipes

thermoplastics having a high dielectric loss factor, for example ABS, PVC and cellulose acetate. The technique uses a high-frequency field usually at 27.12 MHz from a generator similar to a radio transmitter. This is connected to electrodes which hold the sheets together. Pressure applied through the electrode welds the sheets together.

The process is well suited to mass production as it is faster and produces reliable results. One of the largest applications of RF welding is in the stationery industry, making book covers, clear pockets, catalogue binders, credit card holders, etc.

Implant welding

Implant welding is based on the principle of trapping a metallic insert between the two parts to be joined, and then heating the insert by induction or resistance heating.

With induction heating a high-frequency magnetic field (2–30 MHz) is used to induce electric current, which heats the insert and melts the surrounding plastic. The current version of the type of implant is the use of micron size ferromagnetic particles (usually iron oxide), known as Emaweld (developed by Emabond). One of the disadvantages of this technique is the necessity to design a new coil for each application. This is time consuming and expensive.

In resistance implant welding, conductive wire or braid is trapped between the components, in this case the wire is directly resistance heated by the passage of an electric current. The heat causes the surrounding plastic to melt. The main application for this is in the welding of plastic pipes (Fig. 66), and car bumpers.

The advantage of implant welding is its simple process which can be applied to complicated joints in large components. Weld times are short, less than 30 s for large components. However, the presence of the insert after welding reduces the joint strength, in addition the use of such inserts increases the overall cost. Implant welding is a relatively new process, and has been adapted for use with new composite materials.

Select references

C B Bucknall et al., Hotplate welding of plastics: factors affecting weld strength, *Polym. Eng. and Sci.*, **20** (6), 432–40 (1980).

E Pecha, *Hotplate Welding of Thermoplastics & Production Experience*, ed. M N Watson, TWI, Abington, 1988.

M N Watson, Hot gas welding of thermoplastics, *TWI Bulletin*, **30**, May/June (1989).

M N Watson, R M Rivett, K I Johnson, *Plastics – An Industrial and Literature Survey of Joining Techniques*, TWI Research Report 301/1986.

G

Glass transition temperature

D A TOD

In general, polymers are either completely amorphous or have an amorphous phase. These materials are hard, rigid glasses below a temperature known as the glass transition temperature or T_g; above this temperature amorphous polymers are rubbers – soft and flexible. Adhesives share with other polymers the phenomena associated with the glass transition temperature. Some, such as epoxides and acrylics (see **Epoxide adhesives, Acrylic adhesives**) are glassy at room temperature, others (e.g. **Plastisols, Pressure-sensitive adhesives**) are rubbery. All may pass through the glass transition when in use, and this is likely to have a dramatic effect on their properties and performance. For example the elastic modulus of a polymer above the T_g may be a factor of over 1000 lower when compared to the modulus below the T_g.

Crystalline polymers will start to crystallize above the T_g and if there is sufficient crystallization this can suppress the dramatic changes in modulus as exhibited by amorphous polymers. Several other key physical properties change in the region of the T_g; these include the coefficient of thermal expansion, specific heat capacity, refractive index and mechanical damping. The glass transition temperature is very dependent upon the structure of the polymer. Bulky side groups act to increase the T_g of a polymer, whereas hydrocarbon straight chain rubbers have low transitions. The glass transition temperature is sometimes compared with a second-order transition as it is the derivatives of a property such as volume which changes with temperature unlike melting which is a first-order transition where the property itself changes dramatically.[1,2]

Several variables affect the value of the T_g; two of these are external pressure and internal constraints. The application of pressure will cause the T_g of a polymeric system to increase at an approximate rate of $15–35\,°C\,kb^{-1}$. Internal constraints such as higher filler or fibre loadings will also act to inhibit molecular motion and therefore increase the T_g of the material.

Methods of measurement

It is important to appreciate the limits of practical measurement of the T_g; a number of different experimental techniques can be used. Perhaps the commonest are dynamic mechanical spectrometry[3] and differential scanning calorimetry (see **Thermal analysis**). In the former the mechanical response of the material is measured as a function of temperature and in the latter the heat capacity of the sample is measured. Although these different systems give similar trends they will not give exact agreement in the absolute value of the T_g. Within each of these ranges of techniques there are a number of different ways of defining T_g. In the dynamic mechanical tests the peak in the loss modulus is sometimes used or the peak in the loss angle $\tan \delta$ (see **Viscoelasticity**). A further method is to take the point of inflection of the elastic modulus–temperature curve in the glassy and transition regions and to define the T_g at this point. An example of a dynamic mechanical test upon an epoxy adhesive is shown in the article **Testing of adhesives**. The calculations of the T_g from differential scanning calorimetry can be equally difficult.

Several aspects of polymer structure act to compound these differences in measured value for T_g. Normally the molecules in a polymer sample do not have a unique molecular weight, but there is a distribution of molecular weights and a range of different local molecular environments of chain segments. A consequence is that there is a range of values of T_g and this produces a broad peak in the properties measured. A second fundamental effect is that the T_g is sensitive to the rate at which the measurement is made. If the time-scale of the measurement is long then the apparent T_g will be lower than if the measurement were made at a very short time-scale. A rough rule is that the apparent T_g will increase by a value of 3–7 °C for a decade rate of increase in test frequency.

The method of measurement of T_g is often a matter of personal choice or of available equipment. The use to which the values are to be used may also influence the choice of test method. If the T_g is to be used to determine the upper operating temperature of an adhesive then a mechanical method of measurement would be most appropriate. However, to be of most use to other workers the exact experimental details used in determining the T_g must be quoted.

Effect of molecular structure

If a polymer has a stiff backbone and bulky side groups its glass transition will be higher than that of a linear polymer with a flexible main chain.

The rigidity of the total structure will mean that high temperatures will be needed to enable molecular motion to occur. If there is a high degree of attractive forces between adjacent chains and side groups this will also result in a high T_g. The molecular weight of the polymer will influence the T_g in a similar way: for a given polymer, a higher T_g will be exhibited by a sample of higher molecular weight. The degree of cure of a thermosetting polymer will have a major effect upon the T_g. The higher the degree of cross-linking the higher the T_g. When **Epoxide adhesives** cross-link there is a slight increase in density which produces significant stresses in coatings and adhesive layers where contraction is restricted. These stresses are proportional to the temperature difference between the glass transition temperature and service temperature.

Effect of moisture

All organic polymers absorb some moisture and this can act to plasticize the material.[4] This plasticization involves a lowering of the T_g. A useful rule of thumb that can be applied to adhesive systems is that a 1% absorption of water equates to a drop of 20 °C in the T_g. A problem exists with ambient curing structural adhesives as their transitions are not much higher than the ambient temperatures in which the adhesives are used. In time an adhesive can absorb sufficient moisture so that its T_g is depressed and the adhesive changes from being rigid to being rubber-like at its operating temperatures. Once a material is above its T_g then it will absorb more moisture at a faster rate than it does at the lower temperatures. In adhesive joints the effect of plasticization can be complicated. The simple plasticization of the adhesive reduces its effective strength. However, as moisture diffuses into the joint it may attack the adhesive substrate interface causing premature adhesive failure rather than cohesive failure in the bulk (see **Durability – fundamentals, Underwater adhesives**).

The T_g is very important in the **Tack** of adhesive tapes. As the temperature of test is lowered the modulus of the adhesive increases together with the viscosity. The **Wettability** of the adhesive is reduced and the degree of **Tack** is lowered. This loss in tack is associated closely with the T_g, but occurs at a higher temperature.

References

1. R N Howard in *Molecular Behaviour and the Development of Polymeric Materials*, ed. A Ledwith, A M North, Chapman and Hall, 1974, p. 404.
2. G B McKenna in *Comprehensive Polymer Science*, Vol. 2, ed. G Allen, J C Berrington, Pergamon, Oxford, 1989, p. 311.
3. T Murayama, *Dynamic Mechanical Analysis of Polymeric Materials*, Elsevier, Amsterdam, 1978.
4. W W Wright, *Composites*, **12**, 201 (1981).

Good–Girifalco interaction parameter

D E PACKHAM

A significant advance was made in the understanding of interfacial energies when Good and Girifalco[1] defined their interaction parameter ϕ in terms of the work of adhesion between phases 1 and 2 and the geometric mean of the work cohesion of the phases:

$$\phi = W_A / (W_{C1} W_{C2})^{1/2} \qquad [1]$$

By incorporating the definition of W_A and W_C (see **Contact angles and interfacial tension**), Eqn 1 can be rearranged to give the Good–Girifalco equation for interfacial tension γ_{12} in terms of the **Surface energy** values γ_1 and γ_2 of the two phases

$$\gamma_{12} = \gamma_1 + \gamma_2 - 2\phi(\gamma_1\gamma_2)^{1/2} \qquad [2]$$

See **Contact angles and interfacial tension** for the definitions and a note on nomenclature of the surface energy terms. It is important to realize that Eqn 2 depends only on the definitions, and so is univerally valid.

London dispersion forces and evaluation of ϕ

From the definitions of W_A and W_C it can be seen that ϕ is a function of surface energies. If the structure of a material and the molecular potential energy–separation relationships are known, the surface energy can be calculated by evaluating the work required to separate to infinity the material either side of a chosen plane. For a material where the dominant intermolecular forces are dispersion force interactions the Lennard-Jones potential (see **Dispersion forces** and **Polar forces**) will apply, and the calculation is relatively simple.[2] It gives work of cohesion of phase 1

$$W_{C_1} = \frac{\pi n_1^2 A_{11}}{16 r_{11}^2} \qquad [3]$$

and work of adhesion between phases 1 and 2

$$W_A = \frac{\pi n_1 n_2 A_{12}}{16 r_{12}^2} \qquad [4]$$

where n_1 and n_2 are the number of molecules per unit volume in the phases 1 and 2, and A_{11} and A_{12} van der Waals attraction constants and r_{11} and r_{12} intermolecular separations within and between the phases indicated. Substituted into Eqn 1, these expressions give

$$\phi = \frac{A_{12}}{(A_{11}A_{12})^{1/2}} \frac{r_1 r_2}{r_{12}^2} \qquad [5]$$

As discussed in **Dispersion forces**, where substances 1 and 2 are not too dissimilar in structure,

$$A_{12} = (A_{11}A_{22})^{1/2} \qquad [6]$$

The second factor in Eqn 5 will also be close to unity for similar substances, making ϕ itself unity for an 'ideal' case. In general, for dispersion force interfaces between simple liquids, ϕ may be calculated from Eqn 5 using molar volumes to evaluate the 'r' term and explicit expressions for attraction constants (see **Dispersion forces**). The method can be extended to molecules of more complex structure such as polymers, but becomes more difficult.[1]

Where **Polar forces** act across an interface as well as dispersion forces, the Lennard-Jones potential still applies and Eqns 3–6 are still valid. A similar approach to the calculation of ϕ is possible, but the attraction constant terms have to be summed covering the orientation and induction forces involved.[1]

Interfacial tensions and solid surface energies

The Good–Girifalco equation (Eqn 2) can be used directly to calculate the interfacial tension between two phases of known surface energies providing the value of the ϕ is known. It can also be used to estimate solid surface energies from contact angle measurements. This is done by eliminating the interfacial tension γ_{12} between Eqn 2 and Young's equation (see **Wetting and spreading**, Eqn 1). This gives

$$\gamma_{S} = \gamma_{LF}(1 + \cos\theta)^2/4\phi^2 \qquad [7]$$

In deriving this equation, it has been assumed that the spreading pressure is zero: this point is discussed under **Surface characterization by contact angles – polymers**.

Koberstein[3] lists some values of the interaction parameter. A somewhat different treatment of interfacial energies is discussed under **Surface energy components**.

References

1. R J Good, *J. Colloid Interf. Sci.*, **59**, 398 (1977).
2. B W Cherry, *Polymer Surfaces*, Cambridge University Press, 1981.
3. J T Koberstein, *Encyclopedia of Polymer Science and Engineering*, Vol. 8, John Wiley, New York, 1987, p. 255.

H

Health and safety

D C WAIGHT

The aim of Health and Safety Acts, statutory instruments, approved codes of practice, guidance notes, etc. is to ensure that the hazards inherent in adhesives are controlled so that the risks to humans and their environment are at a level acceptable to society.

Safety

This term covers hazards to the person arising from all aspects of the use of adhesives. It includes primary risks (e.g. death from exposure to solvent), secondary risks (e.g. injury from explosive ignition of solvent) and injury by ancillary equipment (nip rollers of a rubber mill).

Health

This term is related to the effects which are not quite so immediate, where the longer-term exposure to the substance or process affects a person's health. Such effects include lung impairment due to exposure to dusts, hearing impairment due to noisy conditions of work, or loss of nervous function due to solvent exposure.

Hazards of adhesives

Hazards of adhesives can be divided into two categories, physico-chemical and toxicological.

Physico-chemical hazards normally found in adhesives include flammability,* but in some cases there may be the risk of exposure to corrosive

---

* 'Flammable' is jargon used in this context. It stands for the English word 'inflammable' which means 'able to be inflamed'.

or oxidizing (e.g. organic peroxide) substances. Hazards from radio-activity and explosive substances are not normally encountered; but vapours given off by flammable substances can explode when mixed with air and ignited, as can the dusts of adhesives.

Toxicological hazards normally found in adhesives can be classified as 'harmful', 'irritant' and 'sensitizing'. The 'harmful' classication covers those products which have the potential to cause limited health risks. Some products may involve toxic risks (where serious, acute or chronic health risks and even death may be encountered), and some may be corrosive. **Cyanoacrylate adhesives** may bond to the skin.

Adhesive safety

Safety and the freedom from risks to health in the manufacture, storage, use and disposal of adhesives, of any type, is achieved by the basic health and safety procedure of establishing a safe system of work. A 'safe system of work' is a formal defined procedure which results from a systematic examination of the task that identifies all the hazards that may be involved.

A safe system involves five steps: (1) assessing the task, (2) identifying the hazards, (3) defining safe methods, (4) implementing the system and (5) monitoring the system.

Hazards from adhesives

There is a wide variety of adhesive types (see **Adhesive classification**, and a corresponding variety of associated hazards). **Solvent-based adhesives** generally retain the hazards associated with the solvent used. All solvents should be considered for their health effects that may be caused by inhalation of the vapours and through contact effects with the body. Adhesives with a flammable solvent will also have fire and explosion risks. It should be noted that solvent vapours are often heavier than air and will therefore concentrate in depressions, as well as travel for considerable distances in search of a source of ignition, extending the hazard well beyond the immediate vicinity of the workplace. In addition thermal degradation products especially of the 'non-flammable' chlorinated solvents may be hazardous. The displacement of oxygen from the air by the solvent vapour is a danger, especially in confined areas.

Water-borne adhesives (see **Emulsion and dispersion adhesives**) are generally less hazardous, but they may contain certain other substances in sufficient quantities to give rise to risk in certain circumstances. For example they may contain fungicides which could be harmful if ingested or inhaled.

Hot melt adhesives are generally free of hazard when solid, but in the molten state there is a risk of burns, from the adhesives and associated equipment. They also degrade to various degrees, especially if heated above recommended temperature, producing vapours which may, like other degradation products from other heated materials, be harmful if inhaled. The build-up of deposits in fume extraction ducts may give rise to fire or explosion risk.

Powder adhesives are of various types: the dusts may be directly hazardous to health and may cause an explosion when finely dispersed in the presence of an ignition source. Additionally some types can cause irritation when in contact with the skin.

Plastisols and curing rubber adhesives (see **Rubber-based adhesives, Rubber to metal bonding**). These product types react by the application of heat. When hot the thermal and vapour risks from them are similar to hot melt adhesives, see above.

Chemically reactive adhesives (see **Reaction setting adhesives**) are a diverse group of adhesives, some of which belong in the other groups listed above. They act by reaction of two or more components, one of which may be atmospheric moisture or oxygen. Generally the hazards of the final adhesive are very much lower than the hazards presented by the component parts which may include chemical substances capable of causing sensitization (see **Polyurethane adhesives**).

Information on adhesives

Under the legal or quasi-legal requirements of many countries the supplier is responsible for ensuring that such steps as are necessary are taken to ensure that persons supplied with the substance are provided with adequate information about any risks to health or safety to which the inherent properties of the substance may give rise (e.g. the UK Health and Safety at Work etc. Act 1974, s. 6 as amended by the Consumer Protection Act 1978 or the USA OSHA Federal Hazard Communication Standard 29CFR 1910.1200 or the EEC General Preparations Directive). This duty is usually performed by the provision of a Material Health and Safety Data Sheet. In the absence of this information the user will need to consider obtaining it in order to assess the hazards prior to use of an adhesive.

Additionally, under the laws or regulations of most countries (such as those derived from the EC Classification, Packaging and Labelling of Dangerous Substances and Preparations Directives, as amended), the supplier is responsible for ensuring that any substance which is subject to the regulations (which covers adhesives as well as other substances with significant toxicological or physico-chemical risks, and also cyanoacrylate adhesives) are properly labelled, in the local language(s), when supplied. In the EC and neighbouring countries the labelling generally includes one or more hazard warning symbols. The hazard labelling has to conform with detailed requirements for colour and size, and should appear on all levels of packaging. In the USA the communication is usually by text only, with a prominent warning on the front panel of the container, with fuller information elsewhere.

This labelling only communicates the principal inherent hazard(s) associated with the substance, and should generally be supported by the separate provision of fuller health and safety information.

Select references

EC Substances Directive 65/547/EEC as amended six times and adapted to technical progress fifteen times.

EC Solvents Directive 73/173/EEC as amended by 80/781/EEC and adapted by 80/1271/EEC and 82/473/EEC.

EC Paints, Varnishes, Printing Inks, Adhesives and Similar Products Directive 77/728/EEC as amended by 83/265/EEC and adapted by 81/916/EEC, 86/508/EEC and 89/451/EEC.

EC Dangerous Preparations Directive 88/379/EEC as amended by 89/178/EEC and 90/492/EEC. See also 90/35/EEC, 90/155/EEC and 91/442/EEC. (This replaced the Solvents and Paints Directive on 7 June 1991.)

Guidance booklets issued by National Adhesive Association, such as the British Adhesive and Sealants Association's *Guide for Users and Safety Officers – Safe Handling of Adhesives and Sealants in Industry*, BASA, 33 Fellowes Way, Stevenage, Herts SG2 8BW. These guides usually contain references to applicable regulations, etc. in the country where the association is based.

High-temperature adhesives

S J SHAW

Introduction

Adhesives are becoming increasingly used in the manufacture of a wide range of engineering components because of the substantial benefits provided by adhesive bonding in comparison with more traditional joining techniques such as riveting and welding.

Over the past decade or so there has been a growing requirement, particularly in the aerospace industry, for adhesives capable of withstanding temperatures in excess of 150 °C for both short- and long-term application. Epoxy resins, which currently form the basis of most structural adhesive systems, are generally formulated to yield high-temperature capabilities of approximately 50–100 °C (see **Epoxide adhesives**). Although modest improvements in this capability are possible by, for example, variation in resin/curative functionality, together with the use of certain filler materials, 150 °C is generally considered the maximum permissible working temperature. The features which impart thermal stability to a polymer are discussed in **High-temperature stability: principles**.

Before going on to consider some of the adhesives capable of exhibiting high-temperature capabilities it is important to consider what is meant by the term 'high-temperature resistance'. This can be viewed in two ways. Firstly there is the need for the adhesive to maintain mechanical properties at the intended service temperature. Second, there is a requirement for the adhesive to maintain its structural integrity, i.e. resist thermal breakdown at elevated temperature. The former is controlled primarily by the **Glass transition temperature**, T_g, while the latter is largely dependent upon the inherent thermal stability of the polymer from which the adhesive system is produced. For short-term applications, i.e. short durations at elevated temperature, whereas the first requirement is of overriding importance, the latter can be, to a large degree, ignored. However, for longer term applications both requirements must be considered important.

Traditional adhesive systems

In consideration of the types of adhesive able to exhibit high-temperature capabilities it is convenient to divide the discussion into two main parts. First, a consideration of what could be regarded as the traditional high-temperature adhesives, i.e. those which have been available for many years, and second a mention of some of the more recent developments in the high-temperature polymer area, developments designed to minimize the disadvantageous characteristics exhibited by the former types.

The traditional **Structural adhesives** capable of operating at temperatures in excess of 150 °C for both short- and long-term applications can be divided into three classes, namely: (1) phenolics, (2) polybenzimidazoles, (3) condensation polyimides (see **Phenolic adhesives, Polybenzimidazoles** and **Polyimide adhesives**).

Phenolics, prepared from a reaction between phenol and formaldehyde, are rarely used alone as **Structural adhesives** due to tendencies towards both high shrinkage during cure and brittleness. Consequently, for structural adhesive applications, phenolics are usually modified with other polymers. One particular combination, the epoxy-phenolics, offer a substantial high-temperature capability, in particular providing good strength retention, for short periods of time, up to about 300 °C. For long-term use, however, a ceiling of 250 °C is advisable due to the onset of severe oxidative degradation at higher temperatures.

Similarly, **Polybenzimidazole** adhesives also offer excellent high-temperature capabilities for short-term use, with the ability to retain 50% of room temperature strength at 450 °C. Unfortunately, as with the epoxy-phenolics, this capability is not maintained under long-term, high-temperature conditions, due to the susceptibility of **Polybenz-imidazoles** to oxidative degradation at temperatures in excess of 250 °C.

Condensation polyimides offer the capability of retaining in the region of 50% room temperature strength at approximately 300 °C. Thermal stability, dependent on a number of factors is such that long-term use at temperatures in the vicinity of 275 °C is feasible.

Recent developments

The three adhesive types discussed above, as mentioned, exhibit substantial high-temperature capability and as such have been, and will probably continue to be employed in various high-temperature applications. However, all three to varying degrees exhibit a serious disadvantage concerning processability which, at the very least, can render them difficult to use. Although essentially discounted for high-temperature use, **Epoxide adhesives** are generally easy to use in comparison to the above-mentioned systems and thus many of the recent developments in the high-temperature polymer field have had epoxy-like processability as a major goal. For convenience, these developments can be divided into the following three broad areas: (1) thermoplastic polyimides, (2) imide prepolymers, (3) other polymers.

In all three areas some considerable success has been achieved in fulfilling the main objective, to the extent that, in some cases, adhesive formulations based on these three categories are at or close to full commercialization. For more detailed accounts of these developments see the articles entitled **Polyimide adhesives, Polyether ether ketone** and **Polyphenylquinoxalines**.

Select references

J P Critchley, G J Knight, W W Wright in *Heat Resistant Polymers*, Plenum Press, New York, 1983.
K L Mittal in *Polyimides – Synthesis, Characterisation and Applications*, Vols 1 and 2, Plenum Press, New York, 1984.
S J Shaw in *Adhesion 10*, ed. K W Allen, Applied Science Publishers, London, 1986, pp. 20–41.
S J Shaw, *Mater. Sci. and Tech.*, **3**, 589 (1987).

High-temperature stability: principles

D E PACKHAM

A serious limitation to the use of organic polymers in general and of adhesives in particular is their poor resistance to thermal degradation. Considerable effort has been put into the development of **High-temperature adhesives** and examples of the materials which have been produced are described in articles on **Polybenzimidazoles, Polyether ether ketone, Polyimide adhesives** and **Polyphenylquinoxalines**. Some of the general principles used in the search for enhanced thermal stability are discussed in this article.

A very wide range of reactions is involved in polymer degradation, depending on the polymer concerned and the environment.[1] Main chain reaction occurs in many polymers: this often involves chain scission, but sometimes cross-linking results. In other polymers a side chain or a substituent may be more vulnerable. Degradation often involves not one, but a series of reactions leading to a complex mixture of degradation products.

Thus generalizations about degradation are few, and there are many exceptions to them. Nevertheless, some molecular features giving thermal stability to a polymer can be described.[2] They are based on the recognition that degradation must start with bond breaking somewhere in the structure, and that this will often initiate a series of reactions leading to a loss of desirable properties. These features are as follows:

1. Only strong bonds should be incorporated in the structure;
2. No easy pathways for rearrangement should be present;
3. Maximum use should be made of delocalization (resonance) stabilization;
4. Where possible polybonding should be used.

The first of these points seeks to avoid initial attack by eliminating thermally labile bonds from the structure, the second suggests one way of limiting the damage once a bond has been broken.

A powerful way of limiting the initial damage is to build stability into the partially degraded molecule: this is the rationale of the third point. The initial attack will often involve homolytic scission of a bond giving a free radical – that is a species with an unpaired electron. Radicals are characteristically reactive (see **Addition polymerization**), but their reactivity is reduced if the unpaired electron is not localized on a single atom. The presence of conjugated structures – structures with a number of alternating single and double bonds – allows such delocalization. The phenyl groups – the C_6H_5 benzene ring – is a common example. If phenyl groups are linked together delocalization may be possible over many aromatic rings, giving even greater stability.

The fourth feature mentioned, polybonding, refers to structures where the main chain consists of more than one string of atoms, like a ladder. If one bond is broken the chain is still held by the second, so the rate of degradation of properties is reduced.

The structures given in the articles cited on high-temperature adhesives illustrate the application of these principles. They show how frequently conjugation is used to provide radical stabilization.

References

1. I C McNeill in *Comprehensive Polymer Science*, Vol. 6, ed. G Allen, J C Bevington, Pergamon, Oxford, 1989, p. 451.
2. W W Wright in *Degradation and Stabilization of Polymers*, ed. G Geuskine, Applied Science Publishers, London, 1975, p. 43.

Honeycomb structure

R D ADAMS

This configuration applies to flat and curved panels which have high bending strength and stiffness per unit weight. Two skins are separated by a core to make a structure which in bending is analogous to an I-beam. The skins take the tensile and compressive loads which carry the bending moment, while the core transmits the shear force from one skin to the other.

The term 'honeycomb' comes from the use in high-strength applications of a core which is similar to the comb used in a beehive. In one form, a series of thin sheets of aluminium are bonded in alternate strips so that when the system is pulled sideways, it opens out to a hexagonal shape as shown in Fig. 67. In addition to aluminium, other metals such as titanium can be used, but this is only for specialist applications. Even woven fibre composites such as glass or Kevlar have been used to make honeycomb.

A commonly used alternative today is paper or polyamide coated and bonded with a resin to form a non-metallic honeycomb. Being non-metallic, these are corrosion resistant and have excellent dielectric and radar-transmission properties.

The shear stiffness of the honeycomb is a function of its depth, the thickness of the strip used to form it, the material used and the size of the honeycomb cell. In order to manufacture a honeycomb sandwich panel, skins have to be bonded to the core. A film or paste adhesive is

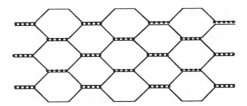

Fig. 67. Cross-section of honeycomb core

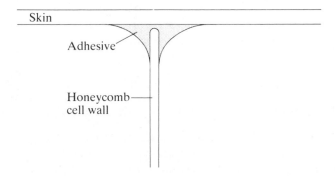

Fig. 68. Formation of adhesive bond between skin and core

used which has a lower cure temperature than that of the bonded core (otherwise the core falls to bits!) Care has to be taken to get a bead with a fillet so that a good joint is formed (see Fig. 68).

See also **Aerospace applications, Repair methods**.

Select reference

H G Allen, *Analysis and Design of Structural Sandwich Panels*, Pergamon, Oxford, 1969.

Hot melt adhesives

D M BREWIS

Introduction

Hot melt adhesives are thermoplastics which are applied molten to the substrates. Depending upon the masses and temperatures of the substrates, a melt adhesive may cool very rapidly (see **Internal stresses**). Setting is normally achieved in a few seconds or less and for this reason melt adhesives are attractive for a number of applications.

Composition

Melt adhesives are based on thermoplastics, but usually contain a number of other components. The most commonly used melt adhesives are based on **Ethylene–vinyl acetate** (EVA) copolymers, but polyethylene, polyesters, polyamides and thermoplastic rubbers, e.g. styrene–butadiene block copolymers, are also used. Depending upon the properties required, the vinyl acetate content in the EVA copolymer may be varied and varying quantities of tackifying resins (e.g. rosin esters) and waxes are incorporated. Tackifying resins will affect the adhesion at elevated temperatures, the time the adhesive takes to harden and the final adhesion achieved. The main purpose of the waxes is to reduce the melt viscosity and thereby improve the wetting of the substrates.

Advantages of melt adhesives

The advantages of hot melt adhesives include the following:

1. Rapid setting leading to very high production speeds;
2. Much-reduced space requirements due to (a) lower warehouse volume, (b) reduced compression sections and (c) avoidance of drying areas;
3. Indefinite shelf life;
4. Low toxicity and fire hazards providing fuming is minimized (see **Health and safety**).

Limitations of melt adhesives

The main limitations of hot melt adhesives are their modest upper service temperature (although polyester and polyamide systems are usually much superior to EVA systems) and modest load-bearing ability.

Application methods

A wide range of application techniques is available, selection of which will be determined by assembly and performance requirements and cost. The methods include the following:

1. *Wheel system.* A wheel rotates in a reservoir of adhesive and applies a line of adhesive to a substrate. The thickness of adhesive is controlled by a doctor blade.
2. *Dauber.* Adhesive is transferred from a reservoir by a bar to particular areas of the substrate. Melt viscosity must be low to avoid stringing of the adhesive.

3. *Spring ball valve*. Adhesive is released when the substrate is pressed into contact with a sprung ball situated in the orifice of an extrusion head.
4. *Extrusion* (see **Extrusion coating**). Adhesive is forced through a die on to the substrate and may be applied as continuous or intermittent lines or bands.

Uses of melt adhesives

The widespread use of hot melt adhesives includes bookbinding (soft cover binding of books and magazines) and component assembly in electronics, consumer durables and automotive fields (see **Automotive applications**). Specialized articles discuss applications in the **Packaging industry** (e.g. carton sealing, bag making, labels), in footwear manufacture (counters and toe puffs, see **Footwear applications of adhesives**), and in woodworking (**Wood adhesives – edgebanding**). They find employment as **Adhesives in the textile industry**.

Some test methods are listed in **Appendix 1**.

Select references

D L Bateman, *Hot Melt Adhesives*, 3rd edn, Noyes Data Corporation, Park Ridge, New Jersey, 1978.
R D Deeheimer, L R Vertink in *Adhesives in Manufacture*, ed. G L Schneberger, Marcel Dekker, New York, 1983.

Humidity

J COMYN

Humidity is a measure of the amount of water vapour in the atmosphere, and is usually quantified as relative humidity (RH), which is defined as

$$RH(\%) = \frac{\text{Partial vapour pressure of water in a sample of air}}{\text{Saturated vapour pressure of water at the same temperature}} \times 100$$

The importance of humidity in adhesive bonding is that water in the atmosphere diffuses into adhesive bond lines and causes weakening of joints (see **Durability**). All adhesives are based on polymers and all polymers are permeable to water. There is a consensus of opinion that the mechanism of weakening is by water attacking the interface,[1] and a large body of practical experience to show that weakening can be minimized by the correct application of surface **Pretreatments**.

There are many reports in the literature[1] where joints which have been exposed at high RHs have weakened with time, but otherwise identical specimens stored at typical ambient laboratory humidities retain their strength. This led Gledhill, Kinloch and Shaw[2] to propose that there is a critical RH below which environmental weakening does not occur.

In the laboratory RH in relatively small containers (typically a desiccator) can be controlled by the use of saturated salt solutions. Some data over a wide range of humidities at temperatures of 22.8, 30.0 and 37.8 °C have been given by Wink and Sears.[3]

References

1. J Comyn in *Durability of Structural Adhesives*, ed. A J Kinloch, Applied Science Publishers, London, 1983. Ch. 3.
2. R A Gledhill, A J Kinloch, S J Shaw, *J. Adhesion*, **11**, 3 (1980).
3. W A Wink, G R Sears, *TAPPI*, **33**(9), 968 (1950).

Hydrogen bonding

D BRIGGS

Hydrogen bonding is a very important mechanism for intermolecular interaction and therefore adhesion. Hydrogen bonding is due to the strong interaction of hydrogen attached to one atom (such as O, N, C) by a polar covalent bond with an adjacent atom of high electronegativity (such as O, N and halogens); it is denoted by a dotted line. Thus dimerization of carboxylic acids and interaction between molecules of PVDF and PMMA both occur by hydrogen bonding.

The strength of this interaction (8–35 kJ mol^{-1}) falls between that of van der Waals forces (4–8 kJ mol^{-1}) and full covalent bonding (40–400 kJ mol^{-1}).

Fowkes (following the earlier work of Drago and co-workers) has argued strongly that intermolecular interactions are due only to either van der Waals **Dispersion forces** or **Acid–base interactions**. Hydrogen bonding is an important component of A–B interactions, often loosely referred to as specific interactions. Liquids and polymer surfaces can have one of the following three types of hydrogen-bonding capability (see **Acids**):

1. Proton acceptor (electron donor or basic) such as esters, ketones, ethers or aromatics which include such polymers as poly(methyl methacrylate), polystyrene, **Ethylene–vinyl acetate copolymers**, polycarbonate;
2. Proton donor (electron acceptor or acidic) such as partially halogenated molecules, including polymers such as poly(vinyl chloride), chlorinated polyethylenes or polypropylenes, poly(vinylidene fluoride) and ethylene–acrylic acid copolymers;
3. Both proton acceptor and proton donor molecules such as amides, amines and alcohols where the polyamides, polyimides and poly(vinyl alcohol) are included.

The work of adhesion between surfaces is given by

$$W_A = W_A^d + W_A^{ab}$$

Where W_A^d and W_A^{ab} refer to the additive contributions from dispersion force and acid–base interactions respectively (see **Acid–base interactions, Contact angles and interfacial tension**). Practical bond strengths can therefore be increased by maximizing W_A^{ab} and this is achieved in polymer surface pretreatments (e.g. corona (see **Corona discharge treatment**), and **Flame treatment** of polyolefin surfaces). These result in the incorporation of highly polar groups with electron accepting or donating, and particularly H-bonding, capacity. Thus in the corona discharge treatment of polyolefins enhanced adhesive characteristics result from the incorporation of $-CH_2-C = O$ groups which can enolize to $-CH = C-OH$, the OH group of which is a strong H-bonder. Likewise, treatment of poly(ethylene terephthalate) produces H-bonding phenolic–OH groups.

Select references

F M Fowkes in *Polymer Science and Technology*, Vol. 12A, *Adhesion and Adsorption of Polymers*, ed. L-H Lee, Plenum Press, New York, 1980, p. 43.

D Briggs in *Practical Surface Analysis*, 2nd edn, Vol. 1, *Auger and X-ray Photoelectron Spectroscopy*, eds D Briggs, M P Seah, John Wiley, Chichester, 1990, Ch. 9.

I

Impact resistance

A MADDISON

The performance of bonded joints under impact conditions has particular relevance in certain applications. Automotive body structures, for example must, if involved in a collision, collapse in a controlled manner to afford maximum protection to the occupants (see **Automobile applications**). It is therefore essential that joint integrity is maintained to permit the necessary adsorption of energy by the large-scale deformation of the structure.

It has been shown that in comparison with spot welds, rivets and a variety of self-piercing mechanical fastenings, toughened structural adhesives may yield single lap shear joints of superior energy absorption when used to bond aluminium alloy sheet, findings confirmed in tests on full-size bonded vehicles (see **Toughened adhesives, Structural adhesives**).

Standard test methods for the determination of impact resistance (ASTM D 950-82 and BS 5350, Part C4) specify the specimen geometry shown in Fig. 69, see also Appendix 2. While providing a simple basis for comparisons between adhesives, a number of disadvantages are associated with this configuration, for example it does not reproduce the differential strain effects often present in real joints and it does not readily permit the evaluation of bonds to many important coated substrates.

Alternative techniques have therefore been devised to more adequately address specific impact dependent variables. For example, studies employing bulk adhesive specimens have indicated increases in shear modulus and yield stress, and decreases in strain to failure, at high strain rates.[1]

The development of high-performance structural adhesives has also been undertaken using instrumented drop-weight and pendulum machines used in conjunction with a **Peel test** specimen which is cleaved by a wedge arrangement.[2] Adhesive formulation effects have also been demonstrated by testing bonded panels in an instrumented guillotine-type machine.[3]

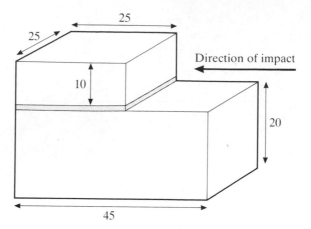

Fig. 69.   Impact resistance test-piece (BS 5350 part C4): dimensions in mm

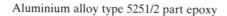

Aluminium alloy type 5251/2 part epoxy

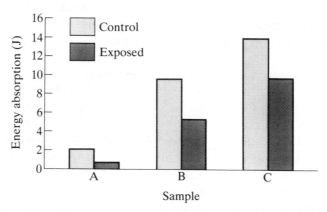

Fig. 70.   Effect of pretreatment of aluminium and environmental exposure
on the energy absorption of single lap shear joints.[4] *Environment*:
deionized water 50°C. *Pretreatments*: A Degreased in acetone then
cleaned in alkali; B As A then anodized in 10 vol percent
orthophosphoric acid at 60 V r.m.s. AC for 10 s; C As B but
anodized for 60 s

An important consideration in bonded structure design is the tolerance
of joints to impact **Fatigue** conditions. Recent studies suggest that a crack,
once initiated, may propagate through the adhesive at low levels of applied
energy.

Box team sections have been extensively employed in the evaluation of
bonding and weld-bonding systems for possible application in the
transport industry. Of particular value in section design and the

optimization of spot-weld distribution, they provide useful indications of initial performance in more relevant geometries.

In the study of bond **Durability** under impact conditions, **Pretreatment** and environmental effects have been evaluated using short diffusion path specimens derived from thin sheet adherend materials.[4] The results (see Fig. 70) serve to emphasize the critical role of surface treatment in the formation of strong, environmentally stable adhesive bonds.

Unfortunately, values of energy absorption obtained are invariably influenced by a number of factors, principally the properties of the adhesives and adherends, test joint geometry and the characteristics of the test machine. Given this degree of sensitivity to test conditions, appropriate experimental controls and careful interpretation of results are essential.

References

1. J A Harris, R D Adams, *Proc. Instn. Mech. Engrs*, **199** (C2), 121–31 (1985).
2. W F Marwick, J H Powell, *Conf. Proc. ASE*, Brighton, Plastics and Rubber Institute, 1988, p. 144.
3. M Jordan, *Int. J. Adhesion and Adhesives*, **8** (1) 39–46 (Jan. 1988).
4. C W Critchlow, A Maddison, *Surface and Interface Anal.*, **17** (7), 539 (1991).

Industrial applications of adhesives

C WATSON

The advantages that adhesives can offer in terms of design flexibility and ease of manufacture over other methods of joining, mean that adhesives find many applications throughout industry (see **Engineering practice, Engineering design with adhesives**). In this article it is intended to give an idea of the scope of adhesives in general industrial practice, and to supplement the information to be found in more specialized articles (see **Aerospace applications, Automotive applications, Footwear applications, Packaging industry, Printing ink adhesion, Rubber to metal bonding applications**).

Bonding cylindrical components

The fixing of cylindrical components is a commonly occurring requirement in industrial manufacture. In recent years adhesives, especially **Anaerobic adhesives**, have found many applications that range across many industries replacing traditional methods. Adhesives have been used to replace brazing or welding in fixing tubes to carburettor housings, in fixing roller ends on Gestetner machines saving £15 000 per annum. They are also used to retain bearings into housings, for instance the spherical bearings in the flight controls of the Lynx helicopter and bearings in electric motor end caps, for instance, on Invicta vibrator motors.

Gear to shaft bonding Larger-scale applications include the fitting of gears to shafts in a French steel mill. Figure 71 shows the assembly, the shaft through the gear being the same size as the exposed diameter. Originally the components were shrink fitted and could only transmit a torque of 700 kN m: when an anaerobic adhesive was employed the torque was increased to 2800 kN m.

On a more modest scale, an increasingly important application is in car and truck gearboxes, for example in bonding the ring gear to the differential housing (see **Joint design**).

Major benefits of this type of application are as follows: weight reduction; augmentation of mechanical fit; reduction of mechanical fit with subsequent reduction of hoop stress; ease of machining of gear teeth of separate components as opposed to machining cluster; cheaper capital outlay, e.g. adhesive application assembly machine £100 000 compared to electron beam welding machine £500 000; easier replacement of gears.

Cylinder liner retention Thin-walled cylindrical components such as bushes and liners can be interference fitted into bores. This more often than not results in distortion, necessitating post machining to bring bush or liner bore back to size. Perkins Engines use an anaerobic adhesive when they manufacture the Prima diesel, for augmenting the light press fit when assembling liners to engine blocks.

Originally in the auto industry, liners were lipped, the block recessed. The liners were assembled with a heavy interference fit. The bore was distorted and subsequently had to be honed back to size. Using adhesive technology, both these stages are eliminated.

Non-metallic components

Many light-weight non-metallic components are bonded with **Cyano-acrylate adhesives**. Figure 72 shows some applications on a typical car body.

Increasing use is made of bonding for parts traditionally joined by other means. Bonding can replace stitching in attaching decorative bows to slippers (see **Footwear applications**); it can replace ultrasonic welding (see **Fusion welding**) for ABS torch bodies; solid rubber tyres can now be bonded to carriage wheels. Figure 73 shows a neoprene sponge rubber being bonded to a PVC tube to produce a paint roller which gives a stippled effect.

Fig. 71. A gear in a steel mill bonded adhesively to its shaft

The applications described are but a few of the many thousands of joints that are assembled using adhesives. They illustrate the varied applications and some detail of the benefits of using adhesives compared to other joining methods.

The adhesives industry continues to increase the awareness of the use of adhesives so that design and manufacturing engineers may at least consider adhesives at design stage.

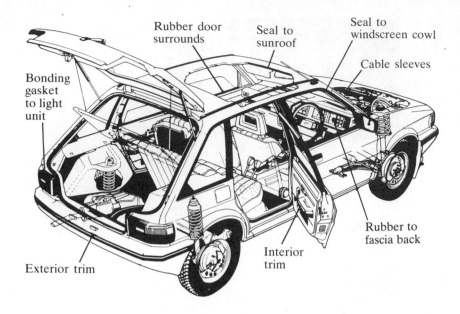

Fig. 72. Some typical applications of cyanoacrylate adhesives in a car body

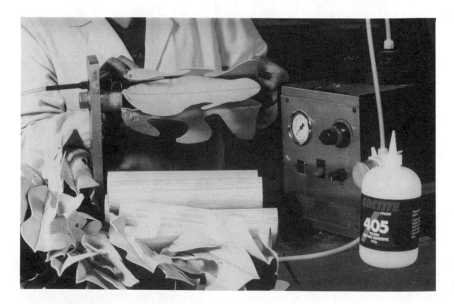

Fig. 73. Neoprene sponge rubber bonded to a PVC tube to form a paint roller

General references

H N Watson, *Joining Plastics in Production*, The Welding Institute, Abington, 1988, Ch. 12.

M M Sadek, *Industrial Applications of Adhesive Bondings*, Elsevier Applied Science Publishers, London, 1987, Chs 1, 4–6.

G Haviland, *Machinery Adhesives for Locking, Retaining and Sealing*, Marcel Dekker, New York, 1986, Chs 1, 6, 7.

G Defrayne, *High Performance Adhesive Bonding*, Society of Manufacturing Engineers, Dearborn, Michigan, 1983.

Inelastic electron tunnelling spectroscopy (IETS)

J COMYN

Inelastic electron tunnelling spectroscopy (IETS) was discovered by Jaklevic and Lambe in 1960[1] and it is a technique which records the vibrational spectrum of a minute quantity of organic material which is placed in a four-layered sandwich (metal–metal oxide–organic material–metal) in which the metal oxide and organic layers are very thin. The first three layers of this sandwich bear some resemblance to the structure of an adhesive to metal interface. A recent review by Brown[2] took IETS as its subject.

Principles of IETS

When an electrical bias V is applied across two metals which are separated by an insulator, the Fermi levels become separated in energy (Fig. 74). Electrons may now tunnel elastically from one metal to the other; the elastic current obeys Ohm's law. When a monolayer of organic material is adsorbed on the insulator, a tunnelling electron may excite vibrational modes in the adsorbate and lose energy in the process; such an inelastic process will require a minimum of energy which corresponds to a minimum bias (V_{min}) such that

$$eV_{min} = h\nu$$

Here e is the electronic charge, h is Planck's constant and ν the vibrational frequency of the excited mode.

Inelastic processes are relatively uncommon, typically $<1\%$ of the electrons tunnel inelastically. However, they may be distinguished from the elastic background by examining the peaks in the second derivative d^2I/dV^2; here I is current (Fig. 75). The second derivative is usually measured by superimposing a small modulation voltage on a slowly increasing bias, and measuring the level of the second harmonic generated by non-linearity.

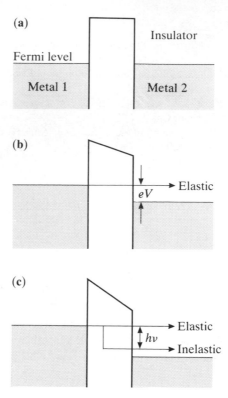

Fig. 74.   The occurrence of elastic and inelastic electron tunnelling through an insulator between two metals, as the bias is increased

Performance

The experimental data on selection rules for IET spectra suggests that there is no evidence for selection on grounds other than the orientation of groups relative to the direction of electron tunnelling. Both infra-red (IR) and Raman modes can be detected by IETS (see **Infra-red spectroscopy of surfaces**). To obtain adequate resolution experiments are usually carried out at liquid helium temperature (4.2 K). The thermal broadening function has a width of $5.4\,kT$ at half-height, which corresponds to $16\,\text{cm}^{-1}$ at 4.2 K.

Resolution also depends on the modulation voltage used to measure the second derivative. This has a smearing function of width $1.22\,eV_2$ at half height where V_M is the r.m.s. modulation voltage. At $V_M = 1\,\text{mV}$ the smearing is $10\,\text{cm}^{-1}$. Resolution may be improved by using a lower modulation voltage, but the time to scan the spectrum then increases. Resolution is also improved by the use of a superconducting metal for the top electrode.

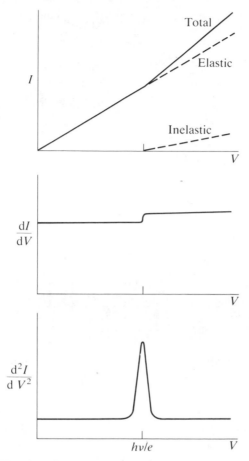

Fig. 75. Principle of the detection of inelastic tunnelling.

Experimental methods

Junctions of IET are prepared by vacuum evaporating the metals on to an inert substrate such as glass. The geometry of the deposits can be controlled by the use of masks (Fig. 76). Oxidation of the bottom electrode can be by plasma oxidation in the vacuum chamber or by simply venting the apparatus to the atmosphere. There are a number of ways by which the organic dopant can be added. One is simply to expose the oxidized metal to the vapour, while involatile materials can be applied by spin casting from a dilute solution. Other techniques are plasma polymerization and infusion doping through the porous grain-boundaries in the lead electrode.

A diagram of an IET spectrometer appears in Fig. 77. It is housed in a room totally screened with aluminium sheets with mains electricity

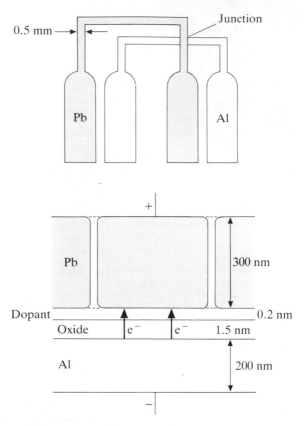

Fig. 76. A plan and section (not to scale) of an IET junction. Distances are only approximate

supplied through a filter. A lock-in amplifier detects the second harmonic voltage and plots it on the y-axis of an x–y recorder. The bias voltage is recorded on the x-axis.

IET studies of adhesion
Aluminium is the most widely used substrate in IETS, and it is also the most widely studied metal for adhesive bondings. IETS has been used to study the interaction of **Epoxide adhesives**, polymers with ester groups and silane **Coupling agents** with aluminium oxide, and the salient features of these studies are as follows.

Epoxides show a peak at 939 cm^{-1} in IET spectra (Fig. 78)[3] which is stronger than the corresponding IR (930 cm^{-1}) and Raman peaks (925 cm^{-1}). Attempts to remove this peak by adding an excess of amine hardener and subjecting the junction to conditions of heating much in

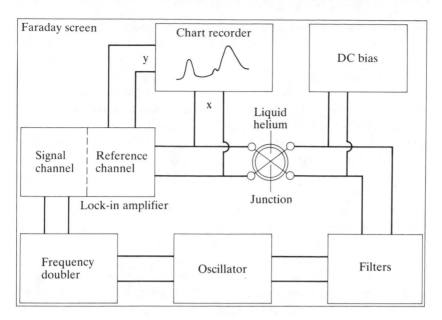

Fig. 77. Diagram of an IET spectrometer

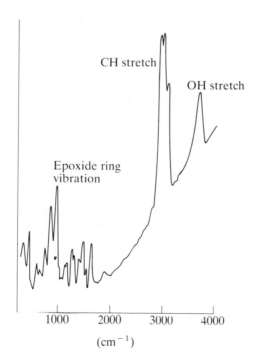

Fig. 78. IET spectrum of the diglycidylether of bisphenol-A on aluminium oxide, prepared by doping from solution in benzene

excess of the normal cure requirements for these adhesives failed to remove the IET peak. The likely reason for this is adsorption of the amine on the metal oxide.

Polymethylmethacrylate has been studied, using both solvent and plasma doping, as a representative of **Acrylic adhesives**. The most significant feature of its IET spectra is the presence of symmetrical and asymmetrical stretching vibrations of the carboxylate anion at 1460 cm^{-1} and 1621 cm^{-1} respectively.[4] This indicates that the ester groups have been saponified; one result of this is that there are strong ion-pair interactions at the adhesive interface. The effect is demonstrated by other ester polymers.

The IET spectrum[5] of triethoxysilane $(HSi(OC_2H_5)_3$ shows a sharp and intense peak at 2191 cm^{-1} due to the Si–H stretching mode. In consideration of the orientational selection rule, this demonstrates that the molecule is adsorbed tripod-like on the metal oxide surface by the alkoxy groups, with the Si–H bond lying perpendicular to the surface.

Other techniques for studying the chemical nature of surfaces are described in articles on **Electron probe microanalysis, Infra-red spectroscopy of surfaces, Secondary ion mass spectroscopy, Surface analysis** and **X-ray photo-electron spectroscopy.**

References

1. R C Jaklevic, J Lambe, *Phys. Rev. Letters*, **17**, 1139 (1966).
2. N M D Brown in *Spectroscopy at Surfaces*, ed R J H Clark, R E Hester, John Wiley, 1988, Ch. 5.
3. J Comyn, C C Horley, D P Oxley, R G Pritchard, J L Tegg, *J. Adhesion*, **12**, 171 (1981).
4. R R Mallik, R G Pritchard, C C Horley, J Comyn, *Polymer*, **26**, 551 (1985).
5. D M Brewis, J Comyn, D P Oxley, R G Pritchard, S Reynolds, C R Werrett, A J Kinloch, *Surf. Interface Anal.*, **6**, 60 (1984).

Infra-red spectroscopy of surfaces

J COMYN

Infra-red (IR) spectroscopy is a means of detecting and quantifying chemical bonds. It depends on the fact that different types of chemical bonds absorb IR radiation at characteristic wavelengths.[1] An example is the $C=O$ group in saturated aliphatic ketones, which absorbs at wavelengths in the region 5.80–5.86 μm (1725–1715 waves per centimetre or wavenumbers, usually given as cm^{-1}). Similarly the range for C—Cl bonds is 13.1–19.8 μm (760–505 cm^{-1}), with the exact position depending on the neighbouring groups.

There are two basic types of IR spectrophotometer. Dispersive instruments have been available for a long time, but in recent years a rapid rise in the development of Fourier transform (FT) machines has taken place.

The most common sampling technique in IR spectroscopy is that of transmission, in which an IR beam is monitored after it has passed through a thin film of the material. Here the principal concern is sampling methods that can be applied to surfaces and adsorbed layers.

ATR and MIR

A technique which has been established for some time is that of attenuated total reflectance (ATR) spectroscopy or multiple internal reflectance (MIR). The difference between these two techniques is that ATR may only involve one reflection and MIR involves many, e.g. 5, 9 or 25 in commercially available attachments (see Fig. 79). The sample is placed in contact with a prism of either germanium or thallium bromide-iodide, which can be placed in either a dispersive or a FT spectrophotometer. The spectra obtained closely resemble conventional transmission IR spectra.

The sampling depth is given by the Herrick equation:

$$d_{\mathrm{p}} = \frac{\lambda}{2\pi n_1 (\sin^2 \theta - (n_2/n_1)^2)^{1/2}}$$

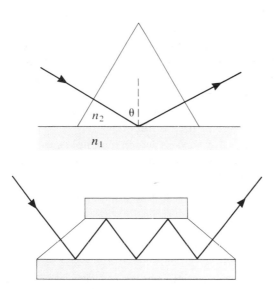

Fig. 79. Attenuated total reflectance ATR (top) and multiple internal reflectance MIR (bottom)

Here d_p is the distance below the surface at which the electric vector has fallen to $1/e$ of its value at the surface, θ the angle of incidence, n_1 the refractive incidence of the sample and n_2 that of the prism ($n_2 = 2.4$ for TlBrI and 4.0 for Ge). Except for the wavelength λ, all the parameters in the Herrick equation have values roughly in the range 0.5–5.0, which means that $d_p \approx \lambda$. The sampling depth is thus about 2–50 μm which is many thousands of molecular layers; it is thus much less surface sensitive than **X-ray photoelectron spectroscopy** and static **Secondary ion mass spectrometry**.

It is important that the sample makes good optical contact with the prism, and for this reason it is easier to work with rubbery polymers. However where rubbers are filled with carbon black no spectrum is obtained unless FTIR is used. Here MIR has been applied to the oxidative surface treatment of polyolefins,[2] where it shows the formation of carbonyl groups. However, because of the sampling depth the samples are probably greatly overtreated.

Fourier transform IR spectroscopy[3]

A conventional IR spectrophotometer employs a prism or diffraction grating to disperse the radiation, which is then passed one wavelength at a time to the sample. A FT instrument is fundamentally different from this, although there is no difference in the way in which the radiation interacts with the sample. It uses a Michelson interferometer (Fig. 80) where radiation from a polychromatic source passes to a beam splitter which forms two perpendicular arms. One is transmitted to a stationary mirror and the other to a moving mirror. The reflected beams return to the splitter and then pass to the detector via the sample. The reunited beams will interfere in either a constructive or destructive manner, depending on wavelength and path difference. The signal at the detector is a polychromatic interferogram which is converted into a spectrum by a computer. The scanning time is about 1.5 s.

There are a number of ways in which a sample can be exposed to the beam in an FTIR instrument. These include transmission and ATR/MIR, and two further methods of value to adhesion science are diffuse reflectance and reflection–absorption (Fig. 81). In diffuse reflectance in FT (DRIFT), light which is reflected from the sample is collected by a set of mirrors and passed to the detector. It is particularly useful in that powdered samples can be examined without further preparation; the method has been extensively used by Koenig to study silane **Coupling agents** on powdered silica. With reflection–absorption IR (RAIR) a beam which is polarized parallel to the plane of incidence is reflected from a polished metal mirror at a high angle of incidence (about 80°). It has been employed

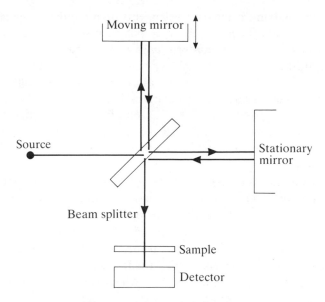

Fig. 80. Michelson interferometer

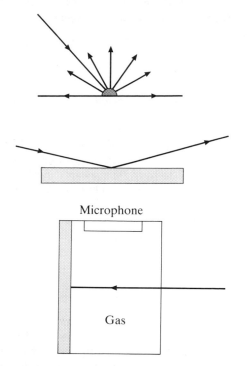

Fig. 81. Techniques in FTIR. DRIFT (top), RAIR (centre) and PAS (bottom)

by Boerio in studying silane coupling agents on a number of metals. The work of Koenig and Boerio in this area has been reviewed.[4] See also the detailed article on **Fourier transform infra-red spectroscopy**.

Photoacoustic spectroscopy (PAS) is a detection method of rising importance. The solid sample is contained in a chamber filled with a coupling gas such as air, helium or argon. The surface of the sample is heated by a small amount ($<0.001\,°C$) on absorbing and this causes a small increase in gas pressure. The IR beam, the temperature rise and gas pressure are all modulated, and the modulation of gas pressure is detected by a microphone.

Great advantage is conferred on the FTIR spectrophotometer by the incorporation of a computer. These include multiple scanning to improve signal-to-noise ratio, and the ability to produce difference spectra. This has been used by Koenig and co-workers to study the adsorption of 3-aminopropyldimethylethoxysilane on titanium and aluminium oxides. The difference spectra (metal oxide treated with silane–metal oxide) revealed weak bands at 950 and 963 cm^{-1} which were respectively assigned to Si–O–Ti and Si–O–Al groups.

Other techniques for studying the chemical nature of surfaces are described in articles on **Electron probe microanalysis, Inelastic electron tunnelling spectroscopy, Secondary ion mass spectroscopy, Surface analysis** and **X-ray photoelectron spectroscopy**.

References

1. G Socrates, *Infrared Characteristic Group Frequencies*, John Wiley, New York, 1980.
2. D Briggs in *Surface Analysis and Pretreatment of Plastics and Metals*, ed. D M Brewis, Applied Science Publishers, London, 1982, Ch. 4.
3. S Naviroj, J L Koenig, H Ishida, *J. Adhesion*, **18**, 93 (1985).
4. J Comyn in *Structural Adhesives; Developments in Resins and Primers*, ed. A J Kinloch, Elsevier Applied Science Publishers, London, 1986, Ch. 8.

Internal stress

K KENDALL

Adhesive joints often contain internal stress, that is, stress which is caused by internal movements as the joint is made, not by externally applied forces. These stresses are also called residual stresses because they remain in the joint after all external stresses have been removed. Such internal stresses are important because they cause premature failure, or even spontaneous fracture of the joint. However, in particular circumstances, an adhesive joint can be strengthened by inserting appropriate internal stress, just as glass is strengthened by blowing cold air on its surface during solidification, to promote residual compressive surface stress.

The most common form of internal stress arises from shrinkage of the adhesive. For example, a resin adhesive may shrink by several volume per cent during polymerization, while the adherends remain constant in size. Obviously, this shrinkage has to be accommodated in some way, usually by the appearance of internal stress within the adhesive, and also within the surrounding adherends. The internal stress depends on the amount of shrinkage, the geometry of the joint and the relative elasticity of adhesive and adherends. The internal stress may be high enough to rupture the joint without applying any external load.

Internal stress can also be produced with **Hot melt adhesives** where the adhesive shrinks more on cooling than the adherend. In this case there may also be a difference in thermal expansion coefficient between the two adherends, and this difference causes additional internal stress in the joint. Such internal stress is seen most dramatically in bimetallic strips used in thermostats, where the internal stress causes large bending of the joined materials. **Solvent-based adhesives** also produce internal stress when shrinkage occurs during solvent evaporation.

The simplest way to visualize internal stresses is by observing the distortions which accompany them, as in the bimetallic strip example above. For example, coating a resin adhesive on to a strip of plastic film (Fig. 82(a)), then polymerizing it, causes bending of the composite strip (Fig. 82(b)) as a result of polymerization shrinkage in the adhesive layer.

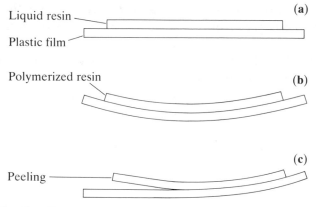

Fig. 82. Visualizing the effects of internal stress caused by polymerization shrinkage of the resin adhesive

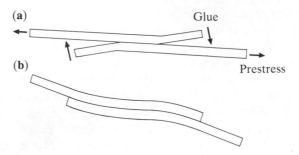

Fig. 83. Strengthening a lap joint by introducing internal stress during formation of the joint: (a) prestressing and glueing of laps; (b) final shape of joint after release of prestress

This bending shows that internal stresses exist both in the adhesive and the adherend, and allows calculation of the stress levels once the dimensions and elastic moduli of the materials are known. If the distortion is sufficiently large, then failure of the adhesive interface may be observed by the peeling mechanism shown in Fig. 82(c) even without the application of any external force. It has been demonstrated by a **Fracture mechanics** argument[1] that peeling is always assisted by internal stress, whether tensile or compressive, because the stored elastic energy released by splitting the joint can drive the crack through the interface. The strength reduction is proportional to the adhesive thickness, to its elastic modulus and to the square of shrinkage strain.

This theory allows the design of adhesive joints with resistance to internal stress. The most important objective is to minimize shrinkage by appropriate choice of polymer. An alternative approach is to add filler to the polymer to reduce shrinkage (see **Filled polymers**). Use of a compliant or rubbery adhesive reduces the internal stress. Weakening of the joint is minimized by using the smallest possible volume of adhesive when making the joint.

Although peeling failure is always assisted by internal stress, lap joint failure may be inhibited by judicious insertion of internal stress.[2] This is most simply achieved by prestressing the sheets before gluing the laps in place (Fig. 83). A factor 3 improvement in lap joint strength is possible by this method.

Paint service properties and adhesion includes a discussion of internal stress in paint films.

References

1. K Kendall, *J. Phys. D: Appl. Phys.*, **6**, 1782 (1973).
2. K Kendall, *J. Phys. D: Appl. Phys.*, **8**, 1722 (1975).

J

Joint assembly

R D ADAMS

Adhesives are available in the form of liquids, solids, pastes and thin sheets. They may be dispersed by hand or machine and there are many methods depending on the structural demands on the adhesive, the manufacturing resource available and the cost of the component.

In mass production, such as assembling motor-car bodies (see **Automotive applications**), pastes or highly viscous liquids are used as these can be dispersed by using a robot and will stay in place on curved surfaces until the joint can be formed. The gap size in the joint is often variable and rarely critical. On the other hand, bonded joints in aircraft are designed to much more stringent safety factors and the tolerances are necessarily tighter. Often, the adhesives are applied in the form of a thin sheet of solid adhesive, sometimes supported by a fabric carrier, so as to give careful control of bond-line thickness.

In all cases, jigging or holding of the parts in some way is necessary. As the adhesive cure state is approached, the material becomes liquid and may run out of the joint, resulting in starvation and low joint strength. Pressure is necessary to hold the parts together, but excess pressure may squeeze the adhesive out.

It is impossible to generalize on assembly matters, except that the advice of the adhesive manufacturer should be sought before committing extensive resources to a manufacturing procedure.

See also **Dispensing of adhesives, Engineering design with adhesives, Engineering practice**.

Select reference

W A Lees, *Adhesives in Engineering Design*, Design Council, London, 1984.

Joint design — cylindrical joints

C WATSON

Most adhesive joint configurations are either 'flat' or 'cylindrical': this article is concerned with the latter type. Broader aspects of design are discussed in **Joint design — general**.

Cylindrical joints

Some typical cylindrical joints are gear to shaft, rotor to shaft, bearing into housing, tube into casting, cylinder liner into engine block, pulley to shaft, fan to shaft, trunnions into rollers and bushings into housings.

Traditional methods of holding cylindrical components together, apart from adhesives, are mechanical fit (press fit or shrink fit), keys, circlips, set screws, splines, welding and bolting. As far as the designer is concerned the traditional methods work — most of the time. However, traditional methods may be heavier and cost more. Also, mechanical fits are not always strong enough and, ingress of moisture into a joint can lead to fretting corrosion.

Adhesive bonding can eliminate extra parts, reduce machining operations and ease manufacture in some instances.

Designing cylindrical joints

The following calculations apply to **Anaerobic adhesives** and use of a computer-aided design program called RETCALC.[1] The program could be modified for application to other adhesives provided that appropriate data for their properties were inserted.

Anaerobic adhesives can be used to retain cylindrical components in conjunction with a clearance fit or an interference fit.

Using anaerobic adhesives with a clearance fit[1-3] The axial force to break a joint can be expressed as

$$F = A\tau f \qquad [1]$$

where F is the breaking force, A the surface area of bond, τ the shear strength of adhesive and f the product of a number of correction factors, determined empirically, to take into account deviations from the simple theory on which the equation is based.

The operating temperature will affect the shear strength of the adhesive: this can be allowed for by use of a suitable factor. Different substrates have different mechanical properties: the substrate factor is unity for

Table 20. Factor f_7: assembly factor

Type of assembly	Clearance fit	Press fit	Shrink fit
Assembly factor	1.0	0.5	1.2
In clearance fit	Adhesive fully fills joint		
In press fit	Some adhesive is scraped off during assembly		
In shrink fit	Adhesive fills joint and is maintained under compressive load		

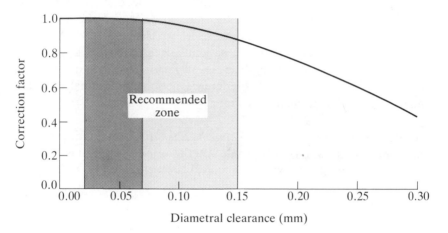

Fig. 84. Diametrical clearance factor f_2 used in the design of cylindrical joints (see Eqn 1 and text, standard = 0.05 mm)

mild steel, but varies from 0.3 to 0.8 for aluminium, according to the alloy.

Three further factors are introduced to allow for joints of differing dimensions. The 'geometry factor' varies according to the length to diameter ratio and the bonding area. A 'diametrical clearance' factor falls below unity as the clearance increases above about 0.07 mm (Fig. 84). The type of assembly – whether a clearance fit, press fit or shrink fit – is allowed for via an 'assembly factor' (Table 20). Data are available for certain types of adhesives.[1,2]

Using anaerobic adhesive with an interference fit The theory for a clearance fit can simply be modified by adding a frictional force term to the term for ultimate load of the adhesive in shear. Thus the expression for the axial force to break the joint is

$$F = A(\tau f + P\lambda) \tag{2}$$

where P is the radial contact pressure between the faces and λ the coefficient of friction; λ is difficult to estimate realistically, but a value of 0.2 for steel on steel is usually found to give a realistic estimate of strength. The maximum torque that the joint can transmit can be obtained by multiplying F by half the diameter of the inner component.

From the derivation of the expression for F, it is clear that they represent an estimate of the ultimate strength of the joint resulting from the application of a steadily applied load. Under dynamic or **Fatigue** conditions, the strength of the joint would be considerably lower than the static value calculated. In order to determine the magnitude of the reduction, appropriate fatigue tests, establishing the 'S/N' curve, would have to be undertaken. These joints usually show a fatigue limit, so it is often possible to apply a safety factor to the static strength calculated. Thus for bonded interference fit joints, torsionally loaded, a factor of 0.35 is normally adequate.

Bonding a ring gear to a differential housing. In fitting a ring gear to differential housing on the Renault R9, the original fitting method was to use a mounting plate with mechanical fasteners (see Fig. 85(a)).

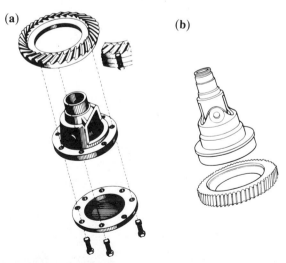

Fig. 85. Fitting of ring gear to differential housing on Renault R9: (a) old design with mechanical fasteners; (b) new design with adhesive bond

The Renault designers wanted to reduce weight and cost. They changed the design (see Fig. 85(b)) to transverse drive and shrink fitted the gear to the differential with a 0.25 mm interference. This, however, did not give the required dynamic strength of 3420 N m. Increasing the interference would have overstressed the gear. They therefore chose to use an anaerobic adhesive with a reduced shrink fit of 0.80 mm nominal, which gave the required strength. The calculation is given below.

Correction factors:

$$f_1 = 0.8 \text{ (steel on cast iron)}$$

$$f_2 = 1.2 \text{ (shrink fit, Table 20)}$$

$$f_3 = 0.6 \text{ (geometry factor)}$$

$$f_4 = 0.9 \text{ (temperature } 120 \, ^\circ\text{C)}$$

$$f = f_1 \times f_2 \times f_3 \times f_4 = 0.52$$

$$\tau = 25 \text{ N mm}^{-2} \text{ for Loctite adhesive}$$

$$\text{Adhesive strength} = 25 \times 0.52 = 13 \text{ N mm}^{-2}$$

$$\text{Mechanical strength} = P \times \lambda = 2.3 \text{ N mm}^{-2}$$

$$\text{Diameter } 140 \text{ mm}$$

$$\text{Length} = 24 \text{ mm}$$

Substituting these values in the formula gives for static torque

$$3.14 \times 140 \times 24 \times \frac{140}{2} (25 \times 0.52 + 2.3) \text{ N m} = 11\,300 \text{ N m}$$

$$\text{Dynamic fatigue factor} = 0.35$$

$$\text{Estimated dynamic torque} = 11\,300 \times 0.35 \text{ N m}$$

$$= 3955 \text{ N m}$$

$$\text{Actual required} = 3420 \text{ N m}$$

Other **Automotive applications** of adhesives discussed elsewhere.

Conclusion

Designers should contact adhesives manufacturers who will have data for their own adhesives and can advise on joint design for flat and cylindrical joints.

References

1. IBM PC-compatible software package *Retcalc*, available from Loctite Ltd, Welwyn Garden City, Herts.
2. Anon., *Design Manual, Module 1, Retailing Cylindrical Assemblies with Adhesives*, Loctite Ltd, Welwyn, undated.
3. G Haviland, *Machinery Adhesives for Locking, Retaining and Sealing*, Marcel Dekker, New York, 1986.

Joint design – general

C WATSON

Most adhesive joint configurations can fall into one of two categories: flat joints, cylindrical joints.

For the best possible performance joints should be specifically designed for adhesive bonding. In a few cases only can an adhesive be used on a joint not specifically designed for adhesives – mainly cylindrical joints. Bond stresses, materials, type of adhesive, surface preparation, methods of application and production requirements can then all be considered in relation to each other at the outset. The designer should consider especially the effect of shear, tension, cleavage and peel stresses upon the joint (Fig. 86).

Tensile shear-loaded joints (Fig. 86(a)) have the highest strength (see **Shear tests**). Stress is distributed over the bond area. Joints which have load applied as a shear loading are most resistant to bond failure.

Tensile loading (Fig. 86(b)) will apply stress over the total bond area. However, care should be taken that joint components have the rigidity to maintain this even loading. When deflection under load occurs a cleavage stress could arise resulting in early bond failure (see **Tensile test**).

Cleavage loading (Fig. 86(c)) will concentrate stress at one side of the joint. The bond area will have to be increased to withstand this uneven loading.

Peel strength is usually the weakest property of a joint. A wide joint will be necessary to withstand peel stress (Fig. 86(d)), plus the use of an adhesive with high peel strength (see **Peel test**).

For an adhesive to be used, a joint must allow the easy application of the adhesive, must allow for the adhesive to cure fully and must be designed to give uniform stress. Even in a simple face-to-face joint it must be possible to apply adhesive to one surface and for it to remain there until the two parts are brought together and after that until curing takes place. These requirements highlight the need for a choice of thin, thick or thixotropic adhesives.

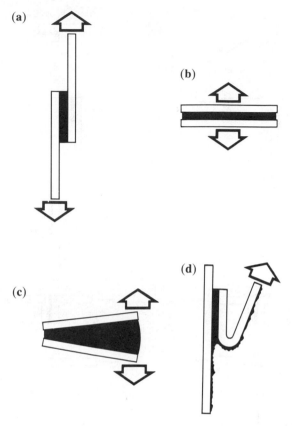

Fig. 86. Types of stress to which an adhesive joint may be subjected:
(a) tensile shear; (b) tensile loading; (c) cleavage; (d) peel

Good and poor designs

Figure 87 shows schematic diagrams[1] of some good and poor joint designs related to the preceding comments. It can clearly be seen that the major difference between good and poor is to restrict peel–cleavage stress of the joint.

The bond line

The gap between the parts and therefore the thickness of the adhesive film has an important bearing on the characteristics of the joint. In terms of simple strength a thick bond line will generally be a weakening feature since the mechanical strength of the unsupported resin film is likely to be less than that of the substrates.

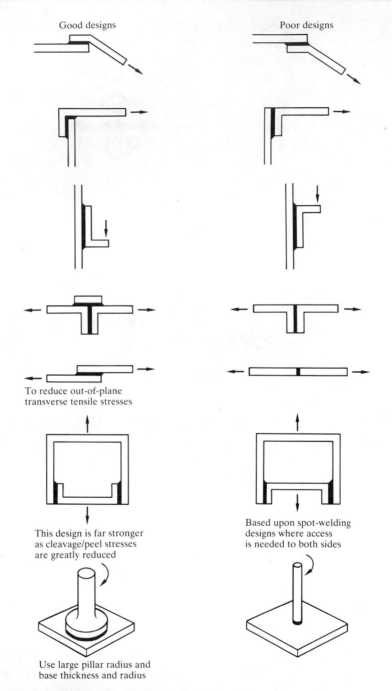

Good designs

Poor designs

To reduce out-of-plane
transverse tensile stresses

This design is far stronger
as cleavage/peel stresses
are greatly reduced

Based upon spot-welding
designs where access
is needed to both sides

Use large pillar radius and
base thickness and radius

Fig. 87.   Representations of good and bad joint designs. (Reproduced with
permission from A J Kinloch, *Adhesion and Adhesives: Science
and Technology*, Chapman and Hall, London, 1987)

A thick bond line can, however, confer advantages. The adhesive is generally more flexible than the adherents or substrates. This is particularly so in most engineering applications where metals or other rigid materials can be bonded. Because of this, a thick bond line can offer a capacity to absorb some impact energy, thus increasing the strength of the bond under this type of loading.

Again, if dissimilar materials are being bonded and there is a differential change in dimensions of the two surfaces, due for example to differential expansion by heating, then the thick film may be better able to accommodate this without being stressed to failure. Consideration of bond-line thickness leads immediately to the question of environmental resistance.

Adhesive bonds will always be susceptible to environmental attack and it is essential that any such attack should not reduce the strength of the bond to an unacceptable level. The most important factor here is the correct choice of adhesive, but design of the joint can make a significant difference. Thus a thick bond line offers a ready path for access by moisture or other solvents which might be able to diffuse through the cured adhesive (see **Durability**).

General references

A J Kinloch, *Adhesion and Adhesives, Science and Technology*, Chapman and Hall, London, 1987.

M M Sadek, *Industrial Applications of Adhesive Bonding*, Elsevier Applied Science Publishers, London, 1987, Ch. 7.

L

Laminating

R J ASHLEY

For some industries and applications, for instance in the **Packaging industry**, the use of a single web of material may not satisfy all the properties demanded. It is therefore of benefit to be able to combine the properties of several and differing materials in order to achieve the performance. A particularly useful means of tailoring exact specifications is to resort to lamination technology. This is essentially based on reel-fed operations although some laminates may be produced by compressing sheets of material together in a hot press (e.g. decorative laminates). The methods described here rely on adhesive components coated on to a substrate forming an interface for chemical bonding to take place. The use of functionally active adhesive materials enables dissimilar substrates to be bonded satisfactorily.

There are three basic categories of adhesive bonding used, each requiring specific equipment. Selection of a particular technique depends on the nature of the substrates used and final application. There are a considerable number of adhesive laminating materials and techniques of variants upon the theme, but it can be reduced to the three basic types of material, aqueous based (see **Emulsion and dispersion adhesives**), **Solvent-based** and **Hot melt adhesives**, and to the two basic techniques, wet or dry lamination.

Wet laminations are those in which the adhesive, based either on water or organic solvents, remains in the laminate during lamination and is dried later. Hence the process is suited to applications where one or more of the substrates is porous to facilitate drying of the adhesive so that solvent may permeate out of the laminate or absorb into one of the substrates. Adhesive is applied, usually by roller coating or air knife, to one of the substrates and while still wet a second substrate is combined at a nip roller. The laminate may be left to air dry or passed through a heated oven to remove solvent and build up bond strength. In general this method is not suited to applications using plastic films, but confined to areas where paper and board form part of the laminate, e.g. paper/foil, paper/paper. Achievement of adequate bond strength is usually indicated by fibre-tearing characteristics. The types of adhesives used are mainly the water-based natural products such as starch and dextrin, polyvinyl acetate or latexes.

Dry lamination can be used with a wider range of substrates such as coated papers, foils and plastic films. The adhesives may be water or solvent based and generally use cross-linking agents to achieve full bond strength. The adhesive film is coated on to one of the substrates generally by a roller system and the coated web passed through a drying oven to remove the solvent leaving a slightly tacky surface. A second web may then be combined at a high temperature and pressure nip to enable good contact and flow of the adhesive. Initial tack or 'green strength' should be sufficient to hold the plies together and resist forces due to relaxation of webs and full strength generally develops over about 24 h. Further plies can be built up by successive passes through the process or several adhesive operations may be combined by using multi-head machines. The adhesives are generally solvent based although considerable development has taken place in this area to reduce the type of solvent and content. Typically the adhesive is a two-part **Polyurethane adhesive** to achieve highest temperature and product resistance. A wide range of adhesive compositions is available to satisfy particular bonding situations.

The method for hot melt adhesive lamination consists of heating the melt to a closely controlled temperature and applying to the most temperature-resistant substrate by roller or similar applicator to obtain a smooth layer. Combination with a second web is achieved at a temperature-controlled nip unit followed by cooling. A pre-made film of hot melt material may be interleaved between two substrates at a high-temperature nip to achieve lamination (thermal bonding). The adhesives are generally complex formulations typically based on **Ethylene–vinyl acetate** copolymers although many other polymers may be used. Cross-linkable grades have been developed to improve heat resistance of laminates.

In many laminating operations plastic films are used as part of the composite. Since many of these will have poor surface-wetting characteristics, especially to water, it is necessary to use surface modification techniques in line with the laminating operation to enhance adhesion (see **Wetting and spreading**). One system commonly used is **Corona discharge treatment**.

A common procedure used to assess adhesion or bond strength in laminated materials is the **Peel test**.

Select reference

R J Ashley et al., *Industrial Adhesion Problems*, ed. D M Brewis, D Briggs, Orbital Press, Oxford, 1985, Ch. 8.

Literature on adhesion

D E PACKHAM

This article gives guidance both to those seeking general background information and to those embarking upon research and development involving adhesion, adhesives and related technologies. It will be of help to students, practising engineers and scientists and to technical service staff. The reference numbers in this article refer to the list in Appendix 3 at the end of the book.

Books

With ready computer access to catalogues of major libraries and lists of books in print, it is relatively easy to search for books with adhesion and similar words in their titles and to obtain a comprehensive list. Such lists are of great length, but of limited value: they cannot offer a critical insight. It seemed better to provide a list of relatively recently published books which seem to make a useful contribution to the areas that they address. This list is provided in Appendix 3. It is idiosyncratic, being limited by my experience and judgement, and should not be taken to include all that are worth while, or even to represent the best published. I must apologize to the authors of what must be many fine works which I have overlooked.

There has been an increasing number of books of broad scope on adhesion published in the last 30 years. Some particularly useful ones are given in the first part of Appendix 3.

My preference for a general text covering the science and technology in a thorough and well-written way is Kinloch.[3] It starts with a consideration of interfacial contact and theories of adhesion, then moves on to surface treatments, setting of adhesives and joint mechanics and concludes with a discussion of the service life of adhesive joints. It is authoritative in its treatment and includes comprehensive references to the original work, making it of value to the research worker as well as to the scientist and engineer needing a textbook.

Mention should also be made of a much larger work, the ASM *Handbook* edited by Brinson[4] which runs to nearly 900 000 words. This, too, is particularly valuable in being comprehensive and up to date. It is written to guide the practising engineer to an understanding of the scope of adhesives and sealants and of how to use them. It comprises about 100 specialized articles. Many of these directly address topics of industrial engineering interest – applications and markets, design, manufacturing technology, non-destructive inspection and repair of bonded structures – but these are placed in a firm scientific context. There is supporting treatment of the materials science of adhesives and sealants including their environmental degradation, the analysis of surfaces, the mechanics and testing of joints. It is a book to be consulted for reference rather than read from cover to cover.

Most of the general texts deal adequately with testing of adhesion, and detailed methods are given in **Standards for adhesion and adhesives**. The monograph of Anderson, Bennett and De Vries[5] on testing gives a valuable survey of the most important tests, placing them in the context of the theory of the mechanics and chemistry which underlie them.

Adams and Wake's book[7] on structural adhesives represents fruitful collaboration between a mechanical engineer and a chemist: it is somewhat unusual in its integration of the treatment of properties of adhesives with that of the mechanics of the joints.

A common practical problem is the **Selection of adhesives** and of **Pretreatments**. Monternot,[10] Shields[12] and Skeist[14] all give a great deal of relevant information on service properties and practical guidance including selection charts.

Periodicals

Adhesion is an interdisciplinary subject. Fundamental sciences such as physics and chemistry and mechanics are involved as well as many branches of engineering and technology including metal finishing, polymer processing and composite materials. As a consequence, original research relevant to adhesion may be published in a very wide range of periodicals. This makes the task of the research worker who needs to be conversant with the literature a difficult one. The journals, shown in Appendix 3, which are devoted to adhesion are an obvious starting-point. A list is also given of some of the other journals in which important articles are likely to appear. The list cannot be complete, but might well form the basis of periodicals to be monitored through *Current Contents*.

Abstracting services are of course indispensable. It is worth pointing out that the scope of *Chemical Abstracts* in this field is good. *Adhesives Abstracts* (published by RAPRA) is more specialized.

Further information

The bibliographies in the works cited in Appendix 3 form an obvious source of further references. The ASM *Handbook*[4] has an 8500-word article by A H Landrock on information sources on adhesives and sealants. It includes lists and short descriptions of periodicals, trade literature and some 70 books and monographs. A useful list of books in the background areas of surfaces and surface analysis, polymer science and stress and fracture analysis is given by Kinloch[3] at the end of Chapter 1.

Locus of failure

D M BREWIS

There are several possible zones where an adhesive joint may fail. It may fail within the adhesive layer (cohesive failure), at the interface between the substrate and adhesive (interfacial or adhesive failure) or within the substrate (material failure). Likewise with a coated substrate, interfacial, cohesive or material failure may occur.

To the naked eye, failure sometimes appears to be interfacial (apparent interfacial failure), but examination by **Optical microscopy** or **Electron microscopy** or by one of the **Surface analysis techniques** such as **X-ray photoelectron spectroscopy** (XPS) may reveal a thin layer of an adhesive or coating perhaps only a few nanometres thick; the true failure mode is therefore cohesive. Examples of different failure types are discussed in **Scanning electron microscopy, Electron probe microanalysis, Auger electron spectroscopy** and **X-ray photoelectron spectroscopy**.

Determining the true locus of failure can be very useful in understanding the reason for an adhesive problem. An example of this is given in the **Weak boundary layer** section. This describes the case where an adhesive bond between untreated PTFE and an **Epoxide adhesive** was broken. Failure appeared to be interfacial, but XPS showed that material failure had occurred within a region of low cohesive strength at the near surface of the PTFE.

Select references

D Briggs in *Industrial Adhesion Problems*, ed. D M Brewis, D Briggs, Orbital Press, Oxford, 1985, Ch. 2.

W J van Ooij, *Industrial Adhesion Problems*, ed D M Brewis, D Briggs, Orbital Press, Oxford, 1985, Ch. 4.

A J Kinloch, *Adhesion and Adhesives*, Chapman and Hall, London, 1987, Ch. 8.

M

Mechanical theory of adhesion

K W ALLEN

Introduction

One of the first clear statements of a mechanical explanation for adhesive phenomena is due to McBain and Hopkins who in 1926 concluded that joints might be of two types. In one of these a mechanical type of mere embedding existed. Thus the idea grew of an interlocking of the adhesive with the texture of the surface and of the advantage of a rough surface. It was a statement of the simple and intuitive explanation which practical men had evolved for themselves over a considerable period. It sufficed so long as the materials being joined were simple fibrous materials such as wood or leather, although it had shortcomings even for these. Once adhesives were used, with even modest success, for smooth impervious surfaces it was clearly inadequate, at least for these materials. Nevertheless it persisted (and persists now) to a surprising extent. It is now apparent that this concept has to be considered on two different scales.

Interlocking on a macro scale

While it is now clear that a mechanical interlocking cannot be an adequate explanation of all adhesion, there are several instances where it is of major importance on a macroscopic scale.

The classical work of Borroff and Wake in 1949[1] on the adhesion between the textile cords and the rubber casing in automobile tyres demonstrated that the only significant factor was the penetration of the fibre-ends into the rubber. These fibre-ends originated from the natural textile fibre (usually cotton), and it was their absence in synthetic fibres (e.g. nylon monofilament) which caused the difficulty in using these for this product. Any specific interaction between the rubber and the fibre was insignificant and only affected the length of fibre-end which must be embedded. There was no significant penetration of the rubber between the strands of the yarn. The bond strength between rubber and cord depended solely upon the number of fibre-ends involved and the depth to which they were embedded.

Rather similar is the case of adhesion to leather, which is of particular importance in the footwear industry. It has been shown that the fibres of the surface must be separated and raised so that they become embedded in the adhesive layer if a satisfactory bond is to be achieved.

A quite different example is the process known as 'electroless' plating of plastics. There are a number of plastic materials which may be coated with a thin layer of metal by this process. The base materials are usually either high-impact polystyrene or ABS (acrylonitrile–butadiene–styrene), both of which have a continuous phase of glassy polymer with an elastomer dispersed in it. The process involves first an etching with chromic acid which oxidizes and removes the elastomer from the surface layer to leave a porous, spongy structure. Then an initial metal layer (usually copper) is deposited by chemical reduction. Once this has been established it is built up and an appropriate metallic surface layer is added by conventional electroplating. A very extensive study by Perrins and Pettett[2] showed that there are two mechanisms simultaneously involved in this adhesion: one a chemical relationship between the metal and the plastic, the other a mechanical interlocking between the metal and the porous surface which is controlled by the topography of the plastic surface.

Reference should also be made to **Flame-sprayed coatings** and **Wood adhesives – basic principles**.

Interlocking on a micro scale

For a considerable period it was common practice to dismiss mechanical interlocking as a 'hook and eye' conception and irrelevant; however, that view has changed more recently. The beginning of this change can be traced to Packham's work[3,4] on the adhesion of molten polyethylene to aluminium. He was exploring the effects of various treatments of the aluminium surface on the adhesion, and particularly the effects of various oxidation processes. It was well known that by **Anodizing** aluminium in an acidic electrolyte the oxide film produced consisted of a dense layer next to the metal covered by a porous layer. The pores were regular in size and shape and were orientated with their axis normal to the surface of the metal. These pores had diameters in the range 120–330 Å and their size and number was governed by the conditions of anodizing. Packham showed a direct relationship between the adhesive bond strength and the size and surface density of these pores. Further, he obtained electron micrographs of polymer surfaces which had been in contact with the aluminium oxide which showed clusters of tufts. These tufts were 500–2000 Å in diameter and each consisted of a cluster of whiskers which had aggregated together. These individual whiskers had originally been inside the pores in the oxide film. Evidently the adhesion involved the penetration of polymer into the pores to give a mechanical interlocking.

More recently, as much more sophisticated techniques of **Electron microscopy** have been developed for examining the features of surfaces, Venables and his colleagues[5] have revealed the detailed morphology of aluminium surfaces which have been prepared by various well-established techniques for adhesive bonding (see **Anodizing, FPL etch, Pretreatments of aluminium**). These surfaces all had whiskers of oxide on the outermost parts; and in some cases, depending upon the details of the treatment, there were also pores similar to those already described. These whiskers were 100–400 Å long and 50–100 Å in diameter and the pores were about 400 Å in diameter. Their work quite clearly implies that there is a double interlocking, with the whiskers being embedded in adhesive and with adhesive penetrating the pores. Thus the region between metal and adhesive has the character of a composite. The mechanism of failure depends upon a viscoelastic deformation of the polymer–adhesive together with a rupture of the other adhesive bonds which have been formed.

It is now quite clear that mechanical interlocking has a significant role in the adhesive process, but all the features involved are on a very much smaller scale than was once imagined. While on a macroscopic scale interlocking may be of little relevance, on a microscopic scale it is vital.

A comparison of different theories is given under **Theories of adhesion**.

References

1. E M Borroff, W C Wake, *Trans. Inst. Rubber Inst.*, **25**, 199, 210 (1949).
2. L E Perrins, K Pettett, *Plast. and Polym.*, **39**, 391 (1971).
3. K Bright, B W Malpas, D E Packham, *Nature* **223**, 1360 (1969).
4. D E Packham in *Aspects of Adhesion 7*, ed. D J Alner, K W Allen, Transcripta, London, Applied Science, London, 1973, p. 51.
5. J D Venables in *Adhesion 7*, ed. K W Allen, Applied Science Publishers, 1982, Ch. 4; *J. Mater. Sci.*, **19**, 2431 (1984).

Microfibrous surfaces

D E PACKHAM

Some form of substrate **Pretreatment** is almost always essential before adhesive bonding. The extent to which pretreatments increase the

Roughness of surfaces and the connection between this and adhesion are complex issues which have interested scientists for more than 50 years (see **Mechanical theory**). The advent of **Scanning electron microscopy** (SEM) in the late 1960s made much more detail of surface topography available. Some surfaces of metals can be seen in SEM to be covered with acicular projections with heights of the order of 1 μm (Fig. 88). Such surfaces have been termed 'microfibrous'.

Preparation

The common feature of microfibrous surfaces is their topography, rather than their chemical nature. It is well known that certain oxidation conditions produce a whisker or blade-like oxide growth on a metal surface, rather than a film of uniform thickness. Similarly, certain conditions of electrodeposition give dendritic crystals of the deposited metal, not a smooth coating. These are ways of preparing microfibrous surfaces. Details of some typical methods will be given.[1]

Needles of copper(II) oxide can be prepared on copper by oxidation of the metal at 90 °C for 20 min in a solution of 3 g l^{-1} sodium chlorite, 10 g l^{-1} trisodium phosphate and 5 g l^{-1} sodium hydroxide. The oxidized copper has a matt black appearance (see also **Pretreatment of copper**).

A blade-like covering of oxide can be produced on mild steel by oxidation for 4.5–5 h at 450 °C in a nitrogen stream which has been bubbled through water at 40 °C (see Fig. 88(b)).

Electrodeposition of zinc from potassium zincate solution (e.g. 100 g l^{-1} potassium hydroxide plus 0.5 g l^{-1} of zinc oxide) on to 'flat' zinc (e.g. at 130 A m^{-2}) gives dendritic crystals on zinc on the surface of the base metal. A micrograph of such a surface is given in **Scanning electron microscope,** Fig. 118.

Adhesion

Microfibrous surfaces typically have a dull appearance as they scatter light. Provided the adhesion of the microfibres to the base metal is adequate, such surfaces have potential as substrates for adhesive bonding. Examples with different polymers will be described.

Polyethylene can be applied as a hot melt coating to metals. Poor adhesion has been found with conventional **Pretreatment of metals** when the polymer was stabilized with antioxidant. By contrast, adhesion to microfibrous surfaces is good even in the presence of antioxidant (Table 21).[1]

The adhesion has been studied of a series of **Ethylene–vinyl acetate copolymers** applied to metals as **Hot melt adhesives**. Much higher adhesion

Fig. 88. Scanning electron micrographs of microfibrous oxides on (a) copper and (b) steel

Table 21. Effect of antioxidant concentration on the adhesion of polyethylene to 'flat' and microfibrous surfaces

Metal	Topography	Antioxidant (ppm)	Peel strength ($N\,mm^{-1}$)
Steel	'Flat'	Nil	2.13
		2000	0.28
Oxidized steel	Microfibrous	Nil	2.73
		2000	2.16
Oxidized copper	Microfibrous	Nil	1.47
		2000	1.44

Table 22. Fracture energy (G_C) for zinc-epoxy single edge notched (SEN) test specimens ($J\,m^{-2}$)

Zinc surface	Percentage rubber reinforcement		
	0	7	15
'Flat'	105	300	700
Dendritic	670	2210	2550

was found to copper with a microfibrous oxide surface than to conventionally prepared copper or steel.

In contrast to polyethylene and EVAs, **Epoxide adhesives** are thermosets and are much stiffer and less ductile. The adhesion fracture energy (G_C, see **Fracture mechanics**) between both unmodified and rubber-toughened epoxies (see **Toughened adhesives**) and several metals has generally been found to be much higher when microfibrous surfaces were involved (Table 22).[1]

It is clear from an investigation of the **Locus of failure** and analysis of the fracture surfaces by **X-ray photoelectron spectroscopy** that the microfibrous surfaces lead to considerable plastic deformation of the polymer. They appear to increase the adhesion by enhanced energy dissipation during fracture[2] (see **Rheological theory, Peel tests**).

References

1. D E Packham, *Int. J. Adhesion and Adhesives*, **6**, 225 (1986).
2. D E Packham in *Developments in Adhesives – 2*, ed A J Kinloch, Applied Science Publishers, London, 1981, p. 315.

Microstructure of joints

D E PACKHAM

The significance of microstructure

The measured adhesion of a joint depends not only on the forces which act at the interface but on the mechanical properties of the bulk phases, adhesive and substrate and the way they interact through the interface (see **Adhesion**). These concepts are developed in the articles **Rheological theory, Adhesion – fundamental and practical** and **Peel tests**). The microstructure of a bulk phase will usually affect its properties, so knowledge of the microstructure of the adhesive, substrate and substrate surface is relevant to an understanding of the strength and performance of an adhesive joint. In many cases the microstructure can be altered when making the joint by altering the process variables. The advantage of this is that it gives the practitioner scope for controlling the performance of an adhesive joint between given materials; the disadvantage is that careless or poorly researched bonding procedures can give rise to adhesive bonds of far from optimum properties.

Consideration of all aspects of joint microstructure would require a scope comprising most of materials science. In this article attention will be directed particularly to microstructural features of adhesive and of substrate surface which can be altered to affect the performance of an adhesive joint.

Adhesive microstructure

A way of significantly altering the properties of a polymer is to incorporate fillers or fibres (see **Filled polymers, Fibre composites – introduction**). The stiffness and often the toughness of the polymers are increased.

Some polymers used as adhesives, e.g. polyethylene, and **Ethylene–vinyl acetate copolymers**, are crystalline and their properties will be altered by altering the degree of crystallinity or the crystalline morphology.[1] Cooling rates, annealing treatments and the use of nucleating agents are process variables which it may be possible to exploit to determine adhesive joint performance.

Where the adhesive is cross-linked (see **Reaction setting adhesives, Epoxide adhesives, Phenolic adhesives, Rubber-based adhesives** and **Rubber to metal bonding**) it is often easy to control properties by controlling the cross-link density.

Some important **Structural adhesives** are two-phase materials comprising dispersed rubbery particles in a glassy matrix (see **Toughened adhesives**). Optimum performance depends upon achieving the 'correct' phase structure which is achieved only by careful attention to the procedure for mixing, applying and curing the adhesive.

The presence of voids in an adhesive might be regarded as a feature of its microstructure. These are often incorporated as a result of careless mixing and exert a deleterious effect. In some circumstances beneficial effects of voids have been found, and they can be introduced deliberately by the use of blowing agents.[2]

Microstructural features of the substrate

The microstructure of the bulk substrate, such as the grain structure of a metal or the porosity of a ceramic, will affect substrate properties, and thence the joint performance. Of special importance to an adhesive bond is the 'microstructure' of the substrate surface.

The surface structure can be altered by **Abrasion treatment** and by many other **Pretreatments**. These pretreatments usually have a crucial effect on the subsequent adhesion. The mode of action of any particular pretreatment is often a matter of speculation. Control of the chemical nature of the substrate surface, of its **Surface energy** and through this the **Wetting and spreading** of the adhesive are usually important; in some cases, however, the **Roughness of surfaces** has a significant influence (see **Mechanical theory**). **Anodizing** of aluminium and titanium can produce a fine porous microstructure which gives particularly good adhesion and **Durability** (see **Pretreatment of aluminium, Pretreatment of titanium**). **Microfibrous surfaces** are another category where the surface structure exerts a strong effect on the adhesion.

Influence of substrate on the adhesive microstructure

Some subtle effects occur when the adhesive microstructure is altered through the influence of the substrate.

During the cooling of a crystallizing **Hot melt adhesive** the presence of the substrate can lead to surface nucleation giving a columnal 'transcrystalline' polymer morphology in the interfacial region. There have been reports that such layers influence measured adhesion, but conclusive demonstration has proved elusive.[3]

Many metals, especially transition metals, are well known for their catalytic activity. Some cases of such chemical influence of a metallic substrate on a curing reaction of an adhesive have been demonstrated. A well-known example which has a considerable effect on adhesion, is the influence of brass or copper on the vulcanization of rubber (see **Rubber to metal bonding – basic techniques**).

Conclusion

The total control of microstructure which is used to influence the properties of materials in general, gives a valuable means of altering the performance of adhesive joints. A more extended discussion of the subject can be found in ref. 3.

References

1. A S Vaughan, D C Bassett in *Comprehensive Polymer Science*, Vol. 2, ed. G Allen, J C Bevington, Pergamon, 1989.
2. T Adam, J R G Evans, D E Packham, *J. Adhesion*, **10**, 279 (1980).
3. D E Packham in Adhesives and sealants in *Engineered Materials Handbook*, Vol. 3, ed. H F Brinson, ASM, Ohio, 1990, p. 406.

N

Napkin ring test

D E PACKHAM

In conventional **Shear tests** there is considerable non-uniformity of stress throughout the joint. The 'napkin ring' test was introduced to provide a test where the variation of shear stress was minimal. It consists of two thin-walled tubes joined end to end by a thin layer of adhesive. The torque required to break the joint is recorded.

Equations have been derived relating the shear strength τ to the torque T and the inner and outer radii of the tubes, r_1 and r_2 respectively.[1,2] Equation 1 applies to elastic failure.

$$\tau = 2Tr_2/\pi(r_2^4 - r_1^4) \qquad [1]$$

When significant plastic deformation occurs during failure the shear strength is given by

$$\tau = 3T/2\pi(r_2^3 - r_1^3) \qquad [2]$$

Provided the mean radius of the cylinders R is much greater than their thickness t, both these expressions reduce to

$$\tau = T/2\pi R^2 t \qquad [3]$$

Foulkes et al.[1] have characterized the test, studying the influence of area bonded, adhesive thickness and failure mode. They discuss the ranges of validity of the different equations for calculating the shear strength. A variation of the test has been adopted as ASTM E 229 (see Appendix 2).

Despite relative uniformity of stress, some stress concentration may occur at the edges of the adhesive (see **Shear tests**). A modification, involving rounding of edges of the tubes has been suggested to reduce this effect.[3]

References

1. H Foulkes, J Shields, W C Wake, *J Adhesion*, **2**, 254 (1970).
2. A J Kinloch, *Adhesion and Adhesives, Science and Technology*, Chapman and Hall, 1987, p. 242.
3. D W Schmueser, Adhesives and sealants in *Engineered Materials Handbook*, Vol. 3, H F Brinson, ASM, Ohio, 1990, p. 441.

Non-destructive testing of adhesively bonded structures

G J CURTIS

Introduction

Testing plays a crucial role in adhesion research and development. **Tests of adhesion**, described elsewhere, involve destruction of the joint: the question arises as to whether the bond strength can be measured without sacrificing the joint. In fact it has long been the basic aim of non-destructive testing adhesively bonded structures to determine the overall shear strength. Regrettably this is still unattainable, principally because there is no intrinsic parameter which relates directly to strength which can be measured non-destructively. There are, however, techniques which depend upon parameters, for example elastic modulus, which relate indirectly to strength, under well-regulated conditions and the most valuable will be illustrated here.

Factors which control bond integrity

Strength, unlike elastic modulus, is not even theoretically a readily determinable quantity. Overall elastic–plastic deformation in a structural adhesive might be describable in terms of intermolecular forces and models of viscous flow, but not at the discontinuous moment of fracture. In fact 'overall' behaviour loses sight of the fact that it is normally isolated phenomena which control the magnitude of joint strength and the **Locus of failure**. The term 'isolated phenomena' refers to voids, cracks, second phase material, etc. which can act as stress concentrators. Clearly it would be unwise to suggest that an adhesive bond tester should merely locate and size voids and cracks, as whether or not such a defect is active depends upon where it lies in the working stress pattern of the structure.

It is useful to consider a bond to be composed of two regions; the interfacial region between the adhesive and the adherend and the region of the bulk of the adhesive, i.e. the adhesive and cohesive regions. There is at least one school of thought which maintains that failure cannot occur at the true interface, and that where it apparently does there is a weak layer between the two phases (see **Rheological theory**). From the non-destructive testing point of view, there is no alternative to aiming tests at both the interface and the bulk of the adhesive. To this end two testing regimes need to be considered; manufacture and in-service use.

Non-destructive test methods

In the search for a 'strength tester' almost every joint parameter which can be determined non-destructively has been studied for a correlation with bond strength. Almost every means or probing the joint has been investigated, for example elastic waves, microwaves, thermal waves, X-rays and neutrons. All have useful features which need to be considered. Without doubt the most useful techniques are those which make use of acoustic (elastic) waves in the joint.

An acoustic wave in a joint has four basic parameters which can be varied by the investigator to produce desired effects and which are modified by structural features in the joint; these are amplitude, velocity, wavelength and phase. The kind of structural features which effect them are elastic modulus (e.g. shear), microporosity, macroporosity, cracks, voids and gross corrosion. Depending upon what features are to be assessed, the wavelength of the interrogating sound can be of the order of the joint thickness or, at the other extreme, of the order of the microporosity present in the adhesive. When the joint is excited into thickness resonance, for example, factors which effect the cohesive character of the joint become evident, e.g. state of cure and glue-line thickness. When the joint is interrogated with a fine pencil beam of acoustic waves, factors like fabrication voids or in-service-induced cracks/non-bonds/corrosion become evident.

Principal devices for production and in-service testing

Resonance testers A number of devices are available. In Europe, the Fokker bond tester is commonly used. This was developed by the Fokker Aircraft Company in order to test laminates and honeycombs (see **Honeycomb structure**). Over 40 years of use and development has shown its strengths and weaknesses. Under conditions of strict fabrication control (or where they have been applied) it is capable of determining the presence of features which effect the cohesive strength of the joint. It yields 'cohesive bond strength' by calibration of either the resonant frequency or amplitude of the joint with shear strength. Where the prospective failure locus is at the interface it is less successful, and in the production regime it is supplemented by surface treatment quality-control tests. In the in-service regime, where gross corrosion of the interface has occurred, it can successfully detect this.

Figure 89 shows an example of the degree of correlation that can be achieved between strengths predicted by the Fokker bond tester and those determined destructively for a bonded laminate. Figure 90 shows a similar correlation for bonded **honeycomb structures**.

Small or large areas of laminates or honeycombs can be tested by using single or multiple acoustic probes mounted upon appropriate scanning mechanics and feeding data process and presentation equipment.

Pulse-echo testers A vast range of general flaw detectors are available which can be used specifically for the location of voids in the glue line or

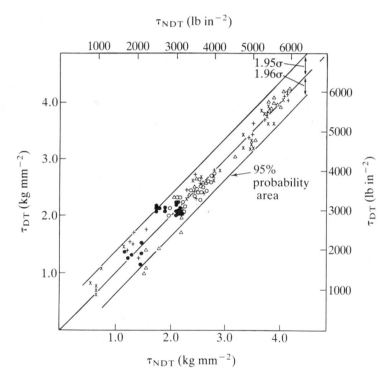

Fig. 89. Comparison of destructive shear strength tests on laminates with the predicted strengths using a Fokker bond tester. A series of laminates is represented where the adherend thickness t varies as follows: (●), $t = 0.6$ mm; (○), $t = 0.8$ mm; (×), $t = 1.0$ mm; (+), $t = 1.2$ mm; (△), $t = 1.5$ mm. To construct this standard plot 120 specimens of Dural 2024-T3 were used; 95% of all results are within the range ± 0.36 kg mm^{-2}. (From R J Schliekelmann, Nondestructive testing of adhesively bonded joints, in *Adhesion, Fundamentals and Practice*, McClaren, London, 1966)

at the interface. Depending upon the relative size of the acoustic beam employer and its wavelength to the size of the defect being sought, the amplitude of the beam is attenuated by their presence. Where the defects are small compared to the beam width, the overall wave amplitude is attenuated by scattering induced by the defects (to an extent depending upon their number and size relative to the wavelength). The relationship between microvoid volume fraction and wave amplitude (also wave velocity), can be determined via careful calibration. The extent of macrovoid size can be determined by scanning the beam across the joint. Figure 91 demonstrates the principle of macrovoid detection and sizing by pulse-echo testing. An acoustic transducer operating at a frequency of ~ 1 MHz is coupled to the specimen by water or grease and scans

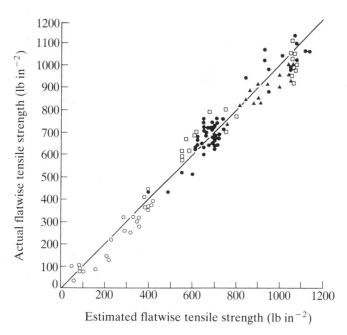

Estimated flatwise tensile strength (lb in^{-2})

Fig. 90. Comparison of destructive tensile strength tests on honeycomb sandwich panels with the predicted strengths using the Fokker bond tester. A series of honeycomb constructions is represented where the foil thicknesses and the cell sizes are both varied: (●), foil thickness 0.007 in (0.177 mm), cell size 0.125 in (3.175 mm); (▢), foil thickness 0.003 in (0.076 mm), cell size 0.250 in (6.25 mm); (▲), foil thickness 0.002 in (0.05 mm), cell size 0.1875 in (4.762 mm); (○), foil thickness 0.001 in (0.025 mm), cell size 0.250 in (6.35 mm). (From R E Clemens, Paper presented at the American Society for Nondestructive Testing Technical Meeting, California, February 1962)

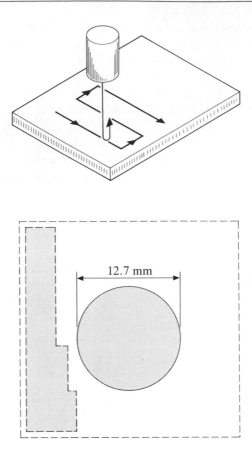

Fig. 91. An ultrasonic C-scan image of two deliberate bonding defects
shown by the dark areas between the dotted boundaries, in the
bond area of a step lap joint. (From J L Rose, P A Meyer,
Ultrasonic procedures for predicting adhesive bond strength,
Mater. Evaluation, **31** (6), 109, (1973))

rectilinearly across the bonded structure. It produces a short pulse of
ultrasound which refracts and reflects at the interfaces of the joint. A
non-bond reflects back over 90% of the incident sound. The magnitude
of the reflected pulse amplitude is registered on a chart recorder as a shade
of grey or as colour and yields a plan view of the void area in the joint.
Relatively large areas of laminates and honeycombs can be studied
depending upon the scale of the scanning mechanics and coupling facilities
that can be made available. In the case of honeycombs or multiple
laminates, the scan can be adjusted to show all defects anywhere in the
depth of the structure or at selected depths.

Select references

G J Curtis, Bonded structure testing, Report No. R6098 (1968), Atomic Energy Research Establishment, UK.

G J Curtis, Adhesively bonded structure nondestructive testing with acoustic methods in *Ultrasonic Testing*, ed. J Szilard, John Wiley, 1980.

G J Curtis, A critique of acoustic methods of adhesive bond strength determination in *Adhesion – 4*, ed. K W Allen, Applied Science Publishers, London, 1979.

C C H Guyott, P Cawley, R D Adams, The non-destructive testing of adhesively bonded structures; a review, *J. Adhesion*, **20** (2), 129–59 (1986).

G M Light, H Kwun, *Nondestructive Evaluation of Adhesive Bond Quality: State of the Art Review*, Report No. AD-A210 0511, Nondestructive Testing Information Analysis Centre, San Antonio, Tex., USA, June 1989.

A Fahr, S Tanary, Nondestructive evaluation of adhesively bonded joints, *Agard Conference Proc.*, No. 462 (1990).

O

Optical microscopy

B C COPE

The average human eye is capable of resolving two points 0.1 mm apart. In order to discern or resolve features smaller than this some form of microscopy must be used.

In the compound optical microscope contrast is produced by the transmission through, or reflection from matter, of visible light. A positive objective lens is used to produce a real image that is viewed through and magnified by a negative eyepiece. The resolution produced by such a microscope is maximized by a reduction in the wavelength of the illuminating light to the minimum visible and a decrease in the focal length of the objective lens. The limits to the latter are set by the physical proximity of the lens and the specimen and the very shallow depth of field obtainable with very short focal length (and hence high magnification) objectives. Overall the maximum useful magnification obtainable with a 'dry' objective is around $700 \times$, giving a resolution of 0.4 μm, whereas an 'oil immersion' objective, used with a refractive oil filling in the space between the specimen and the objective, is capable of $1400 \times$ magnification and a resolution of 0.2 μm.

In transmission mode thin specimens (a few micrometres) are essential. These may be cut using a steel or glass microtome and are mounted between a glass slide and a glass cover slip. The microscope is especially versatile in this mode as polarization may be employed to reveal contrast in birefringence generated by differences in crystallinity or strain level. Sections through adhesive joints may yield useful information, for example, on adhesive penetration into a porous substrate, when viewed in transmission (see **Roughness of surfaces**). However, the sectioning may be difficult, and may introduce artefacts (i.e. visible features originating in the sample-preparation method rather than in the inherent nature of the sample).

This difficulty in producing reliable sections is probably the single factor that most limits the scope of optical microscopy in adhesive studies. Adhesives and substrates that are flexible tend to deform rather than cut and may need to be supported in a cast block of epoxide or other resin, or may need to be cooled down in liquid nitrogen prior to cutting. At the other extreme very hard samples, such as many thermosets and liquid crystal polymers, are too hard to form coherent sections by cutting and must be reduced by grinding in the same way as geological or metallurgical specimens.

Phase contrast illumination employs special condensers and objectives that improve the visible differences between phases of similar optical characteristics. The technique may be useful in showing phase segregation in adhesive films.

The reflection technique is more limited in resolution and magnification than is transmission. Reflection optical microscopy of adhered surfaces is often revealing, but the very limited depth of field is a handicap. Nevertheless, examination of surfaces after testing or failure is a simple, but important first step in determining **Locus of failure**. Surface reflectivity of specimens is often improved by deposition of a layer of metal. The reflection microscope is especially useful when operated in 'dark-field' mode where topographical differences are accentuated.

The Nomarski differential interference contrast technique accentuates changes in specimen thickness and refractive index in transmission without giving the haloes round fine features that distract from the usefulness of the phase-contrast technique. In reflection images are obtained that strongly accentuate the topographical features present.

Scanning electron microscopy offers vastly improved depth of field and magnification in the study of topography, but the simplicity of sample preparation, the freedom from vacuum and beam damage and the relatively modest cost of the optical microscope are attractive.

Select references

S Bradbury, *An Introduction to Optical Microscopy*, Oxford University Press, Oxford, 1984.
L C Sawyer, D Grubb, *Polymer Microscopy*, Chapman and Hall, London, 1987.
B C Cope in *Surface Analysis and Pretreatment of Plastics and Metals*, ed. D M Brewis, Applied Science Publishers, London, 1982.

P

Packaging industry

R J ASHLEY

In the UK around 40% of the adhesive market is consumed by the packaging industry. The construction of packaging materials involves the use of a wide spectrum of adhesive types and covers many methods of assembly. Some of the applications include coatings which act as adhesives for bonding to other forms of packaging, for example in the form of labels or tapes (gummed and pressure sensitive), while some adhesives may be used to assemble a package as in the case of a carton or prepare sophisticated laminates.

Materials

In general the adhesives used fall into the categories of solvent-borne resins, water-based dispersions and solutions and 100% solids hot melts. Apart from coatings the methods of **Laminating** used for producing multilayer complexes include adhesive bonding, **Extrusion coating, Coextrusion** and thermal bonding. Some of the adhesive materials used may be broken down as follows (see **Adhesive classification**):

1. Solvent-based dispersions such as polyvinyl acetate, polyesters, polyethers, acrylic copolymers, rubbers with tackifiers and plasticizers.
2. Water-based dispersions or emulsions such as polyvinyl acetate, acrylics, polyvinyl chloride and polyvinyl alcohol with plasticizers and tackifiers. In addition this range can include urea formaldehyde and **Phenolic adhesives**, resins, natural adhesives produced from starch, dextrin, casein, animal glues (see **Animal glues and technical gelatins**) and rubber latex (see **Emulsion and dispersion adhesives**).
3. **Hot melt adhesives** include **Ethylene–vinyl acetate copolymers**, polyolefins, polyamides, polyesters with tackifiers and waxes. More recent additions include cross-linkable systems.

Trends over the last few years have been for a reduction in the use of **Solvent-based adhesives** in favour of water-based or 100% solids. The use of hot melt adhesives has gained importance within the industry for use as carton assembly and labelling or tapes. The **Acrylic adhesives** have been considerably developed and have found increased use in lamination, especially in view of the advantages offered by UV or electron beam curing systems for rapid cross-linking (see **Radiation cured adhesives**). Some of the natural adhesives such as starch and dextrin have been replaced by the synthetic polyvinyl acetate emulsions for improved tack.

The packaging industry is diverse in the range of materials used and areas where adhesives are required for bonding. The materials include cardboard, paper, metal sheets and foils, glass, metallized films, plastic containers and films. To satisfy some of the properties required of a package such as a barrier, no single material may offer suitable performance so use is made of composite structures where the properties of several materials are combined. In the form of laminates the substrates to be bonded include paper, metal foils and most types of polymer as films. Specific uses of adhesives cover paper making, fibreboard construction (cases and corrugated boards), carton assembly, adhesive tapes, labels, composite tube winding, cold seals for films and envelopes and flexible packaging laminates. In addition to these areas adhesion situations arise in the application of a coating or lacquer to a substrate, printing inks (see **Printing ink adhesion**), varnishing, metallizing or heat seal applications. The nature of the surface to be bonded and the interfacial properties are of prime importance to achieve satisfactory performance. The technology may therefore also include surface modification and application of primers to enhance adhesion (see **Primers for adhesive bonding**).

Service requirements

Packaging applications serve a wide range of industries from heavy-duty wrappings or sacks to the containing and preservation of foodstuffs and medical products. The final materials may be expected to perform under hostile environments such as high or low temperature, for example deep-freeze packs and cook-in pouches or extremes of humidity. Some of the products packaged can have detrimental effects on adhesive bonds such as solvent-based polishes, cleaning agents, shampoos, essential oils and spices. Adhesive bonds can also be affected by the migration of additives (see **Rheological theory, Compatibility**) used in some plastic films leading to failure of the package and spoilage of the product (especially foodstuffs). Some packages, for example those used in medical applications may be required to resist steam sterilization and irradiation.

Apart from adequately bonding various substrates together and retaining the desired shelf life for the package, care must be taken with the selection of adhesives to suit the application involved. Problems can arise due to interaction effects between product and pack, causing odour or taint. In some cases strict codes of practice and regulations limit the types of adhesives that may be used. The two most common reference sources are the Food and Drug Administration (FDA) Register of the USA and Bundesgesundheitsamt (BGA) of Germany. In addition local standards of good manufacturing practice apply to each country. These have important implications in the case of goods manufactured in one country and used for packaging of a product exported and sold elsewhere.

Residual solvents in laminates is a common source of odour and taint in products. A solvent may also act as a carrier to aid migration of additives to surfaces leading to adhesion problems such as loss of bonding or heat-seal strength.

See also **Laminating, Extrusion coating, Coextrusion**.

Select references

R J Ashley, *Adhesion 7*, ed. K W Allen, Elsevier Applied Science Publishers, London, 1983, pp. 221–51.

Paint constitution and adhesion

J L PROSSER

Introduction

Paints are used for protective purposes (e.g. to prevent corrosion of metals, to protect wood against the ingress of moisture), for cosmetic purposes (enhancement of appearance of a surface, improving gloss, conferring colour on a drab surface) and for functional purposes (maintaining a cool temperature by IR-reflective paints, increasing emission from a radiating body, use in non-IR reflecting camouflage schemes). It is an essential requirement for all end uses that the film remain firmly adherent to the surface under degradative conditions for as long as an acceptable life cycle. The adhesion of a paint film, immediately on formation and cure, and particularly after submission to degradative conditions, such as rain, hail, sea and water immersion, sunlight, atmospheric pollution, abrasive conditions, etc. is therefore of primary importance.

Coatings (paints and clear finishes) consist of a continuous medium, commonly called the binder, which is invariably a polymer, copolymer or a macromolecule (in the sense that a simple primary monomeric building block cannot be identified – as in alkyds). This may be modified with a lower-molecular-weight compound intended to change the physical properties of the binder and called a plasticizer, either used in a solvent selected for solubility properties but also for ease of removal from the deposited film or as a dispersive (water or other non-solvent) or as a powder in a powder coating. If the coating is a paint, then additionally it will carry a content of finely divided powder (generally a micrometre or less in size) either for opacifying the film conferring colour or certain desirable physical properties. There may also be secondary constituents such as wetting and dispersing agents which may affect adhesion.

Mechanisms of adhesion

The adhesion of the film is primarily determined by the binder and its chemical composition is important. Adhesion to a clean metal (metal oxide) can be by a mix of all possible mechanisms, e.g. **Dispersion forces, Acid–base interaction, Hydrogen bonding, Covalent bond** formation and other specific chemical interactions such as chelation and also electrostatic charge separation; all of these explanations have been advanced to explain the adhesion of paint films and are sometimes supported by experimental evidence for the particular effect without, however, affording a complete explanation. Apart from these molecular interactions across an interface, the microroughness of the substrate plays a part by mechanical interlocking (see **Mechanical theory**); additionally, if the substrate is itself organic (plastic or another paint film) the possibility exists of diffusion of the chain ends of the overcoating polymer into the substrate; in this latter case, mutual compatibility of the binders is of crucial importance (see **Theories of adhesion**).

Binders and plasticizers

A large variety of macromolecular types are used as binders in coatings, including polyvinyl acetate and copolymers, polyvinyl butyral, polyesters, acrylic polymers, epoxides and polyurethanes (see **Polyurethane adhesives**), polyvinyl chloride, polyvinylidene fluoride, oxygen-convertible media containing polyalcohol esters of long-chain unsaturated acids, such as epoxy esters and alkyds. All these compounds can adhere through dispersion forces (which are probably weak), and many contain chemical groups which can be envisaged as participating in specific interactions of the kind listed above. Hydroxyl and carboxyl groups have been shown to increase adhesion for an initially poorly adhering film, but too high a content of such groups can lead to undesired water sensitivity; in multicoat systems, also, too high a hydroxyl content can lead to incompatibility between coats, and the **Locus of failure** shifts from primer/substrate to primer/overcoat.

Chain polymers need to be plasticized for effective adhesion, presumably because plasticization extends the chain (more points of contact with the substrate) and renders it more mobile for a longer period as solvent evaporates. Study of the exact configuration of polymer molecules at a surface in a film is an area of uncertainty, but is of primary importance in elucidating the mechanism of adhesion in individual cases; an initial attachment by, say, an hydroxyl group probably reduces the mobility of the rest of the chain to conform to the contour of the surface and so maximize adhesion; in any case, the configuration is posited to consist of a vicinal run alternating with a loop, where the loops contribute nothing to adhesion and micro-roughness may cause bridging of the chain across protuberances.

Spectroscopic methods have been used to study configuration and packing of polymer molecules in the interface, **Fourier transform infra-red – FTIR**, principally, but also newer ones such as **Inelastic tunnelling spectroscopy**. These have conferred validity on the concepts of absorbance at specific sites, e.g. for (Lewis) **Acid–base interactions**. Support for such concepts is also found empirically in adhesion tests, e.g. polymethylmethacrylate (acid) adheres well to (basic) glass but not to acid-washed glass; it is known that basic polymers adhere best to Fe and acidic ones to Al; oxide surfaces can be characterized by their isoelectric surface potential.

Importance of solvent

When films are deposited from solution or dispersion at room temperature, it is usually observed that raising the system temperature increases the adhesion; one explanation is that the mobility of the chain is increased by the reduction in viscosity and consequently better surface wetting (more points of contact) occurs; additionally, a reaction may be promoted at the original points of contact; it has been shown, for example, that heating a polyethylene film on metal in the presence of oxygen improves adhesion markedly; in the absence of oxygen, the improvement is negligible.

Solvents used in coatings are hydrocarbons for alkyds, aromatics in admixture with ketones and other powerful solvents for most other media; some highly insoluble polymers like polyvinyl chloride and polyvinylidene fluoride are applied as dispersions and after deposition the film must be heated both to fuse it to a continuous material and to remove the dispersant. Selection of a wrongly formulated solvent mixture can have an adverse effect on the adhesion of a film formed at room temperature – a non-solvent diluent for example, or a slow evaporating powerful solvent like one of the higher ketones which may be trapped at the film–substrate interface, especially in thick films of high-build coatings; poor ventilation when a chemically reacting coating like an epoxy or amine is used can produce film detachment (especially in corners) by reabsorption of ambient solvent held in a stagnant atmosphere by a hardened surface with a soft interior – this causes swelling stresses followed by blistering and detachment. Too high a content of poorly evaporating diluent in a solvent mixture may lead to formation of a weak, poorly adherent film.

Poor adhesion can occur if the substrate is contaminated with an adherent greasy film, as from a rolling lubricant in the finishing of sheet metal, parting agents applied to tinplate, mould release agents on plastics (see **Rheological theory, Weak boundary layers**).

Further information on the adhesion of paints can be found in the articles **Paint service properties and adhesion, Primers for paints** and **Pretreatment of metals prior to painting**.

Paint primers

J L PROSSER

Paint primers are the first coat to be applied to a surface in a multicoat system. On wood, the function of the primer is to satisfy the suction of the wood, to adhere equally well to summer and winter wood and to produce a uniform surface for following coats.

On metals, promotion of adhesion of the multicoat system is important, and additionally there is usually a requirement that the paint suppress or control the corrosion of the metal; this is effected by incorporating pigments which are anodic to the metal substrate or inhibit corrosion by some other mechanism; such pigments are, among others, red lead, metallic zinc, zinc chrome and zinc phosphate. The binder itself may inhibit corrosion by imposing a high resistance between cathodic and anodic areas on the metal surface.

Good adhesion to the surface is a prime requirement of a primer, especially in water-wet conditions, and this is achieved by ensuring cleanliness of the metal and selection of binders carrying chemical groups favouring adhesion.

Specialist primers are **Adhesion promoters, Autophoretic primers** and **Etch primers**; these last react with the metal.

Select reference

S R Finn, *Convertible Coatings*, Pt 3 of Paint Technology Manuals, Oil and Colour Chemists Association, Chapman and Hall, London, 1965.

Paint service properties and adhesion

J L PROSSER

Introduction

A discussion of mechanisms of adhesion and a consideration of the influence of paint constitution on adhesion are given in **Paint constitution and adhesion**. In this article aspects of paint adhesion are considered which affect the paint film once formed.

Internal stresses

Once the film has formed, its physical properties are important for the retention of adhesion. All films possess an **Internal stress** by reason of the manner of their formation; thus solvent-deposited films have a shrinkage restraint imposed by evaporation of solvent continued beyond the point at which the bulk film has a certain rigidity; thermally curing films acquire a stress by cooling through their T_g; additionally, because organic coatings have coefficients of thermal expansion two to three times greater than that of metals, stresses arise from differential contraction on cooling a painted metal, either from a stoving temperature to ambient, or from ambient to sub-zero temperatures in a cold climate. Then again, repeated repainting can produce a film so thick that the cumulative stress from each coat eventually overcomes the adhesion of the first coat and the system pulls away; weakening by water penetration may be a contributory factor. Stress can also be applied by impact, as by flying stones in the chipping phenomenon so common on vehicles; a compromise between elasticity and hardness is required here.

The importance of water penetration cannot be too strongly stressed. This occurs by activated diffusion through the binder network, or through submicroscopic cracks in the film (creep around the periphery of hydrophilic pigment particles is possible); arriving at the film–substrate interface, water accumulates there, either by occupying unwetted areas or by swelling of water-soluble constituents; development of an osmotic pressure can then force the film off (see **Durability**).

Testing paint films

Many tests have been devised for the measurement of adhesion after the film has formed; the relationship of such a measurement to the forces operative at the interface is at best tenuous, if only because the tensile properties of the film constitute an interface (see **Tests of adhesion**).

If the metal surface is examined by any of the current spectroscopic methods after test, it is invariably found that a film (possibly only of monolayer thickness) remains firmly adherent, at least if high and commercially acceptable orders of adhesion are in question. Obviously, one can only say in such a case that the true adhesion is greater than the tensile strength of the film; pigmentation may increase the tensile strength and give an appearance of higher adhesion in a test, but adhesion is an interface phenomenon and is not affected by bulk properties. However, these tests are useful in confirming the unsuitability of a surface for painting – because, for example of contamination; the progressive effect of weathering can also be followed, and the tests may be employed (using reference coupons, small sheets painted for test purposes) to confirm the maintenance of a specified standard (see **Weathering of adhesive joints, Durability**).

Weathering

Exposure of a painted surface to the weather, especially, but not exclusively if the paint is an oxygen-convertible one, usually leads to a progressive reduction in adhesion. A recent innovation to prolong retention of adhesion is the use of adhesion promoters (see **Coupling agents**). These are effectively films applied very thinly and which have the property of reacting with the hydroxyl of the metal surface to form strong covalent bonds, the rest of the adhesion-promoting molecule being miscible with or also reacting with the binder. The commonest, but not the only, members of this group are substituted silanes; these, being of low molecular weight make many points of contact with the surface in a way that extensive or large-volume binder molecules cannot do. It has also been reported that very thin films of the binder itself, as a primer before the application of the main film, act as adhesion promoters; perhaps the low concentration of solution necessary allows greater mobility to the binder molecule permitting better surface conformation.

Select references

K Mittal (ed.), *Adhesion Aspects of Polymer Coatings*, Plenum Press, New York, 1983.

J C Bolger, A S Michaels, Molecular structure and electrostatic interactions at polymer–solid interfaces in *Interface Conversion for Polymer Coatings*, eds P Weiss, G S Cheever, Elsevier, New York, 1968.

D M Brewis, D Briggs (eds), *Industrial Adhesion Problems*, Orbital Press, Oxford, 1985, Chs 2, 4, 7.

K W Allen, Review of contemporary views of theories of adhesion, *Proc. Adhesion International*, Williamsburg, Va, 1987, pp. 85–101.

Peel tests

D E PACKHAM

A peel test, as its name suggests, is a test in which the force required to peel a flexible member is recorded and gives a measure of adhesion. There are many varieties of the test; some are illustrated in Fig. 92. Often the substrate is rigid and the flexible member is peeled at a defined angle, but where both materials bonded are flexible, such as laminated plastic film, a T-peel can be done (see also **Pressure-sensitive adhesives – adhesion properties**). It is possible to use the **Climbing drum peel test** for relatively rigid materials. It is evident that the 'peel angle', 90 or 180° or whatever is a formal angle between the lines of action on substrate and peeled strip, cannot be the actual angle at the point of fracture: for any given formal angle this will vary according to the bending stiffness of the peeled member (Fig. 93).

The choice of an appropriate form of the test will often be dictated by practical considerations of the adhesive bond of interest. As with all **Tests of adhesion** it is important to standardize the test-piece dimensions and

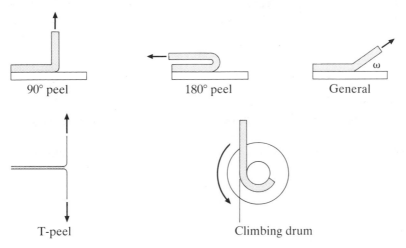

90° peel   180° peel   General

T-peel   Climbing drum

Fig. 92. Some different types of peel test

Fig. 93. Difference between the formal and actual peel angle

details of the test method in order to ensure that results are comparable. It is usual to express the results as a 'peel strength', the average peel force per unit width of the strip peeled. Even with this normalization, it is unwise to alter the width peeled without checking whether there is an effect on the normalized peel strength.

In many adhesion tests, each piece tested gives only a single estimate of strength such as stress at failure in **Shear tests** and **Tensile tests**. An advantage of a peel test is that each strip peeled yields a trace which shows how the force varies along the whole distance peeled; in taking an average the deviation from the mean and any systematic variation along the sample can easily be seen.

Peel force and peel energy

Although for many straightforward comparative purposes it is often adequate to record the results as a peel force/width, it is easy to derive an expression for the peel energy from basic mechanical principles by equating the work done by the test machine to the work done on the sample. The result forms a basis for understanding and interpreting peel tests (see also **Fracture mechanics test specimens**).

Consider a sample peeling a distance AB ($=x$) at angle ω by the application of a force F (Fig. 94). The point of application of the force will have moved

$$l = x(1 - \cos \omega) + \Delta x \qquad [1]$$

where the first term is a result of freeing a length x of strip and the second (Δx) represents the extension of the feed length x caused by the force F; Δx could be measured in a separate experiment or calculated if the tensile properties of the material of the strip were known.

The work done by the machine is then Fl, and can conveniently be written as

$$Fl = Fx(\lambda - \cos \omega) \qquad [2]$$

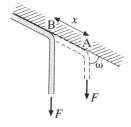

Fig. 94. Simplified peel theory: strip peels a distance AB $= x$

where the expression for l has been simplified by introduction of the extension ratio λ (extended length/original length).

The work done on the sample can be expressed as the sum of two terms. The first is the peel energy P which we take to be the energy (per unit area of peeled substrate surface) dissipated in the broad region of the peel front. The second is the work done in stretching the freed strip, which will be the strain energy density W_λ for extension to λ. Here W_λ can be calculated if an expression, such as Hooke's law, for the tensile stress–strain relationship for the material is known for extensions up to λ, otherwise it can be evaluated from the work done in an appropriate tensile test (Fig. 95). This can be used whether or not the yield point has been exceeded. Thus

$$\text{Work done on the sample} = Pbx + W_\lambda bxt \qquad [3]$$

It should be noted that as W_λ is expressed per unit volume it has to be multiplied by the volume of peeled strip, t is its thickness.

By equating 2 with 3 we obtain, for the peel energy,

$$P = \frac{F}{b}(\lambda - \cos \omega) - W_\lambda t \qquad [4]$$

The experimental conditions selected may well enable Eqn 4 to be simplified. If the extension of the peeled strip is negligible, as is often the case, λ is unity and W_λ is zero; 90 and 180° are commonly chosen peel angles. Under these circumstances Eqn 4 reduces to

$$P_{\pi/2} = F/b \qquad \text{at } 90° \qquad [5]$$

and

$$P_\pi = 2F/b \qquad \text{at } 180° \qquad [6]$$

These suggest that the peel load at 90° should be twice that at 180°. This point is discussed below.

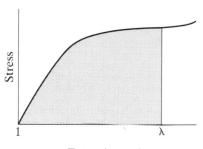

Extension ratio

Fig. 95. Strain-energy density is given by the area under a stress–extension ratio curve

Peel energy

An advantage of the analysis given is that it concentrates attention on
the peel energy. Here P is the energy (per unit area) dissipated by all the
energy-dissipating processes involved in the broad region associated with
the peel front. Fracture occurs and new surfaces are created, so there will
be a thermodynamic term, work of adhesion W_A or work of cohesion W_C
(see **Contact angles and interfacial tension**), depending on whether failure
is adhesive or cohesive (see **Locus of failure, Rheological theory**). To this
must be added other terms according to the materials and circumstances
of the experiment. These may include terms for plastic deformation of the
adhesive close to the fracture surface, for viscoelastic dissipation as the
peel front advances causing adhesive to be stressed and then relaxed (see
Viscoelasticity), for losses (plastic and/or viscoelastic) in bending the freed
strip through the peel angle. Thus P may be written as the sum of various
terms:

$$P = W_A \text{ (or } W_C) + \psi_{\text{plast}} + \psi_{\text{v/e}} + \psi_{\text{bend}} + \cdots \qquad [7]$$

Factors affecting peel energy

Values of the thermodynamic terms (W_A and W_C) in Eqn 7 can usually
be obtained from **Contact angle measurement** (see **Good–Girifalco
interaction parameter** and **Surface energy components**. Sometimes esti-
mates can be made of some of the other terms.[1] In most practical
circumstances the thermodynamic terms are orders of magnitude smaller
than the others: the magnitude of peel energy is largely determined by
the extent of dissipation within the materials of the adhesive bond.

   The thickness of the adhesive, and of any backing used, will affect the
peel strength in several ways. It directly enters the strain energy density
term ($W_\lambda t$), and may also alter some of the dissipation terms in Eqn 7
by changing the actual angle at the peel front (see Fig. 93) or by altering
the volume of polymer in which plastic or viscoelastic dissipation occurs.
Many of the terms in Eqn 7 will be temperature and rate dependent, so
the peel strength will also depend on these variables.

Peel angle variation

In as much as Eqn 4 is only based on definitions and principles of
mechanics, it must be correct for a given peel angle. It will predict the
variation of P with peel angle if the peel energy itself is a term independent
of angle of peel. With some simple rubbery adhesives the equation is quite
well followed, and the predictions of Eqns 5 and 6 about 90 and 180°
peel strengths are observed.[2] In other cases somewhat different angle

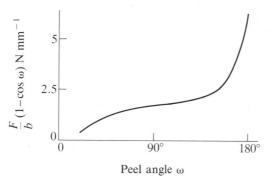

Fig. 96. Adhesion of polyethylene to aluminium: results[3] for variation of peel force with peel angle plotted to test equation [4] as λ is unity and W_λ is 0, deviation from a horizontal line indicates variation of peel energy with angle, ω

dependence is found. In Fig. 96 results[3] for the variation of peel strength (F/b) of polyethylene peeled from sulphochromated aluminium (see **FPL etch**) are compared with the prediction of Eqn 4, assuming that P is independent of peel angle. As the peel angle changes bending losses will vary and the balance between shear and cleavage forces at the peel front will change, affecting energy losses at the peel front: all of these may exert a significant influence on P and will therefore lead to deviations of the angle dependence predicted by Eqn 4.

References

1. T A Hatzinikolaou, D E Packham in *Adhesion – 9*, ed. K W Allen, Elsevier Applied Science Publishers, London, 1984, p. 33.
2. K Kendall, *J. Adhesion*, **5**, 179 (1973).
3. D E Packham in *Aspects of Adhesion – 6*, University of London Press, 1971, p. 127.

Phenolic adhesives

S TREDWELL

Resoles and novolaks

Although two types of phenolic resins exist (resole and novolak, Figs 97 and 98), **Structural adhesives** are usually based on the alkylated (methylated or butylated methylol group) resole variety. The main modifying resins (which reduce shrinkage and provide stress relief) most frequently copolymerized with resole phenolics are linear, high-molecular-

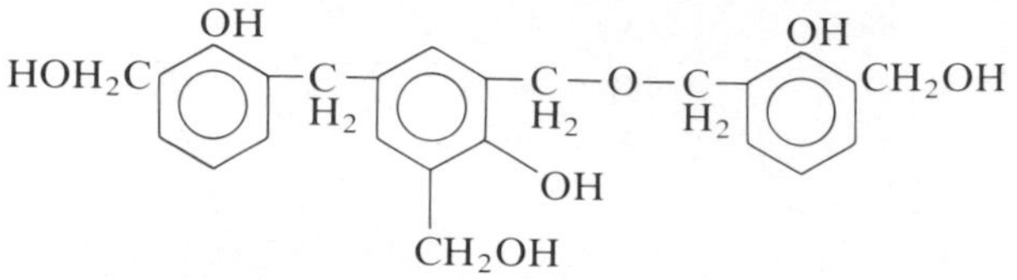

Fig. 97. An example of a resole structure

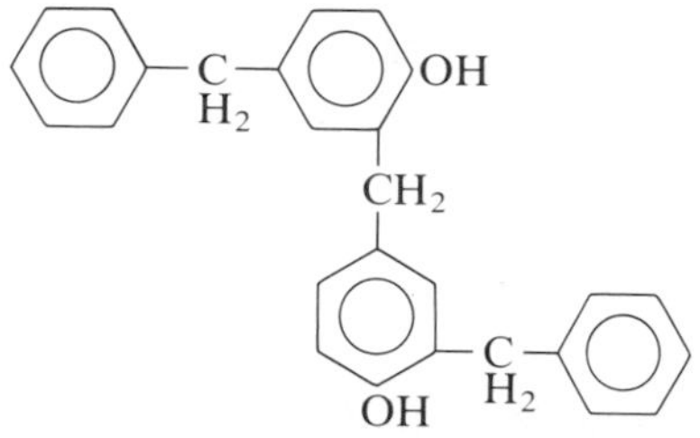

Fig. 98. An example of a novolak structure

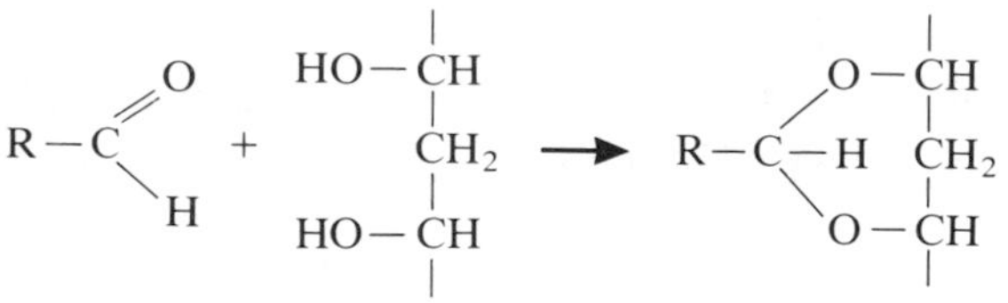

Fig. 99. Polyvinyl acetal is the generic name for condensation products of aldehydes with polyvinyl alcohol. $R = H$ poly(vinyl formal); $R = CH_3$ poly(vinyl acetal); $R = C_3H_7$ poly(vinyl butyral)

weight polymers with recurring hydroxyl groups along the polymer backbone. Reaction then proceeds between these hydroxyl groups and the phenolic methylol group (see **Condensation polymerization**). Examples would be epoxy–phenolic and vinyl–phenolic adhesives. In the vinyl–phenolic adhesives, the actual resins used are poly(vinyl formal) (PVF) or poly(vinyl butyral) (PVB) resins (Fig. 99) which can average as many as 46 hydroxyl groups per chain.

Resoles are produced by reaction of excess formaldehyde with phenol under basic conditions. On heating to 130–160 °C cross-linking occurs between methylol groups and phenol rings. A typical structure is given in Fig. 97.

If excess phenol reacts with formaldehyde, under acidic conditions, then the product is a novolak. An example of this structure is given in Fig. 98. The absence of methylol groups means that novolaks will not themselves react further on heating. A cross-linking agent, such as hexamethyltetramine, is then used. Long-chain alkane groups, instead of reactive methylol groups, must be present in a phenolic resin to be compatible with the nitrile rubber.

Nitrile–phenolic and vinyl–phenolic adhesives

A phenolic novolak resin, when blended with nitrile rubber, produces a nitrile–phenolic structural adhesive. The ratio of phenolic resin to nitrile rubber may be varied over a broad range and determines the properties of the adhesive. An increase of the phenolic resin component improves the elevated temperature strength of the adhesive, but, at the same time, decreases the impact resistance. A typical formulation contains equal amounts of both rubber and resin.

Tough structural adhesives with good impact strength, resistance to oil and aromatic fuels, and good salt spray and **Weathering** resistance can be produced when phenolic resins are combined with either vinyl formal, vinyl acetal or vinyl butyral (Fig. 99). As with nitrile–phenolic adhesives, variations in performance can be obtained by changing the amount of phenolic resin incorporated into the adhesive. The proportions of phenolic: vinyl in these mixtures may vary from 1:10 to 2:1 with a ratio of 1:1 providing a reasonable compromise between good high-temperature strength and low-temperature impact resistance and peel strength (see **Peel tests**). A range of 50–125 parts of phenolic per 100 parts of vinyl resin is the range in which most commercial formulations fall. A good detailed study on the chemistry and application of phenolic resins has been written by Knop and Scheib.[1]

It was a vinyl–phenolic system that, during the Second World War, was developed as the first structural adhesive. The technique used, although crude, provided much higher strength and durability than had been obtained previously with any synthetic resin metal-bonding adhesive. Phenolic-cured adhesives had, and continue to have, one major performance advantage, i.e. bond **Durability**, that is still not matched by the newest epoxide systems (see **Epoxide adhesives**). Work over the years[2,3] has documented the exceptional capability of the best vinyl–phenolic, nitrile–phenolic, and epoxy–phenolic adhesives to retain bond strength on aluminium or other metals after extended exposure to combinations of moisture, salt, stress and impact.

This superior durability of phenolic-based adhesives may, in part, involve the presence of covalent bonds across the interface (see **Adsorption**

theory). Such bonds can be formed by the coordination of phenolic oxygens to aluminium, as has been observed by the formation of complexes between aluminium and 8-hydroxyquinoline or alizavin dyes.[4] It has also been agreed that interfacial ion pairs could account for the partial weakening of joints in the presence of water, with recovery when the joints are dried.[5] Such ion pairs can arise in with phenolic adhesives by phenol groups being absorbed as the phenolate ion.

Uses

Although the vinyl–phenolic and nitrile–phenolic adhesives continue to be used for non-aerospace bonding (e.g. brake shoes), they have largely been displaced for aircraft assembly by the newer toughened **Epoxide adhesives**. The major application disadvantages of the phenolic systems are slow rate of cure, high cure temperature and high pressure needed during cure to prevent void formation.

Vinyl–phenolic and nitrile–phenolic adhesives show their highest strengths near room temperature, but much lower strengths at temperatures above about 80 °C, although the decrease is less marked for nitrile–phenolics. These adhesives are unsuitable for bonding non-perforated **Honeycomb structures** because they evolve volatiles when they are cured.

As the size and complexity of bonded assembles increased, airframe manufacturers began to demand the lowest possible cure time, temperature and pressure to minimize the required numbers and size of their autoclaves. In addition, although the performance properties of these systems could be varied by adjusting the ratio of phenolic resin to modifying polymer (e.g. lowering the phenolic content and giving up some hot strength and increased flexural properties), the total available 'trade-off envelope' did not yield the higher peel strengths demanded by designers and ultimately provided by the faster, lower pressure curing nylon–epoxy and nitrile–epoxy systems.

Further information can be found in other articles, especially, **Aerospace applications, Durability, Structural adhesives** and **Weathering of adhesive joints**.

References

1. A Knop, W Scheib, *Chemistry & Application of Phenolic Resins*, Springer-Verlag, Berlin, 1979.
2. J L Cotter in *Developments in Adhesives – 1*, ed. W C Wake, Applied Science Publishers, London, 1977.

3. J D Minford, *Int. J. Adhesion and Adhesives*, **2**, 25 (Jan. 1982).
4. W Brockman, *Natl. SAMPE Symp. Exhib.*, **21**, 383 (1976).
5. J Comyn, D M Brewis, S T Tredwell, *J. Adhesion*, **21**, 59 (1987).

Plasma treatment

D BRIGGS

Strictly, plasma treatment encompasses any form of surface treatment in which a material is exposed to an ionized gas (plasma). This would include **Flame treatment** and **Corona discharge treatment**, but 'plasma treatment' generally refers to low-pressure treatments requiring vacuum equipment. Such processes are also often referred to as glow discharge treatment.

In simple versions of such a process power is coupled into the discharge either capacitively or inductively; in the former the electrodes are usually inside the vacuum system with the sample resting on one of them, while in the latter the coil is external to the system. The a.c. power is either in the radio frequency (r.f.) band (almost always at the fixed frequency of 13.56 MHz) or in the microwave band (frequencies in GHz).

Depending on the gas used to generate the plasma and the plasma parameters, essentially three different forms of 'treatment' are possible.

The first involves noble gas plasmas and has previously been known as 'casing'. Here the principal effect is one of surface cross-linking and elimination of potential **Weak boundary layers**, but a very useful effect is the generation of a barrier layer to additive diffusion.

The second involves simple 'reactive' gas plasmas (e.g. O_2, N_2, SO_2, CF_4). Here the principal effect is the introduction of functional groups containing the elements present in the gas molecules. In either of these two types of treatment a prolonged exposure will lead to ablation of material (etching) and surface texturing.

The third involves 'polymerizable' gas plasmas capable of depositing a film on to the substrate. These molecules do not have to be polymerizable in the conventional sense; examples are CH_4, C_6F_6, CH_3OH. The highly cross-linked 'polymer' laid down on the substrate is pinhole free even at thicknesses below 100 Å and adheres tenaciously to the substrate. Depositions using mixed gases can build up 'graded' interface structures. Clearly the scope for fine tuning surface properties is immense with this technology. Many gases can lead to either modification or deposition depending on the plasma parameters chosen (pressure, say 0.1–10 torr, flow rates, power, substrate temperature etc.).

Plasma treatments have been used for some time to change the adhesive properties of irregular-shaped articles by batch processing, but more recently reel-to-reel systems for film treatment have become available both out of line (inside a vacuum chamber, as in metallization) and in line (air–vacuum–air).

Select references

J R Hollahan, A Bell (eds), *Techniques and Applications of Plasma Chemistry*, John Wiley, New York, 1974.
H Yasuda, *Plasma Polymerisation*, Academic Press, Orlando, Fla, 1985.

Plastisols

J PRITCHARD

What are they?

Plastisols form a unique class of materials composed of particles of solid polymer dispersed or suspended in high-boiling-point liquids known as plasticizers. In practice, the major polymer used is polyvinyl chloride with possible additions of copolymers of vinyl chloride and vinyl acetate.

The particles of polymer have to be able to pack closely in the liquid medium, but must leave enough space for adequate mobility. Hence they are usually spherical, between 0.5 and 2.5 μm in diameter, with a fairly wide distribution of particle sizes within this range. The liquid phase consists of mixtures of organic or inorganic esters – the most important being phthalate and phosphate esters.

How do they function?

Polyvinyl chloride is a hard, tough material at normal temperatures, because the polymer chains are so packed they have little freedom of movement. It was discovered in the 1930s that molecules of certain organic liquids could be interposed between the polyvinyl chloride polymer chains giving them more freedom to move. This phenomenon, known as plasticization, is able to change the original hard polyvinyl chloride into a flexible rubbery or leathery material.

At ambient temperatures below 30 °C, polymer particles are suspended in the plasticizer forming a paste-like plastisol which remains stable and unchanged for long periods, providing it has been formulated correctly. On warming, the solubility of the polymer particles in the plasticizer begins to increase, producing absorption of plasticizer and increased viscosity. This process continues until all plasticizer has been absorbed into the particles; however, it is not until temperatures in excess of 140 °C have been reached that the PVC becomes truly plasticized by the incorporation of the plasticizer between the polymer chains.

This unique physical process means that plastisols can wet out large areas easily at room temperature, and then by the application of heat the paste is converted into a tough material without the problem of having to get rid of solvent by evaporation.

Manufacture and compounding

The preparation of plastisols is carried out using any suitable internal paddle or anchor mixer. All the powdered resin is added to the equipment and worked up to a stiff paste with an optimum amount of plasticizer. This ensures sufficient shear to promote satisfactory dispersion of the powder. When this stage is smooth and lump free, any remaining plasticizer is added gradually, ensuring that a smooth paste is maintained the whole time.

If fillers are part of the formulation, they are added to the dry resin powder at the beginning. Small amounts of critical materials such as stabilizers or pigments are added in the form of predispersed pastes. The pigments or stabilizer will have been ground into an amount of plasticizer either by ball mill or paint mill. These predispersions are normally bought from specialist suppliers.

Choice of polymer

The choice of polymer depends on the required properties of the paste and the final application. The main variables are molecular weight, whether homopolymer or copolymer and particle size and distribution.

Higher-molecular-weight grades of polymer give tougher finished product and greater viscosity stability of the plastisol; however, they require higher processing temperatures. Copolymers with vinyl acetate and vinylidene chloride are common, 5–10% of the second component being typical values. The effect of the second monomer is to loosen the interchain attraction thereby reducing gelling and processing temperatures; however, it also leads to increasing viscosity of the plastisol on ageing. The particle size, shape and distribution will control the rheology of the plastisol – whether high or low viscosity, whether dilatant or thixotropic.

Choice of plasticizer

The number of plasticizers in common use is relatively small. The two main groups consist of phthalate esters and phosphate esters followed by the less important sebacates, adipates, chlorinated paraffin wax and epoxidized vegetable oils.

Phthalates are general-purpose plasticizers, giving good all-round properties, reasonable low-temperature resistance and good electrical properties. Phosphates are used in conjunction with phthalates, their main advantage being their degree of non-inflammability.

Stabilizers

Polyvinyl chloride polymers will degrade with heat and light, hence stabilizers are added. In general, stabilizers are accepters for hydrochloric acid, and the reaction products should have no deleterious effect on the durability, colour, odour, clarity or water resistance of the compound. Typical stabilizers are lead salts and soaps, calcium stearate and dibutyl tin dilaurate.

Typical proportions

The proportions of ingredients in any formulation depend upon the finished properties required, but in general, the more plasticizer, the softer the finished product. Typical levels of the three major ingredients are as follows:

PVC polymer	100 p.b.w. (parts by weight)
Plasticizer	40–100 p.b.w.
Stabilizer	1–5 p.b.w.

Applications of PVC plastisols

The processing characteristics of PVC plastisols do not allow them to be readily used as a straightforward adhesive. This is because of the high temperatures involved; it requires 150–200 °C to convert a plastisol into a homogeneous plasticized polymer. At these temperatures the material is fluid and behaves like a **Hot melt adhesive**. Unlike conventional hot melts, it cannot be kept at this temperature for long periods because of its tendency to degrade.

Large-scale use is made of plastisols in the **Automotive industry**, where their toughness and good ageing characteristics make them ideally suited for underbody protection and sealing.

Plastisols can be sprayed readily without hazard on to the underside of cars during manufacture and heated to the correct temperatures during the painting cycle. For filling gaps in the bodywork, for example at seams, plastisols can be made to expand during heating by means of flowing agents giving a closed cell foam which seals gaps efficiently.

Another frequent use of plastisols is for laminating fabric to make conveyor belting. Polyvinyl chloride can be compounded to be virtually non-inflammable, hence it is ideally suited for the manufacture of conveyor belting for coal-mines. Plies of fabric are coated easily with plastisols, and passed through gas-heated infra-red ovens to pre-gel the plastisol. The coated plies are then laminated with heat in suitable equipment which enables the hot laminate to cool while still under pressure.

Hazards are mentioned under **Health and safety**.

Select references

European Vinyls Corporation, *Corvic Info*, Note TSO2E, 1988.
Wacker GmbH Chemie, *Paste Making Resins*, No. 4251.1088, 1988.
Shell Tech. Bulletin, *Phthalate Plasticizers*, ICS(X)/76/3, 1980.
W S Penn, *PVC Technology*, McLaren, London, 1966.

Polar forces

K W ALLEN

The attraction between two electrostatic charges of opposite sign (or repulsion between charges of the same sign), originally investigated by Coulomb in 1785, is well known. It is dependent directly upon the magnitude of each charge and inversely upon the square of the distance between them. In considering the interaction of atomic and molecular particles, this force of attraction is of importance in several situations – in materials which are ionic in nature as well as in cases where one or both particles has a dipole moment.

Ionic structures

Consider the simplest case; when there is a sufficiently great difference in electronegativity (attraction for electrons) between two atoms, then a transfer of an electron may occur to give a positively charged cation and a negatively charged anion. These two ions will then attract each other and form an ionic compound. This can be extended to cases where the particles are not simple atoms, but are groups of atoms which give rise to charged radicals.

In the solid this will give infinite lattice with the ions occupying symmetrical positions so that there is overall electrical neutrality. In solution the ions become separated and solvated to give a conducting electrolyte solution. Typical examples are sodium chloride and, slightly more complex, ammonium nitrate.

This type of strictly ionic bonding is not of importance in most examples of adhesion, but it needs to be reviewed here to give a satisfactory appreciation of the situation.

Dipoles

In covalent molecules, where the bonding depends upon a sharing of electrons between atoms, the distribution of electron charge density may not be symmetrical. If the electronegativity of the atoms is significantly different then there will be a bias in this charge density towards the nucleus with the higher electronegativity. Thus, while the molecule will be overall electrically neutral, one end will carry a small negative charge (excess of electron density) and the other will carry a small positive charge (deficit of electron density). This situation is called dipole and the molecule is said to have a 'dipole moment'.

Within the molecule this situation is usually described in terms of the covalent bond having a partial ionic character, and this will affect its properties and behaviour. Between two (or more) molecules with dipole moments there will be additional interactions arising from these electrostatic dipoles.

The dipole moment of a molecule is represented by μ and is expressed in Debyes (D) where $1\,D = 1 \times 10^{-18}$ e.s.u. of charge $\times$ centimetre (e.s.u. = electrostatic unit). If a molecule had a charge equivalent to one electron separated by a distance of 1 Å, then the dipole moment μ would amount to 4.8 D. ($1\,D = 3.336 \times 10^{-30}$ C m.)

Intermolecular interactions arising from dipoles can be divided into two groups: those between two molecules each of which has a dipole moment, and those between two molecules only one of which has any permanent dipole moment.

Dipole–dipole interactions

If two molecules each with a dipole moment are within the vicinity of each other, then there will be a mutual interaction and attraction between them depending upon the alignment between them. This is known as the 'orientation' effect and was first investigated by Keesom in 1912. He derived a relationship for the energy between two molecules, of dipole moment μ_1 and μ_2 separated by a distance r:

$$U = -\frac{2}{3} \cdot \frac{\mu_1^2 \mu_2^2}{r^6} \cdot 1/kT \qquad [1]$$

where k is Boltzmann's constant and T absolute temperature.

Dipole–molecule interactions

Consider the situation when a molecule with a permanent dipole moment comes near to another, different molecule which does not have any dipole moment. The first will induce a dipole in the other by polarization of the electron field in that second molecule. This is known as the 'induction' effect and was extensively investigated by Debye who derived a relationship similar to that by Keesom but including polarizability terms, α_1 and α_2:

$$U = \frac{\alpha_1 \mu_2^2 + \alpha_2 \mu_1^2}{r^6} \qquad [2]$$

This induction effect may also be effective when the second molecule itself has a dipole moment, so that it exists in the interaction between two identical molecules. Calculations indicate that, for example, with HCl which has a dipole moment of 1.03 D the orientation effect amounts to 3.3 and the induction effect to 1.0 kJ mol^{-1}.

It is common for both Eqns 1 and 2 to be written in the abbreviated form

$$U = -A/r^6 \qquad [3]$$

where A is the attraction constant. It is evident that

$$A_{12} = (A_{11}A_{22})^{1/2} \qquad [4]$$

is an exact relationship for orientation forces and an approximate one for induction forces. (The subscripts 11 and 22 refer to interaction between like molecules, 12 to those between unlike molecules.)

The complete potential energy curve for the molecules is given by the Lennard-Jones potential

$$U = A/r^6 + B/r^{12} \qquad [5]$$

where B is a constant. The second term represents the repulsion energy which rises very rapidly with distance once the electron orbitals start to interact. Equations analogous to 3–5 apply for **Dispersion forces**.

Debye and Keesom forces together with London **Dispersion forces** are known collectively as 'van der Waals' forces'. They play a significant role in the **Adsorption theory of adhesion** and in surface phenomena such as **Contact angles** and **Interfacial tension**.

Select references

K W Allen in *Aspects of Adhesion 1*, ed. D J Alner, University of London Press, 1965, pp. 11–21.

J A A Ketelaar, *Chemical Constitution*, 2nd rev. edn, Elsevier, London and New York, 1958, Ch. V.

R J Good in *Treatise on Adhesion and Adhesives*, Vol. 1, ed. R L Patrick, Edward Arnold, London and Marcel Dekker, New York, 1967, Ch. 2.

Polybenzimidazoles

S J SHAW

Initially developed in the late 1950s and early 1960s, polybenzimidazoles are prepared by reaction of tetrafunctional aromatic amines with aromatic esters (Fig. 100). They find application as **High-temperature adhesives**.

Since high-molecular-weight polybenzimidazoles exhibit, depending upon structure, fair to high levels of intractability, adhesive formulations based on them generally employ the polymeric component initially in a low-molecular-weight prepolymer (dimer or trimer) form, with the final polymerization step being conducted within the bond line. Unfortunately this occurs by a polycondensation reaction, resulting in the liberation of both phenol and water which can therefore result in a porous adhesive layer of low mechanical strength (see **Condensation polymerization**). To alleviate such problems polybenzimidazole adhesives require bonding pressures of approximately 1.4 MPa at cure temperatures in the region of 320 °C, together with the venting of condensation products. Even under such circumstances, however, large-area bonding can remain particularly difficult.

In spite of these difficulties it is important to note that polybenzimidazoles offer substantial high-temperature capabilities, particularly for short-duration applications with the not inconsiderable ability to retain about 50% of room temperature strength at approximately 450 °C. In fact they can be regarded as superior to all their high-temperature competitors, including both the condensation polyimides and epoxy–phenolics in this respect. Unfortunately this capability is not maintained under long-term,

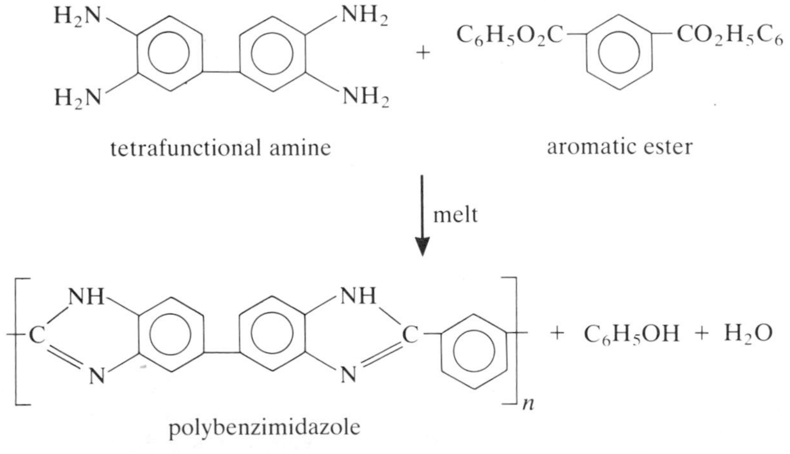

Fig. 100. Reaction scheme/structural formula for polybenzimidazole

high-temperature conditions due to their susceptibility to oxidative degradation at temperatures in excess of 250 °C. For this reason, post-cure operations at about 400 °C, which are often considered necessary, need to be conducted in inert atmospheres, thus adding further complexity and cost to the processing operation.

One further substantial advantage which the polybenzimidazoles enjoy over their competitors is their ability to retain good mechanical properties at temperatures as low as − 190 °C.

Although in certain respects the polybenzimidazoles offer a virtually unique combination of properties, they have not enjoyed the success of other **High-temperature adhesives**. A major reason for this has undoubtedly been due to the monomeric materials required, most notably aromatic tetraamines, being both costly and difficult to obtain in the required purity. In addition doubts concerning carcinogenic activity have also been expressed (see **Health and safety**), which, together with the adverse processability mentioned above has severely restricted acceptance of them. For these reasons, commercial availability has to date been somewhat limited.

Further discussion on the general theme of high-temperature adhesives can be found in the articles entitled **Polyether ether ketones, Polyimide adhesives** and **Polyphenylquinoxalines**.

Select references

S Litrak, *Adhesive Age*, **11** (1), 17 (1968).
J P Critchley, G J Knight, W W Wright in *Heat Resistant Polymers*, ed. J P Critchley, Plenum Press, New York, 1983.

Polyether ether ketone

D A TOD

Polyether ether ketone, poly(aryl-ether-ether-ketone) or PEEK is a high aromatic semi-crystalline thermoplastic (Fig. 101). It is being increasingly used in a number of advanced products such as wire coatings and high-performance mouldings. PEEK is available from ICI Ltd

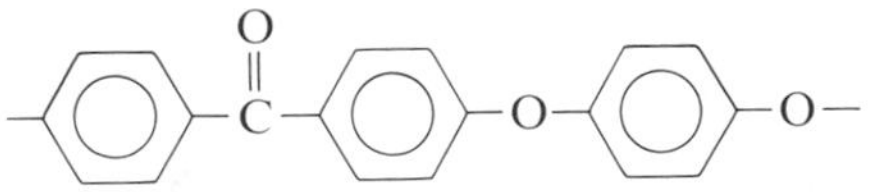

Fig. 101. The structure of the repeat unit in PEEK

unreinforced, reinforced with glass and as the matrix material in a carbon **Fibre composite** known as APC2. The history and development of PEEK within ICI Ltd has been described by Rose[1] and a general review of the subject has been given by Nguyen and Ishida.[2]

PEEK is a relatively expensive polymer, but it does have several attractive properties: high tensile strength (93 MPa at 20 °C, 37 MPa at 150 °C); high stiffness (3.6 GPa at 20 °C, 2.2 GPa at 150 °C); low inflammability; **High-temperature stability**; excellent chemical and radiation resistance; good **Fatigue**, wear and abrasion resistance. The modulus for the unfilled grades drops sharply above the **Glass transition temperature** (143 °C) as the material approaches its melting temperature (334 °C). This reduction in modulus is significantly reduced with the filled grades and almost removed in the continuous **Fibre composite** versions. The crystallinity of thermoplastics has a major influence upon their resultant properties. The crystallites act as reinforcing particles to the amorphous matrix and they also act to make the material solvent resistant. The crystallinity in PEEK can be controlled by the processing conditions and values from 0 to 40% have been reported. The relationship between crystallinity and structure-mechanical properties in PEEK has been reviewed by Medellin-Rodriguez.[3]

Surface pretreatment of PEEK usually involves some degree of mechanical abrasion (see **Abrasion treatment**) to provide a key and then **Degreasing** with a suitable solvent such as 1,1,1-trichloroethane. Etching techniques have been used to good effect in adhesive bonding and a suitable etchant for PEEK has been developed by Olley.[4] This is a permanganic etchant based on orthophosphoric acid. Other pretreatments such as chromic acid etching have proved effective as well as a treatment with a blue (oxidizing) flame.[5] **Epoxide adhesives** have been found to bond well to such surfaces as well as **Anaerobic adhesives**.[5]

PEEK has also been used as a thermoplastic **Hot melt adhesive** for bonding of titanium and to the PEEK composite APC2. Lap shear strengths at room temperature of 20 MPa with titanium and 50 MPa with APC2 were obtained.

References

1. J B Rose in *High Performance Polymers: Their Origin and Development*, ed. R B Seymour, G S Kirshenbaum, Elsevier, New York, 1986.
2. H X Nguyen, H Ishida, *Polym. Compos.*, **8**, 57 (1987).
3. F J Medellin-Rodriguez, P J Phillips, *Polym. Eng. and Sci.*, **30**, 860 (1990).
4. R H Olley, D C Bassett, D J Blundell, *Polym.*, **27**, 344 (1986).
5. A H Landrock, *Adhesives Technology Handbook*, Noyes Publications, NJ, 1985.

Polyimide adhesives

S J SHAW

Introduction

Of the numerous polymers developed since the early 1960s having high-temperature resistance as a primary requirement, the polyimides have by far achieved the greatest commercial success due primarily to their ability to maintain acceptable mechanical properties at elevated temperatures together with a measure of thermal stability sufficient to allow their long-term use at temperatures in excess of 250 °C. Key application areas have included fibre-reinforced composites, surface coatings, electronic devices and, of course, as **High-temperature adhesives**.

Adhesives based on polyimides which have, or appear in the process of achieving a measure of commercial acceptability can be divided into three broad classes, and it is convenient briefly to consider these major types separately. These are:

1. Condensation polyimides;
2. Thermoplastic polyimides;
3. Imide prepolymers.

Condensation polyimides

Proprietary condensation polyimide adhesives are usually based upon starting materials (dianhydrides and diamines) which yield a polyimide structure having a high degree of intractability. Consequently they are processed at an intermediate stage with conversion of an essentially soluble intermediate polymer (known as a polyamic acid) to the polyimide being conducted within the bond line (see Fig. 102). Unfortunately, evolution of water which occurs during this conversion process, together with the need to remove solvent, provides a major processing difficulty with the possibility of porous, mechanically weak bonds (see **Condensation polymerization**). Although both high bonding pressures and venting techniques can alleviate such problems, there are usually quite stringent limitations on the area which can be successfully bonded using condensation polyimide adhesives.

In spite of these difficulties these adhesives can exhibit excellent properties. For short-term applications they can retain in excess of 50% room temperature strength at approximately 300 °C. Table 23 indicates typical joint strength values which can be achieved.

Long-term thermal stability is dependent upon a number of factors including the addition of stabilizers such as arsenic thioarsenate and the nature of the substrate. Studies have shown that, under certain

Table 23. Lap-shear strengths of condensation
polyimide bonded titanium joints

Temperature ($^\circ$C)	Lap-shear strength (MPa)
25	26.5
100	24.5
200	22.2
300	18.5
400	12.5
500	8.5

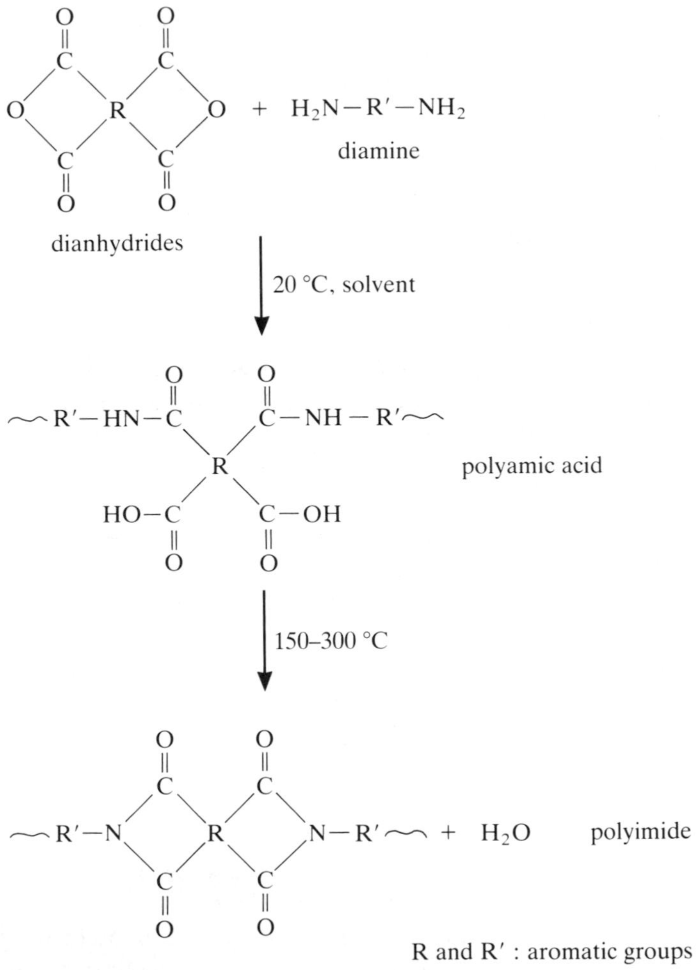

Fig. 102. Reaction scheme/structural formula for condensation polyimide

Table 24. Influence of thermal ageing on adhesive joint strength of condensation polyimide bonded titanium joints

Ageing time at 260 °C (h)	Lap-shear strength (MPa)
0	21.0
1 000	19.5
2 000	19.2
4 000	17.8
40 000	20.7
0*	17.7
1 000*	19.2
2 000*	11.0
3 000*	1.9

* Adhesive not containing arsenic thioarsenate stabilizer.

circumstances, long-term use at temperatures in the vicinity of 275 °C is possible, as indicated in Table 24.

Although condensation polyimides can exhibit remarkable elevated temperature properties, their processing disadvantages have clearly limited the acceptance they would have otherwise deserved.

Thermoplastic polyimides

Investigations aimed at developing polyimide-based adhesives exhibiting substantially improved processing characteristics relative to the condensation polyimides have resulted in two major success areas, one of these being thermoplastic polyimides (Fig. 103). The reaction route is essentially the same as that for condensation polyimides (see Fig. 102); the difference lies in the flexible linkages in the original dianhydride and diamine which give greater chain flexibility in the resulting polyimide.

Many studies have shown that the imposition of molecular flexibility allows the possibility that the flexibilized polyimide could exhibit sufficient thermoplasticity above its **Glass transition temperature** for fabrication by a hot-melt process (see **Hot melt adhesives**). In such circumstances, bonding using a fully imidized film could remove the need for both solvent evaporation and polyamic acid to polyimide conversion within the bond line, thus allowing the bonding of relatively large areas.

One of the most successful thermoplastic polyimides has been LARC-TPI, developed by the National Aeronautics and Space Administration (NASA) in the USA. The polymer is commercially available. Table 25 shows some of the very reasonable properties obtained from this adhesive. As indicated, the ability both to maintain mechanical properties at elevated

Table 25. Typical properties of LARC-TPI
bonded titanium joints (see **Shear tests**)

Property Lap-shear strength (MPa)	Value
20°	36.5
232 °C	13.1
3000 h at 232 °C	20.7
T_g (°C)	250
Decomposition temperature (°C)	520

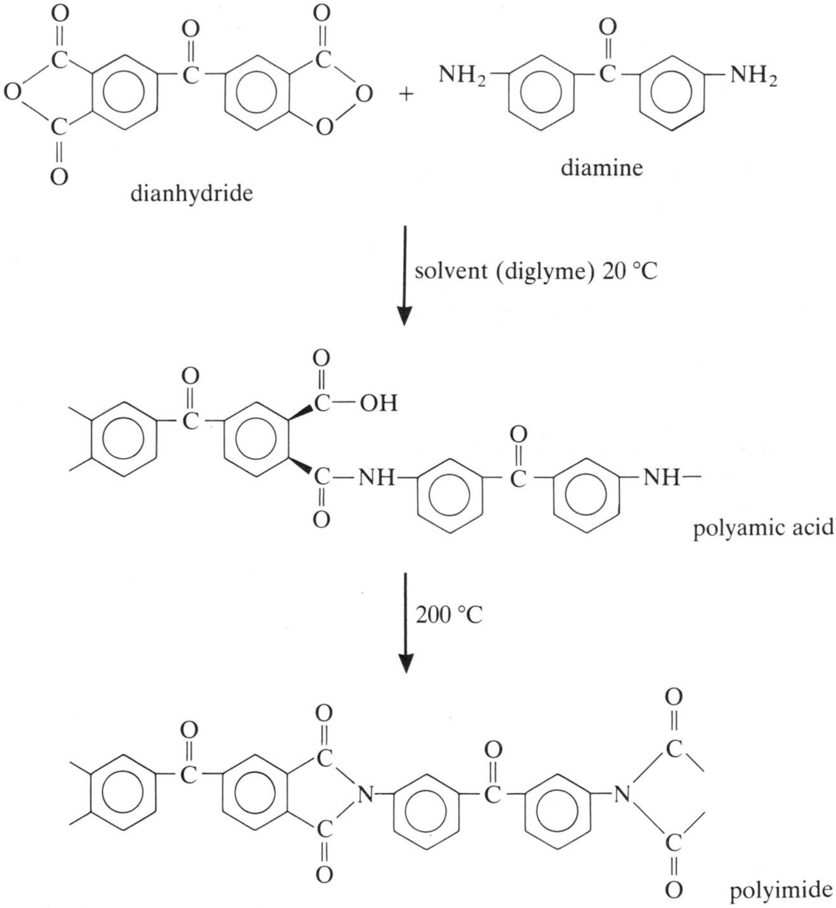

solvent (diglyme) 20 °C

200 °C

(C=O groups act as hinge points
allowing a degree of thermoplasticity)

Fig. 103. Reaction scheme/structural formula for thermoplastic polyimide
(LARC–TPI)

Table 26. Typical properties of LARC-13
bonded titanium joints

Property Lap-shear strength (MPa)	Value
20°	22.1
232 °C	17.9
1000 h at 232 °C	17.9
T_g (°C)	300
Decomposition temperature (°C)	450

temperatures and indeed retain mechanical integrity for long periods of time are clearly apparent.

In addition to LARC-TPI, further recent developments have included the polyimide sulphones, polysiliconeimides and polyetherimides. All three have been evaluated and shown promise for adhesive bonding applications.

Adhesives based on imide prepolymers

The concept of low-molecular-weight imide prepolymers can be viewed as an alternative route to enhanced processability. The development of such systems has been conducted based upon three fundamental requirements. First, the prepolymers should be of low molecular weight allowing for the possibility of a low melting point and low viscosity. Second, imide groups should be present in the prepolymer so as to remove the particularly troublesome polyamic acid to imide conversion process mentioned previously. Third, the prepolymers should have reactive terminal groups capable of reaction by an addition mechanism so as to convert the molten prepolymer to a cross-linked polymer without the harmful evolution of volatiles.

A wide range of imide prepolymers have been developed based on this approach. The various forms differ primarily in the type of terminal reactive group employed so as to convert the prepolymer to a cross-linked product. Three main types have achieved prominence and shown potential as **High-temperature adhesives**, these being prepolymers based upon norbornene, acetylene and maleimide (bismaleimide) functionality.

Studies into the first of these types, the norbornene imides have resulted in a high-temperature adhesive system known as LARC-13 which demonstrated improved processability over condensation polyimides. Some typical results obtained from this adhesive are shown in Table 26, demonstrating its high-temperature capabilities.

Table 27. Lap-shear data for bismaleimide
bonded aluminium joints

Temperature (°C)	Lap-shear strength (MPa)
25	20.0
232	18.6
260	12.7

An alternative approach to imide prepolymer formation using acetylenic termination as the means for further reaction was developed in the 1970s. Although various investigations have provided ample evidence of the excellent properties possible using acetylenic polyimides, current proprietary systems suffer one inherent drawback in having close melting and cure temperatures. The resulting extremely short gel times have resulted in poor lap-shear strengths (see **Shear tests**) at both ambient and elevated temperatures presumably due to poor wetting characteristics. There is little doubt that disappointing results of this kind have resulted in the poor acceptance of these materials to date.

Bismaleimides offer possibly the greatest benefits in terms of enhanced processability, being more akin to the easy-to-use epoxies than any of the systems described above. As a result they have received a great deal of consideration in recent years for many potential applications. Due to these highly promising characteristics attempts have been made to employ bismaleimides in high-temperature adhesive formulations. In particular a bismaleimide modified **Epoxide adhesive** has been developed considered capable of operating at temperatures up to approximately 280 °C as suggested by the data shown in Table 27.

Although bismaleimides offer significant advantages over many other high-temperature systems, they do not exhibit the same degree of thermal stability and hence long-term performance of many other polyimide systems. As a result they are restricted to service applications requiring a 200–300 °C long-term capability. They essentially 'fill the gap' between the epoxies and the condensation and thermoplastic polyimides described earlier.

Further discussion on the general theme of high-temperature polymers and adhesives can be found in the articles entitled **Polyphenylquinoxalines, Polybenzimidazoles** and **Polyether ether ketones**.

Select references

See article on **High-temperature adhesives**.

Polyphenylquinoxalines

S J SHAW

Although the development of polyphenylquinoxalines (Fig. 104), was first reported in the late 1960s, it is only fairly recently that serious attempts have been made to exploit them as **High-temperature adhesives**. One particular adhesive system currently available on a commercial basis exists as a 20 wt% solution of the high-molecular-weight polyphenylquinoxaline polymer in a xylene–cresol solvent mixture. Bonding using this system requires three main stages, as follows:

1. Application of the solution to the substrate or carrier;
2. Evaporation of the solvent;
3. Bonding of the substrates at approximately 400 °C and 1.4 MPa pressure.

Evaporation of the solvent leaves the polymer film which can then be employed as, essentially a 'hot-melt' adhesive, requiring bonding at temperatures substantially in excess of the **Glass transition temperature** (approximately 290 °C), together with high-pressure application so as to induce polymer flow and thus allow a degree of wetting (see **Wetting and spreading**).

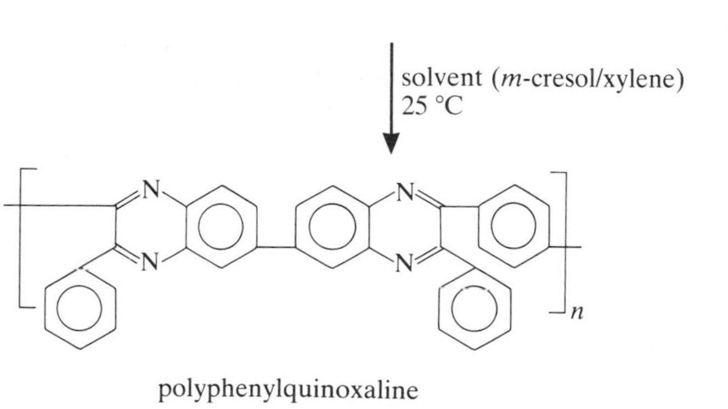

tetrafunctional amine   dibenzil

solvent (*m*-cresol/xylene)
25 °C

polyphenylquinoxaline

Fig. 104. Reaction scheme/structural formula for polyphenylquinoxaline

Table 28. Lap-shear strength values for
polyphenylquinoxaline bonded titanium joints

Test condition	Lap-shear strength (MPa)
20 °C	21.1
177 °C	20.0
232 °C	8.1
177 °C (after 3000 h at 177 °C)	24.1
232 °C (after 3000 h at 232 °C)	13.4

Some results obtained from a proprietary polyphenylquinoxaline adhesive system are given in Table 28. Although these show lap-shear strength (see **Shear test**) deteriorating rapidly as T_g is approached it should be noted that joint strength values have been found to vary substantially from one study to another. For example, with titanium substrates changes in surface pretreatment have been shown to have significant effects with room temperature lap-shear strength varying between 10 and 36 MPa. The values quoted in Table 28 may therefore represent an over-pessimistic view of the capabilities of polyphenylquinoxaline.

As with other thermoplastic-based **High-temperature adhesives**, the main advantage of the polyphenylquinoxalines is undoubtedly their ability to produce large area bonded structures free of significant bond-line porosity. Disadvantages, including the need for solvent removal together with a high-temperature, high-pressure bonding requirement would be seen in some circumstances, however, as being particularly serious. In common with many other thermoplastic polymers, they have also shown susceptibility to gross deformation and failure at temperatures substantially below the limits set by thermal stability.

A reduction in excessive thermoplasticity requires the introduction into the polymer molecule of some means of cross-linking, and much of the research conducted into polyphenyl quinoxalines in recent years has been concerned with this problem. One approach, making use of technology developed initially with the polyimides, has involved the development of phenylquinoxaline prepolymers having acetylene terminal groups. Under the appropriate conditions the acetylene groups react by an addition mechanism to produce a cross-linked structure exhibiting reduced thermoplasticity and hence a greater high-temperature capability. More recent studies have been concerned with the incorporation of pendant cross-linking sites distributed along the polymer chain. Most success has been achieved by the addition of so-called phenylethyl groups which have resulted in substantial improvements in high-temperature adhesive joint strength. Although extremely promising, **High-temperature adhesives** based on these developments are not yet, to the author's knowledge, available commercially. Linear thermoplastic polyphenyl quinoxalines are, however, commercially available, albeit at a high price.

Further discussion on the general theme of high-temperature polymers and adhesives can be found in the articles entitled **Polybenzimidazoles, Polyether ether ketones** and **Polyimide adhesives**.

Select references

P M Hergenrother, *Polym. Eng. and Sci.*, **16**, 303 (1976).
P M Hergenrother, *Polym. Eng. and Sci.*, **21**, 1072 (1981).
C Hendricks, S G Hill, *SAMPE Q.*, **12**, 32 (July 1981).
J P Critchley, G J Knight, W W Wright in *Heat Resistant Polymers*, Plenum Press, New York, 1983.
P M Hergenrother, *Polym. J.*, **19** (1), 73 (1987).

Polyurethane adhesives

G C PARKER

Introduction

A polyurethane adhesive is produced by reacting together two basic raw materials – a polyol and an isocyanate (see **Condensation polymerization**). The two most common polyols are based on polyether and polyester, and the two principal isocyanates used are toluene diisocyanate (TDI) and diphenyl methane diisocyanate (MDI), both of these latter products being extremely reactive.

When a polyol and an isocyanate are mixed together, the isocyanate reacts with the hydroxyl groups in the polyester/polyether polyol.

$$R\text{—}OH + OCNR' \rightarrow ROCONHR'$$

This reaction is exothermic and is stopped by cooling the vessel and introducing a vacuum before all the isocyanate has reacted. Depending on the properties required of the adhesive, between 5 and 35% of the isocyanate is left unreacted or 'free'. This free isocyanate will subsequently react in the presence of moisture, or when mixed with a substance containing free hydroxyl groups, to complete the curing process.

Polyurethane adhesives can be conveniently divided into two types – solvent-free and solvent-based systems.

Solvent-free systems

These adhesives are liquids with a 100% solids content and are available in both 1- and 2-component forms.

The 1-component adhesives typically contain about 15% free isocyanate. After application the curing process is completed by exposure to atmospheric humidity and/or moisture on the substrate surface.

The 2-component adhesives comprise of one part which is 100% polyol and a second part which is a partial reaction of polyol with isocyanate. This second part typically contains approximately 30% free isocyanate. When the two components are mixed the free isocyanate in the second part reacts with the hydroxyl groups in the first part and these two components then polymerize to form an adhesive film. With this type of product it can be appreciated that the cure rate is not therefore moisture dependent.

The 1-component adhesives will typically form a holding bond after 24–36 h and will cure completely in a further 4–5 days. The 2-component adhesives form a holding bond after 24 h and are usually completely cured after 2–3 days.

These products are used primarily for flexible film lamination on purpose-built machines running at speeds of up to 300 m min^{-1}, although they do find certain other applications elsewhere.

Radiation-cured adhesives include acrylated urethanes.

Solvent-based systems

These adhesives are usually 1-component systems in non-flammable solvents and are moisture curing. The solids content of the products used are varied according to the method of application. A 40–60% solids product is typical where a spray application is used, a solids content of 70–80% is common for roller coater applications and a solids content as high as 95% is typical for sealants and mastics.

When solvent-based polyurethanes are spray applied, the solvent will evaporate very rapidly and leave an adhesive film possessing a very aggressive **Tack**. These adhesives are generally applied to one surface only, and each has a finite open time during which the bond must be made if optimum bond strength is to be achieved. The tack of the applied film of adhesive will gradually diminish as the adhesive cures, and a good holding bond may be achieved in around 20 min with the fastest-curing products. Complete curing will typically take a further 24 hours, thereby giving a much faster cure rate than can be expected with the solvent free systems discussed above.

Solvent-based polyurethane adhesives find application particularly in the diverse field of product assembly operations, and are used for bonding materials such as steel, aluminium, timber products and rigid insulation. Products are available, however, which have been formulated to cure to a soft flexible adhesive film for bonding flexible materials such as fabrics and nylon scrim to flexible foams and sponge.

Consequently, major markets for these products are modular floor panels, caravan manufacture, insulated building panels, kitchen scouring pads and sound-insulation materials along with **Automotive applications** such as upholstery, door trims and head linings.

Properties of structural polyurethanes are compared with those of **Epoxide adhesives** and **Toughened acrylic adhesives** in the article on **Structural adhesives**.

Toxicological considerations

The main potential hazard associated with polyurethane adhesives is due to the free isocyanate content, although the potential hazards need to be kept in perspective. All isocyanates are very reactive with moisture and will therefore react on the skin and, if inhaled, the lungs. It is now well established that TDI can cause sensitization of the respiratory tract if inhaled in quantity or over a prolonged period of time.

Compared with TDI, MDI is relatively safer in the working environment because it has a much lower vapour pressure. In practice, free MDI is unlikely to be a hazard at temperatures below 40 °C, especially if adequate extraction is provided at the point of application. If the application temperature is above 40 °C, or if the adhesive is spray applied, a vapour mist will always be formed and exhaust extraction must be used. To ensure a safe working environment, monitoring equipment should always be used to determine the level of isocyanate in the atmosphere, the maximum permitted level for MDI being 0.02 parts per million (0.2 mg m^{-3}). See **Health and safety**.

Select references

J M Buist, H Gudgeon, *Adhesives in Polyurethane Technology*, Maclaren, London, 1968.
J M Buist (ed.), *Developments in Polyurethane*, Elsevier Applied Science Publishers, London and New York, 1978.

Polyvinyl alcohol in adhesives

C A FINCH

Polyvinyl alcohols are a group of water-soluble polymers used in a wide variety of general-purpose and industrial adhesives, to bond porous and cellulosic substrates such as paper and paperboard, wood and textiles and also some smooth surfaces such as metal foils. The polymer may be used alone, in aqueous solution, and is also employed with other polymers to modify the viscosity and rheological properties of formulations. In particular, it is used as the protective colloid component of vinyl acetate polymer and copolymer emulsions used as the base polymers of liquid 'white' glues for bonding wood and paper products.

Polyvinyl alcohol is produced in large quantities by a two-stage process. First vinyl acetate in methanol solution is polymerized to polyvinyl acetate:

$$n\text{CH}_2\text{=CH} \rightarrow \text{-[CH}_2\text{-CH-]}_n$$
$$\underset{\text{OOCCH}_3}{|} \qquad \underset{\text{OOCCH}_3}{|}$$

This reaction is exothermic to the extent of 5.2 kJ mol^{-1} vinyl acetate. The polyvinyl acetate is hydrolysed, usually with alkali, to give polyvinyl alcohol:

$$\text{-[CH}_2\text{-CH]}_n + n\text{NaOH} \rightarrow \text{-[CH}_2\text{-CH]}_n + n\text{CH}_3\text{COONa}$$
$$\underset{\text{OOCCH}_3}{|} \qquad\qquad\qquad \underset{\text{OH}}{|}$$

There are many detailed variations on this process.

The polymer is separated from the residual methanol–methyl acetate solvent mixture (which is recovered), washed to remove residual salts (mainly sodium acetate), and dried to a fine white powder. This can be made in many grades, defined by the degree of polymerization (d.p.) of the original polyvinyl acetate, and by the degree of hydrolysis, usually considered to be 'partly' (87–89%) or 'fully' (97–99%) hydrolysed. Use of polyvinyl alcohol with higher d.p. (up to 2400) results in increased solution viscosity and dry film strength, and increased initial 'grab' to substrate: lower d.p. polymer is easier to dissolve, but has less attractive properties. The water resistance of polymer films with 99% hydrolysis is greater than those with lower values and they are more difficult to dissolve. Such grades of polyvinyl alcohol are used as components of several types of adhesives, since they are compatible in aqueous solution with other polymers, including most starches (e.g. wheat, maize or potato (farina) starches) and dextrins (see **Compatibility**). The polymers are also compatible with some natural gums and alginates, used for remoistenable paper coating adhesives (where low-d.p., low-hydrolysis polyvinyl alcohol is used). Fillers such as wood flour, china clay or whiting can also be added to reduce cost. All these considerations affect the use of polyvinyl alcohols in water-based adhesives for many applications (see **Emulsion and dispersion adhesives**). However, polyvinyl alcohols have only limited compatibility with proteins, such as gelatin, **Animal glue** and casein, and with natural rubber latex (although this can be mixed with some low-hydrolysis grades).

Polyvinyl alcohol is used in several types of **Wood adhesives**. It is added at the condensation stage in the manufacture of modified urea–formaldehyde or urea–melamine–formaldehyde resins used in plywood and particle board, to increase viscosity, and improve initial grab and the ageing characteristics of the adhesive bond. In emulsion-based wood adhesives the polymer is used in production of the emulsion, as a protective colloid, and then further polyvinyl alcohol is added in formulation of the adhesive, to improve both 'open time' during formation of the adhesive bond and the resistance to 'creep' (the cold flow in wood-to-wood adhesive joints under tensile load). Frequently, a mixture of different grades of the polymer is employed, to control the viscosity and rheological properties of the adhesive during application. Adhesion to substrates can be improved by employing a polyvinyl acetate copolymer emulsion with 1–2% of a carboxylate-containing comonomer. Improved moisture resistance in the final adhesive bond can be attained by employing polyvinyl acetate copolymer emulsions which contain cross-linking comonomers (N-methylolacrylamide is most frequently employed).

Adhesives for paper-to-paper applications are made from polyvinyl acetate emulsions, with added plasticizers, or from vinyl acetate copolymers (notably **Ethylene–vinyl acetate copolymers**) with added polyvinyl alcohol to modify viscosity, flow and coating properties, remoistenability and the rate of formation of the adhesive bond. These properties are important for effective use of the adhesive on high-speed application equipment. In water-remoistenable adhesives, partially hydrolysed polyvinyl alcohols with low viscosity are included as a major component, usually with added humectants, such as polyalkylene oxides. A range of adhesives for production of water-resistant corrugated paperboard has been prepared from a high-viscosity fully hydrolysed polyvinyl alcohol in acid solution, slightly gelled with boric acid, used to stabilize a dispersion of finely divided calcium carbonate. Frost resistance during transport may be improved by addition of glycols or other polyols.

Many types of acrylic copolymer emulsions are used, with polyvinyl alcohol as binder and thickener, for building adhesives, bonding concrete to different substrates such as ceramic tiles, polyvinyl flooring tiles, polystyrene insulating panels and hydrophobic films to paper.

Polyvinyl alcohol is added to acrylate copolymer emulsions, which are usually surfactant-stabilized, to alter or improve their adhesive performance. Many emulsions of this type are used as components of water-based pressure-sensitive emulsions – typically, a tacky adhesive for paper can be prepared by dissolving polyvinyl alcohol in a 2-ethylhexyl-acrylate copolymer emulsion.

Polyvinyl alcohol solutions are also used as adhesion binders for non-woven fabrics, screen printing, ferrites, ceramics and some building materials. In powder form, it is also an effective binder for finely divided powders, such as pigments. Mixtures of low-d.p. polyvinyl alcohol with ethylene glycol, however, have low adhesion to substrates, and are used as peelable coatings for metals and plastics, for protection during storage and transport.

Select references

C A Finch, *Polyvinyl Alcohol: Developments*, 2nd edn, John Wiley, Chichester, 1992.
I Skeist (ed.), *Handbook of Adhesives*, 2nd edn, Van Nostrand Reinhold, New York, 1977.

Powder adhesion

K KENDALL

Fine powders stick together tenaciously. Anyone who has tried to dig through dry clay will bear witness to the powerful adhesion developed between the micrometre-diameter clay particles. Like the strength of many other adhesive systems, for example epoxy–aluminium joints, the adhesion of these clay particles is much diminished by the addition of water, and also by the influence of pH, as when lime is added (see **Durability of coatings in water**).

Control of clay particle adhesion was perhaps the first historical example of adhesive technology for human benefit. It was discovered in Babylonian times that wet clay could be plastically shaped when its interparticle adhesion was low, then dried and used as a construction material when the adhesion between grains was raised by drying and firing.

Such understanding and control of particle adhesion is essential to many industrial processes. For example, powder adhesion is useful in making pharmaceutical tablets which are manufactured by compressing the powder in shaped steel dies. Polymer powders made by dispersion polymerization are agglomerated into larger granules in which the fine grains stick together to allow easier handling. On a much larger scale, powdered iron ore is passed through a pelleting process before entering the blast furnace. Similarly, ceramics ranging from building bricks through to ball-bearings are made by compacting fine powders in moulds before sintering the products in a furnace. During this sintering process the particles flow below the melting-point to form extensive interparticle contacts, further enhancing the adhesion.

In contrast, particle adhesion can be a problem in certain areas; powders may refuse to flow out of hoppers, pigments may form intractable sediments in paints and dust grains may wreck electronic micro-circuitry by adhering strongly to the chips. Perhaps the most dramatic consequences of problematic particle adhesion are found during earthquakes or mudslides, when the relatively weak interparticle forces are overcome by vibration or by fluid flow. The theory of powder adhesion attempts to describe and predict such problems from first principles.

The first ever equation used to describe adhesion between solids was devised by Coulomb in 1773 to explain the movement of soils under load. In a shear box test (Fig. 105(a)), a powder is loaded with a normal force N, and sheared with a force F. In general, the plot of F vs N (Fig. 105(b)) does not pass through the origin, but is displaced by an amount A on the horizontal axis. This displacement A represents the adhesive force pulling the particles into contact by **Dispersion forces** or other molecular attractions. Coulomb's equation describing this behaviour may be written

$$F = \mu(A + N) \qquad [1]$$

where μ is the coefficient of **Friction**. This equation is the basis of soil mechanics as described for example by Bolton.[1]

It became apparent in the 1930s that all solid particles must attract each other unless some barrier material is interposed between the grains, the attractive force increasing as the particles approached. For two spheres of diameter D, separated by a gap a, the attractive force A is given by Israelachvili[2] as

$$A = HD/24a^2 \qquad [2]$$

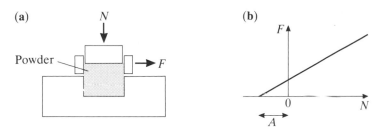

Fig. 105. (a) Shear box test for powder adhesion and friction; (b) plot of shear force versus load for a fine powder

where H is the Hamaker constant, about 10^{-19} J. This equation has been verified by bringing very smooth mica surfaces together, or by studying the contact of fine-pointed probes using the atomic force microscope.

When the particles touch, the force of attraction rises to such a large value that deformation of the particles occurs. This may be elastic deformation as demonstrated by Johnson et al.,[3] or it may be plastic indentation as noted by Krupp.[4] In the elastic case, spherical particles form a small circle of intimate contact diameter d which depends on the elastic modulus E, the Poisson ratio v of the material and the particle diameter D according to

$$d^3 = 9\pi WD^2(1 - v^2)/2E \qquad [3]$$

where W is the work of adhesion of the surfaces (see **Contact angle and interfacial tension**). But the force required to separate the grains does not depend on the elasticity and is given by

$$A = 3\pi DW/8 \qquad [4]$$

In practice it is extremely difficult to measure the adhesion between two small particles, and satisfactory methods have not yet been devised. Generally, experiments are carried out on compacted pellets containing around 10^{12} particles. Attempts to interpret the strength and fracture properties of such gross compacts have had mixed success. Perhaps the best technique for understanding the adhesion of these powders is to measure the elastic modulus of the compacted power. Because this elasticity depends on the number and size of contacts between the individual grains, it is possible to estimate the work of adhesion of powders from such tests.[5]

References

1. M Bolton, *A Guide to Soil Mechanics*, Macmillan, London, 1979, Ch. 4.
2. J N Israelachvili, *Intermolecular and Surface Forces*, Academic Press, London, 1985, Ch. 1.
3. K L Johnson, K Kendall, A D Roberts, *Proc. Roy. Soc. Lond.*, **A324**, 301 (1971).
4. H Krupp, *Adv. Coll. Int. Sci.*, **1**, 111 (1967).
5. K Kendall, Relevance of contact mechanics to powders – elasticity, friction and agglomerate strength, in *Tribology in Particulate Technology*, eds B J Briscoe, M J Adams, Adam Hilger, Bristol, 1987, p. 91.

Pressure-sensitive adhesives

D W AUBREY

Nature of pressure-sensitive adhesives

Pressure-sensitive adhesives are unusual materials in the sense that they are somewhere between the viscous and rubber states at room temperature, i.e. their response to deformation is viscoelastic. They show sufficient liquid-like behaviour to deform or flow into contact with a smooth surface under light contact (the 'bonding' process), yet they show appreciable resistance to flow during a separation or 'debonding' process. The level of bond strength (force or energy of separation) depends on obtaining a rather delicate balance of viscoelastic properties in the adhesive polymer (see also **Viscoelasticity, Tack**), as well as favourable interfacial energetics (see also **Surface energy**).

The main advantages of pressure-sensitive adhesives compared with other types of adhesive are those of convenience of use. There is no storage problem, there is no mixing or activation necessary, no waiting is involved. Often the bond is readily reversible. Disadvantages are that the adhesive strength (both peeling and shearing strengths) are low, they are unsuitable for rough surfaces and they are expensive in terms of cost per unit bond area. Some standard test methods are listed in Appendix 1.

Materials used as pressure-sensitive adhesives

The selection of materials to be used is based not only on the final adhesion properties required, but also on considerations of the economy of the manufacturing process for the particular pressure-sensitive product.

The most well-known type of adhesive consists of natural rubber (NR) blended with an approximately equal amount of a tackifier resin, and a small amount of antioxidant. It is applied to a suitably primed supporting member or backing, such as regenerated cellulose film ('Cellophane'), as a solution in petroleum spirit. A release agent on the reverse side of the backing is commonly applied to facilitate unwinding (see **Release**). Early rosin ester tackifier resins were later largely replaced by better ageing terpene resins (mainly those derived from β-pinene), which in turn have now been largely replaced for economic reasons by cheaper resins obtained by the cationic polymerization of petroleum fractions (C_4, C_5 and C_6), although the terpene resins are still regarded as being excellent technically. Considerable economy of solvent may be achieved by the higher solution concentrations possible from the use of the newer thermoplastic rubbers, especially styrene–isoprene–styrene (SIS), in place of some or all of the natural rubber, and this is quite commonly practised in current solution coatings. Rigid PVC and polypropylene are now also extensively used as backing materials for general-purpose pressure-sensitive tapes, and many other plastic films and papers are used in specialized products including tapes, labels and decals.

Another type of solution coating extensively used is that based on polyacrylate (e.g. poly-2-ethyl hexyl acrylate) or an acrylate–vinyl acetate copolymer. These are generally used in high-quality tapes for their better technical properties – they are paler in colour, much better ageing, and if cross-linked after coating can give very good shear strength. Other adhesive polymers used to a lesser extent in solution coating are polyvinyl alkyl ethers, polyisobutylenes, and silicones.

However, the use of organic solvents for coating is becoming increasingly unpopular and various means have been sought to eliminate the solvent. The use of latex rather than solution systems for rubber/resin adhesives has not resulted in commercial success, because of problems of dispersion, mechanical stability, film homogeneity and cost of drying. However, latex (or 'emulsion') systems based on polyacrylate emulsions have been developed successfully as pressure-sensitive adhesives, and are widely used in sheet materials and labels. These materials do not need added tackifier resins, and are very stable mechanically, so the above problems are largely eliminated. They cannot be formulated, however, to give the same high adhesion and shear strength as the solution polyacrylates, and are therefore not much used in tape form.

The most recent development in pressure-sensitive tape manufacture is the development of the hot-melt coating process. Almost all of the current hot-melt applied adhesives (see also **Hot melt adhesives**) are based on SIS thermoplastic rubber, mixed with hydrocarbon tackifier resins and oils in substantial amounts, and an antioxidant in minor amounts. Usually the materials are mixed in an extruder and fed via a holding tank direct to the coating head. It is therefore a very fast and economical process, but is not very versatile and is most suitable for large-scale production of the same basic tape or sheet. Much of the general-purpose packaging tape currently produced in the UK is made by this method.

A more complete account of the many varieties of pressure-sensitive products commercially available may be found in the first reference (see also **Pressure-sensitive adhesives – adhesion properties**).

Select references

D Satas (ed.), *Handbook of Pressure-sensitive Adhesive Technology*, Van Nostrand Reinhold, New York, 1982.
C A Dahlquist, Pressure-sensitive adhesives in *Treatise on Adhesion and Adhesives*, Vol. 2, ed. R L Patrick, Marcel Dekker, 1969, Ch. 5.

M Toyama, T Ito, Pressure-sensitive adhesives, *Polym. Plast. Technol. Eng.*, **2** (2), 161–229 (1973).

D W Aubrey, Pressure-sensitive adhesives – principles of formulation in *Developments in Adhesives – 1*, ed. W C Wake, Applied Science Publishers, London, 1977, Ch. 5, p. 127.

Pressure-sensitive adhesives – adhesion properties

D W AUBREY

Pressure-sensitive adhesives are normally supported on 'backing' materials and sold as tapes, labels or sheet products. Their adhesion properties are often tested in the form of tapes cut directly from such products. Since most adhesion properties are influenced by the nature and thickness of the adhesives and backing film layers, the results obtained cannot be regarded as intrinsic properties of the pressure-sensitive adhesive, but are in fact properties of the composite tape.

Standard tests used to characterize the adhesion properties of tapes are for the assessment of shear strength (see **Shear tests**) (the ability of a tape joint to resist a load applied in the shear mode), peel strength (see **Peel tests**) (the resistance of a tape joint to peeling under specified conditions) and **Tack** (the ability of a pressure-sensitive adhesive to form a bond immediately on contact with another material). There are many standard test specifications laid down by different authorities to assess these properties and many differences in detail between them (e.g. see Appendix 2). No attempt will be made to describe them comprehensively, but the principles of the tests will be discussed separately.

Shear strength tests

Shear strength of a tape joint is often assessed in the industry by a shear adhesion or holding-power test, in which a weight (e.g. 0.5 kg) is hung on a tape bonded over a known area (e.g. 1 in^2) to a steel test plate (Fig. 106). The time taken (at a given temperature) for the tape either to move a certain distance or to drop from the plate is taken as the indication of shear strength. The test is commonly carried out at room temperature or at 40 °C. A typical test is described in PSTC-7.[1] Although it may be adequate for routine assessment of tape performance under simulated service conditions, it gives little information about intrinsic properties of the adhesive itself. This is partly because there are at least three modes of failure possible in the test (Fig. 106).

First, the adhesive may undergo a true shear failure in which viscous flow is involved; it may be possible to correct this by increasing adhesive

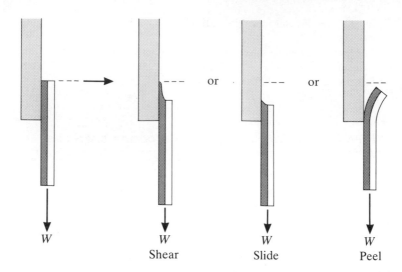

Fig. 106. Modes of failure in the holding power test

molecular weight or cross-linking the adhesive. Second, the tape may appear to slide intact from the plate, a process which probably involves the 'waves of detachment' mechanism well known in the sliding friction of rubbers. Third, peeling may occur from the unloaded end of the tape, due to low adhesion accompanied by a turning moment induced from elastic deformation of the backing, a well-known effect in lap joints generally. The latter two effects involve peeling phenomena and are best corrected by increasing the level of peel strength (see below).

If it is desired to study shear strength of the adhesive itself, without contribution from peeling processes and without backing effects, a somewhat more elaborate shear sandwich creep test is recommended.[2] From the results of such a test the behaviour of the adhesive may be expressed in terms of a fundamental viscoelastic parameter such as the time-dependent creep compliance function[3] (see also **Viscoelasticity**).

Peel strength tests

Peel strength is determined by pulling a tape from a test plate (usually steel, sometimes glass) at a constant rate and temperature in a tensile testing machine. Procedures for plate cleaning, bonding the tape, conditioning before test, etc. are tightly specified and must be rigorously adhered to for reproducible results. Normally a 180° peel angle is used, as in the well-known standard tests PSTC-1[1] and ASTM D 903-78 but sometimes 90° or lower angles are used. Peeling at 90° is subject to less

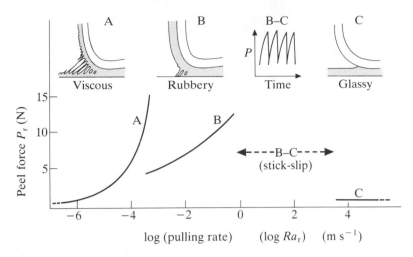

Fig. 107. Typical peel adhesion master curve[4] for polybutyl acrylate supported on polyester film peeled at 90° from glass. Insets show tape behaviour or force-time trace ('stick–slip')

error and is probably more related to practical situations, but requires more complicated test equipment than 180° peeling. Peel strength may be expressed either in force units ($N\ m^{-1}$) or in work units ($J\ m^{-2}$).

Peel strength for a given tape joint is very dependent on temperature and pulling rate and it is found that peel strength measured over a comprehensive range of rates and temperatures may be treated as a viscoelastic function. Application of the time–temperature superposition principle (see **Viscoelasticity – time–temperature superposition**) either by experimental shifting or by use of the WLF equation enables master curves of peel force against pulling rate at a given temperature to be obtained over a very wide (e.g. 10^{12}-fold) range of rates. Such master curves illustrate the change in viscoelastic response of the adhesive as rate and temperature is varied, and regions of steady peeling can be seen corresponding to viscous, rubbery and glassy behaviour of the adhesive (labelled A, B and C, respectively, in Fig. 107). A wide region (labelled B–C) of stick–slip peeling involves regular oscillation of peel force and is associated with a region of (theoretical) negative slope in the master curve. The master curve illustrated in Fig. 107 is for a simple amorphous uncross-linked polymer (poly-*n*-butyl acrylate) peeled at 90° from glass.[4]

Tack tests

The phenomenon of **Tack** is easily understood subjectively as the bond which forms when another surface (e.g. the thumb) is pressed against the adhesive surface for a brief time. There are three kinds of test, reviewed in detail by Hammond[5] which have been most often used to measure tack.

A probe tack test may be regarded as a 'mechanical thumb' in which a disc or hemisphere of standard material (e.g. brass) is brought into contact with the adhesive surface under a fixed load for a specified dwell time (e.g. 1 s), then removed at a specified rate. The maximum force of removal is usually taken as the tack value, although sometimes the work or energy of separation is reported. In such a test it is important to control accurately the variables: probe material and finish, probe diameter and shape, load on probe, thickness of adhesive, dwell time, rate of debonding of probe from adhesive, and test temperature. The effects of independent variation of these parameters have been described by Counsell and Whitehouse.[6] Types of probe-testing devices are described by Hammond,[5] and in ASTM D 2979-77.

The rolling ball test for tack usually involves allowing a stainless steel ball, diameter c. 12 mm, (or sometimes a table-tennis ball) to accelerate down a ramp of fixed (21.5°) inclination on to the horizontal adhesive surface. The distance travelled by the ball on the adhesive surface before arrest is taken as an inverse of tack for that surface. For a given adhesive, the ratio $h^{2/3}d^{-1}$ is often found to be constant, where h = ramp height and d = distance travelled. Although the test is inexpensive and easy to perform, it suffers from greater variability than the probe test and interpretation of results is more difficult. Typical arrangements for the test are described in PSTC-6[1] and ASTM D 3121-73.

The quickstick test[1], sometimes regarded as a measure of tack, involves the peeling of a loop of tape which has been allowed to contact a steel test plate for a brief time under its own weight only. It is clearly a modified form of peel test in which minimal contact between tape and plate has been achieved. As such, it will depend strongly on the stiffness of the tape backing layer, and is therefore not a property characteristic of the adhesive itself. The test is often carried out alongside the conventional peel test, requiring the same equipment. Again, it is difficult to interpret the results of this test in any fundamental way, since the act of peeling the tape will increase the contacting force due to backing leverage[4] and the additional contacting force from this effect will increase with the peeling force.

A more fundamental basis for the phenomenon is discussed under **Tack**.

References

1. *Test Methods for Pressure-sensitive Adhesives*, Pressure-sensitive Tape Council, Glenview, Ill., 7th edn, 1976.
2. D W Aubrey in *Adhesion – 3*, ed. K W Allen, Elsevier Applied Science Publishers, London, 1977, Ch. 12, p. 191; *Adhesion – 8*, ed. K W Allen, Elsevier Applied Science Publishers, London, 1984, Ch. 2, p. 19.

3. C A Dahlquist in *Handbook of Pressure-sensitive Adhesion Technology*, ed. D Satas, Van Nostrand Reinhold, New York, 1982, Ch. 5, p. 78.
4. D W Aubrey in *Developments in Adhesives – 1*, ed. W C Wake, Applied Science Publishers, London, 1977, Ch. 5, p. 129.
5. F H Hammond Jr in *Handbook of Pressure-sensitive Adhesive Technology*, ed. D Satas, Van Nostrand Reinhold, New York, 1982, Ch. 3, p. 32.
6. P J C Counsell, R S Whitehouse in *Developments in Adhesives – 1*, ed. W C Wake, Applied Science Publishers, London, 1977, Ch. 4, p. 99.

Pretreatment of aluminium

D M BREWIS

Introduction

Good adhesion to aluminium is required in several technologies including adhesive bonding, painting, printing and **Extrusion coating**. To achieve satisfactory adhesion, it is usually necessary to carry out a pretreatment. When only moderate initial adhesion is required and when the service conditions do not involve aggressive environments such as high humidity, then solvent **Degreasing** or an **Abrasion treatment**, for example grit-blasting, may be adequate. However, chemical pretreatments will provide higher initial joint strengths and more consistent results.

If bonded aluminium structures are subjected to high **Humidity** or immersion in water, then a chemical pretreatment is essential for good **Durability**. Three such methods have been developed for use in the aerospace industry, namely (1) chromic acid etching (see **FPL etch**), (2) chromic acid **Anodizing** and (3) phosphoric acid anodizing. These treatments are all complex and details do vary to some extent, but typical conditions[1] are given below.

Details of pretreatments

Chromic acid etch

1. Degrease (see **Degreasing**) with trichloroethylene or methyl ethyl ketone.
2. Immerse in mild alkaline degreasing agent for 5 min at 65 °C.
3. Rinse immediately in hot tap water to remove the alkaline cleaner.
4. Immerse in an etching solution of distilled water containing concentrated sulphuric acid (179.5 cm^3 dm^{-3}), chromium trioxide (68.5 $g\,dm^{-3}$), copper sulphate (0.39 $g\,dm^{-3}$) and stearate-free aluminium powder (5 $g\,dm^{-3}$) at 62 °C for 30 min.
5. Wash immediately in cold running tap water for 20 min.

6. Dry in warm air for 20 min.
7. Bond or prime within 3 h of pretreatment.

Chromic acid anodizing Steps (1)–(4) as for chromic acid etching.

5. Immerse in a chromic acid solution containing 50 g dm^{-3} chromium trioxide at 38–40 °C. Over 10 min raise the voltage to 40 V in steps of 4 V; maintain the voltage at 40 V for 40 min with a current density of 0.1–0.6 A dm^{-2}; increase the voltage to 50 V during the next 5 min in steps of 2 V; maintain this voltage for 5 min.
6. Wash in cold running tap water for 20 min.
7. Dry in warm air for 20 min.
8. Bond or prime within 3 h.

Phosphoric acid anodizing Steps (1)–(6) as for chromic acid etching.

7. Anodize in a phosphoric acid solution at 20–25 °C containing 10% by weight of 85% orthophosphoric acid in distilled water. Raise voltage to 10 V in steps over 3 min and maintain full potential for 25 min. Parts should not remain in the phosphoric acid in excess of 2 min after the power is disconnected as the oxide begins to dissolve.
8. Wash in cold running tap water for 20 min.
9. Dry in warm air for 20 min.
10. Bond or prime within 3 h.

Comparison of pretreatments In general, durability under hot wet conditions is found to be in the order: chromic acid etch < chromic acid anodize < phosphoric acid anodize; some typical results are given in Fig. 108.

Although generally inferior to the two anodizing methods, chromic acid etching provides much better durability than solvent or mechanical pretreatments. However, careful control of etching conditions is essential.[2] For example, the presence of small quantities of copper and aluminium (see above formulation) is beneficial, whereas the presence of other ions, especially fluoride, can be highly detrimental to durability. Rinsing conditions are also important, tap water being used to prevent the formation of thick, relatively weak oxide layers.

The Boeing Company concluded that even under optimum conditions, chromic acid etching was inadequate for bonding primary structures. In the late 1960s, Boeing commenced a programme to study phosphoric acid anodizing and this was introduced commercially in 1974. The process is now widely used in **Aerospace applications** in the USA, while in Europe chromic acid anodizing has been used successfully for many years.

The reasons for the superior durability provided by anodizing

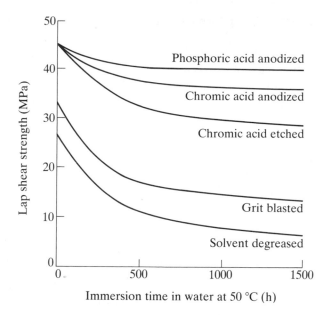

Fig. 108. Effect of surface pretreatment on the durability of aluminium joints bonded with a toughened epoxide adhesive (Redrawn with permission from ref. 3)

pretreatments are not clear. The stability of the interface and the oxide layer and the nature of the topography have been suggested as important factors[3] (see **Anodizing**).

Less complex pretreatments

The methods used in the aerospace industry are complex and not generally acceptable to other industries. There is a clear need for less complex procedures which still provide good durability. Two rapid pretreatments have been developed involving either a **Conversion coating** or the deposition of fine silica particles coated with chromium trioxide on to the aluminium.[4] The pretreated surface can then be covered with a protective press-lubricant which can later be absorbed by the adhesive used. This means that pretreated aluminium with good durability can be supplied to engineering companies. Another useful feature of these pretreatments is that the aluminium can be formed into complex shapes prior to bonding without any serious loss of performance.

Pretreatments prior to painting

As with adhesive bonding, the durability of painted surfaces will depend to a large extent on the quality of the pretreatment. The aerospace industry in the UK uses either a chromate conversion coating or chromic acid anodizing.[5]

Further information, including sources of pretreatment recipes, can be found in **Pretreatment of metals prior to bonding.** See also articles on **Paint adhesion** and **Roughness of surfaces.**

References

1. A C Moloney in *Surface Analysis and Pretreatment of Plastics and Metals*, ed. D M Brewis, Applied Science Publishers, London, 1982, p. 175.
2. D M Brewis, *Mat. Sci. and Tech.*, **2**, 761 (1986).
3. A J Kinloch in *Adhesion and Adhesives*, Chapman and Hall, London, 1987, p. 339.
4. P G Selwood, A Maddison, P G Sheasby, UK Patent Application 2 139 540A (1984).
5. P Poole in *Industrial Adhesion Problems*, eds D M Brewis, D Briggs, Orbital Press, Oxford, 1985, p. 258.

Pretreatment of copper

D E PACKHAM

Careful surface preparation before adhesive application is essential for consistent and successful bonding to most substrates (see **Pretreatment of metals prior to bonding**), and this certainly applies to copper which has a reputation for being difficult to bond. Part of the difficulty is the friability of the black copper(II) oxide which forms on the surface in air at temperatures in the range 200–500 °C. At ambient temperatures a thin layer of copper(I) oxide is present.

Detailed recipes for pretreatment of copper (which are also suitable for some copper alloys) are given by Shields,[1] Cagle,[2] Meynis de Paulin[3] and in ASTM D 2651.[4] The simplest treatment consists of abrasion (see **Abrasion treatment**) with emery cloth followed by **Degreasing** in a chlorinated hydrocarbon solvent. Most of the treatments recommend a more elaborate routine of (1) degrease, (2) etch and (3) dry. Etching solutions for use at ambient temperature include iron(III) chloride in either nitric or hydrochloric acid. Immersion for between 1 and 3 min is advised. Ammonium persulphate solution (1:3 or 1:4 parts by weight) is also used at room temperature: Shields[1] suggests etching times as short as 30 s.

A two-stage etching treatment is also described:[1–3] Shields[1] recommends it for maximum bond strength. The first stage involves immersion in a iron(III) sulphate/sulphuric acid solution for 10 min at 66 °C. After rinsing this is followed by ambient temperature treatment in sodium dichromate/sulphuric acid solution.

A proprietary treatment using 'Ebonol C' is recommended by Cagle[2] and Meynis de Paulin[3] and ASTM D 2651.[4] Cagle describes it as 'one of the better methods' for pretreating copper. It involves treatment in the solution at 98 °C for periods up to 10 min. This method produces a **Microfibrous surface** consisting essentially of a 'floral' array of needle-like copper(II) oxide which does adhere well to the base metal. The alkaline chlorite solution described in the article just cited gives a similar, but not identical, effect.

It is worth mentioning that microfibrous copper(II) oxide surface can also be produced by alkaline **Anodizing** at elevated temperature.[5] Suitable conditions are 4 M sodium hydroxide at 90 °C with a current density of 143 A m^{-2}.

Selection of a pretreatment may well depend on striking a balance between convenience and effectiveness. A proper assessment of convenience and effectiveness will take account of the potential damage to the **Health and safety** of operators and of the general public as well as the environmental impact of disposal of waste materials. Whatever treatment is chosen, the recipes cited (refs 1–4) should be regarded as a starting-point for a series of experiments aimed at optimizing the process variables for the particular application involved.

References

1. J Shields, *Adhesive Handbook*, 3rd edn, Butterworths, 1984.
2. C V Cagle, *Handbook of Adhesive Bonding*, McGraw-Hill, 1973, pp. 21–3.
3. J J Meynis de Paulin, P Cognard, *Les colles et les adhésifs, III collage des métaux et des matières plastiques*, Guy le Prat, Paris, 1979, p. 155.
4. ASTM D 2651 in *Annual Book of ASTM Standards*, American Society for Testing and Materials, Philadelphia, published annually.
5. J R G Evans, D E Packham, *J. Adhesion*, **10**, 177 (1979).

Pretreatment of fluorocarbon polymers

D M BREWIS

The fully-fluorinated polymers, i.e. polytetrafluoroethylene (PTFE) and the copolymer of tetrafluoroethylene and hexafluoroethylene (FEP) are difficult to adhere to and, to obtain satisfactory adhesion, it is usually necessary to pretreat these polymers. With partially fluorinated polymers it is often possible to obtain good adhesion without a pretreatment; for example, good bond strengths have been obtained with poly(vinyl fluoride) and poly(vinylidene fluoride)[1] and polychlorotrifluoroethylene and the copolymer of chlorotrifluoroethylene and vinyl fluoride.[2]

The wide range of pretreatments for fluorinated polymers has been reviewed by Dahm.[3] The methods available include the following:

1. Immersion in a solution of an alkali metal in liquid ammonia;
2. Immersion in a solution of sodium naphthalenide in tetrahydrofuran;
3. Reduction with electrochemically generated tetra(alkylammonium) radical anions;
4. Direct electrochemical reduction of the surface when placed in contact with a metal electrode in a non-aqueous electrolyte;
5. Treatment with alkali metal amalgams;
6. Exposure to an activated inert gas;
7. **Plasma pretreatment** in air or ammonia.

The pretreatment (2) is by far the most widely used method in practice and a number of proprietary solutions are available. Immersion in a 1 M solution of sodium naphthalenide in tetrahydrofuran for 60 s results in a fifteen-fold increase in the adhesion of PTFE to an epoxide[1] (see **Epoxide adhesives**).

There is much evidence that the pretreatments (1)–(4) lead to marked chemical changes in the surfaces of the fluorinated polymers. In particular, **X-ray photoelectron spectroscopy** (XPS) has shown that much defluorination occurs and this is accompanied by the introduction of various functional groups containing oxygen.[4,5] The chemical changes caused by these pretreatments result in a large increase in the **Surface energy** of the polymer.

The reductive treatments (1)–(5) result in the formation of a brown–black carbonaceous layer, the nature of which has been discussed elsewhere.[3] Long exposure to oxygen or UV or treatment with a strong oxidizing agent, may result in the removal of this dark layer and the loss of the benefits of the pretreatment.

See also **Pretreatments of polymers** and **Surface characterization by contact angles: polymers**.

References

1. D M Brewis, *Prog. in Rubber and Plast. Technol.*, **4** (1), 1 (1985).
2. H Schonhorn, L H Sharpe, *Polym. Lett.*, **2**, 719 (1964).
3. R H Dahm in *Surface Analysis and Pretreatment of Plastics and Metals*, ed. D M Brewis, Applied Science Publishers, London, 1982, p. 227.
4. H Brecht, F Mayer, H Binder, *Makromol. Chem.*, **33**, 89 (1973).
5. G C S Collins, A C Lowe, D Nicholas, *Eur. Polym. J.*, **9**, 1173 (1973).

Pretreatment of metals prior to bonding

D M BREWIS

To achieve satisfactory adhesive bonding, it is often first necessary to pretreat a metal. Sometimes solvent cleaning is sufficient, but frequently a mechanical pretreatment, a **Conversion coating**, or a chemical pretreatment is necessary.

Solvent cleaning may be carried out in a number of ways. Vapour **Degreasing** is the most effective, as relatively pure solvent condenses on to the metal. An ultrasonic bath may give satisfactory results provided the solvent is changed when it becomes significantly contaminated. Chlorinated solvents such as trichloroethylene are usually effective.

Mechanical pretreatments include the use of emery paper, wire brushes, abrasive pads and shot or grit-blasting (see **Abrasion treatment**). The latter method is usually preferred as it gives more consistent results. In grit or shot-blasting, the surface is bombarded by hard and usually sharp particles; chilled iron and alumina are commonly used. The effects of shot-blasting are complex. Most of the organic contaminants and also the mill scale will be removed; this should improve the subsequent wetting by the adhesive or primer (provided bonding or priming is carried out shortly after the pretreatment) and remove potential **Weak boundary layers** (see **Wetting and spreading**).

The potential bonding area will be increased and the possibility of mechanical keying will exist (see **Mechanical theory**). However, the latter effect is not thought to be important with the shot-blasting of metals.[1] The random nature of the roughness prevents voids and other defects from propagating rapidly.[1]

Chemical conversion treatments such as chromating and phosphating will also cause several changes which may affect the adhesion. First, the topography may be markedly changed, some coatings resulting in very rough surfaces (see **Roughness of surfaces**). The chemistry of the conversion coating will clearly be very different from the underlying metal oxide, one of the objects being to provide a coating with good durability to water.

Chemical pretreatments including **Anodizing** cause several changes to metal surfaces. They will change the morphology, thickness and detailed chemistry of the metal oxide and also produce marked changes in topography. Evans and Packham[2] have demonstrated clearly the beneficial effects of producing a very rough microporous copper oxide (see **Microfibrous surfaces**). With aluminium and titanium, the changes in topography have been studied in detail. It has been suggested[3] that mechanical interlocking of the primer/adhesive system with anodized aluminium and titanium is a major reason for the good **Durability** observed, but this has been disputed.[1]

Table 29. Possible effects of different types of pretreatments for metals

Pretreatment type	Possible effects of pretreatment
Solvent	Removal of most of organic contamination
Mechanical	Removal of most of organic contamination. Removal of weak or loosely adhering inorganic layers, e.g. mill scale. Change to topography (increase in surface roughness). Change to surface chemistry
Conversion coating	Change to topography (increase in surface roughness). Change to surface chemistry, e.g. the incorporation of a phosphate into the surface layers
Chemical (etching, anodizing)	Removal of organic contamination. Change to topography (increase in surface roughness). Change to surface chemistry. Change in the thickness and morphology of metal oxide

The effects of various pretreatments[4,5] are summarized in Table 29. The **Pretreatments** used for **Aluminium, Copper, Steel** and **Titanium** are described elsewhere in this *Handbook*. Provided care is taken over the pretreatments and suitable primers are used where necessary, good durability can be obtained with these metals.

References

1. A J Kinloch in *Adhesion and Adhesives*, Chapman and Hall, London, 1987.
2. J R Evans, D E Packham, *J. Adhesion*, **10**, 39 (1979).
3. J D Venables, *J. Mater. Sci.*, **19**, 2431 (1984).
4. J Shields, *Adhesives Handbook*, 3rd edn, Butterworths, London, 1984.
5. ASTM D 2651 in *Annual Book of ASTM Standards*, American Society for Testing and Materials, Philadelphia, published annually.

Pretreatment of metals prior to painting

J L PROSSER

Metals (certainly commercial metals) have a surface of oxide, hydroxylated, carrying layers of water and probably also being coated with organic materials. The oxide, if it is not epitaxial to the metal is likely to be friable and liable to detachment; this is certainly true of iron; aluminium has a more compact, adherent oxide which, however, may not be uniform. The oxide may also be cracked and loosely adherent by virtue of a production process, e.g., mill scale from a hot-rolling operation. The organic

Table 30. Epoxy powder coatings: adhesion to treated steel

Pretreatment	Initial adhesion	Appearance	Tape adhesion
Vapour degrease	Very good	Fine blisters	Detachment
Spray Fe phosphate	Very good	Few blisters	Good
Spray Zn phosphate	Very good	Satisfactory	Good
Dip Zn phosphate	Good	Satisfactory	Fair

2000 h Humidity test; BS 3900: F2

contaminant may be adventitious or applied as part of a production process, e.g. cold-rolling lubricants (long-chain alcohols and acids and others), and as protectives for the surface, as for tinplate, sheet steel, galvanized steel, etc. (see **Engineering surfaces of metals**). Such a surface is inimical to the attainment of good adhesion by the following paint system, so some form of pretreatment is necessary to prepare the surface for painting, this produces initially a uniform, clean, grease-free surface which is subsequently covered with a film preventing corrosion of the basic metal if the paint film cracks to allow corrosive agents through.

For iron the commonest pretreatment is a phosphate; for aluminium, zinc, copper and many other non-ferrous metals, chromates. The metal must first be cleaned by an alkali/acid pickling (as appropriate), rinsed and chemically converted in an immersion bath or by spray.

Phosphates may be either a zinc phosphate (from zinc dihydrogen phosphate solutions), or an iron phosphate (from alkali phosphate solutions) and the conversion reaction is promoted by accelerators (depolarizers), e.g. chlorates, bromates, molybdates or sodium meta-nitrobenzoate in alkali phosphate baths, or nitrate/nitrite at 65 °C, or chlorate at 35 °C for the zinc phosphate bath (with Ca or Ni grain-refining additions). An example of the beneficial effect of phosphating is quoted in Table 30 (ref. 1, p. 69, see **Epoxide adhesives**). A recent development is phosphating with an organic phosphate in a hot chlorinated hydrocarbon solution.

Aluminium may be given a green (chromium phosphate) or yellow (chromium chromate) treatment.

Electrochemical processes include anodic oxidation, e.g. of Cu and Al in alkali, or of Ti in NaF/HF, and cathodic oxidation of Cu in sodium bicarbonate. Pretreatments produce a porous surface (**Anodizing** of Al), or needle-like or dendritic oxide structures, to promote adhesion, although the most suitable structure seems to vary with the nature of the following paint.

A complementary discussion is to be found in **Conversion coating**.

References

1. D B Freeman, *Phosphating and Metal Pretreatment*, Woodhead-Faulkner, Cambridge, 1986.
2. D E Packham, Adhesion of polymers to metals; role of surface topology, in *Adhesion Aspects of Polymer Coatings*, ed. K Mittal, Plenum Press, New York, 1983.

Pretreatments of polymers

D M BREWIS

Introduction

Good adhesion to polymers is required in a number of technologies including adhesive bonding, metallizing, painting and printing. To achieve satisfactory adhesion it is often necessary to pretreat the polymer. The wide variety of pretreatments available may be divided into a number of different types.

Solvent pretreatments

The poor adhesion obtained with a polymer may be due to contaminants or additives on the surface. In such a case a wipe or dip with a suitable solvent should be the only pretreatment required. An example is the removal of plasticizer from the surface of poly(vinyl chloride).

Such a treatment is inadequate with polyolefins and some other polymers, although immersion of polyolefins in hot chlorinated solvent vapours for a few seconds can lead to great improvements in adhesion.[1] This vapour treatment is discussed in the article **Pretreatments of polyolefins**.

The properties of some polymers are seriously affected by particular organic solvents and certain polymer–solvent combinations must be avoided.

Mechanical pretreatments

Mechanical treatments such as grit-blasting and the use of abrasive pads are used to improve the adhesion to thermosets, especially those of the fibre-reinforced type (see **Abrasion treatment** and **Fibre-composites – joining**). The improved performance is probably due partly to the removal of contaminants from the surface and partly to the increased roughness.

The latter can be beneficial owing to (1) the increased potential bonding area, (2) the possibility of mechanical keying and (3) increased plastic deformation of the adhesive or coating.[2] It is important to remove the particles of polymer and abrasive from the surface with a solvent wipe or with a short immersion in an ultrasonic bath.

An alternative pretreatment for composites is the tear (peel) ply method. A fabric is incorporated in the composite near the surface. The fabric is pulled away just prior to bonding, exposing a clean surface, provided there is no contamination from the fabric.[3]

Oxidative pretreatments

The growth in the use of low-density polyethylene after the Second World War was restricted by the fact that it was very difficult to print upon or bond. This led to the development of a number of pretreatments including the use of chromic acid, **Flame treatments** and corona discharges. Under the correct conditions, these pretreatments are all highly effective.[3,4]

The **Corona discharge treatment** is especially suitable for the continuous treatment of plastic films, whereas the flame treatment is generally preferred for treating thicker sections such as bottles. Chromic acid is sometimes used prior to the metal plating of plastics and for treating complex shapes. These three methods are discussed in more detail in **Pretreatments of polyolefins**.

Although these treatments were developed for low-density polyethylene, they are now used to treat other polyolefins and other types of plastic. For example, the corona treatment is used with poly(ethylene terephthalate) and chromic acid with acrylonitrile–butadiene–styrene (ABS) and polyoxymethylene.

Plasma treatments

Plasma pretreatment has been the subject of much research with a wide variety of polymers. In this method, power is applied to a gas or a monomer at low pressure (typically 1 torr) and a plasma, consisting of ions, electrons, atoms and free radicals, is formed. Improved adhesion may be due to a variety of mechanisms depending upon the gas or monomer used. With inert gases such as argon, ablation of small molecules or cross-linking may occur. Using gases such as oxygen and ammonia, functional groups such as $\diagdown C{=}O$ and $-NH_2$ will be introduced into the surface. With monomers, such as acrylates, grafting to the polymer surface can occur.

Table 31. Possible effects of pretreatments for polymers

Treatment type	Possible changes to polymer surface
Solvent	1. Removal of contaminants and additives 2. Roughening (e.g. trichloroethylene vapour/polypropylene) 3. Weakening of surface regions if excessive attack by the solvent
Mechanical	1. Removal of contaminants and additives 2. Roughening
Oxidative	1. Removal of contaminants and additives 2. Introduction of functional groups 3. Change in topography (e.g. roughening with chromic acid treatment of polyolefins)
Plasma	1. Removal of contaminants and cross-linking (if inert gas used) 2. Introduction of functional groups if active gases such as oxygen are used 3. Grafting of monomers to polymer surface after activation, e.g. by argon plasma

Summary of changes caused by pretreatments

The possible effects of the various pretreatments are summarized in Table 31.

Methods to study pretreatments

The chemical changes caused by pretreatments may be studied by the use of contact angle measurements (see **Surface characterization by contact angles: polymers**) or preferably with surface techniques such as **X-ray photoelectron spectroscopy** (XPS). The topographical changes are best studied by means of **Scanning electron spectroscopy** (SEM).

Other related articles are as follows: **Contact angles and interfacial tension, Infra-red spectroscopy of surfaces, Microscopy, Pretreatment of fluorocarbon polymers, Primers to improve adhesive bonding, Roughness of surfaces, Surface analysis**.

References

1. E W Garnish, C G Haskins in *Aspects of Adhesion 5*, ed. D J Alner, University of London Press, 1969, p. 259.
2. D E Packham in *Adhesion Aspects of Polymeric Coatings*, ed. K L Mittal, Plenum Press, New York, 1983, p. 19.
3. A J Kinloch in *Adhesion and Adhesives*, Chapman and Hall, London, 1987, p. 101.
4. D M Brewis, *Progr. in Rubber and Plast. Technol.*, **1** (4), 1 (1985).

Pretreatments of polyolefins

D M BREWIS

Introduction

After the Second World War, it soon became apparent that low-density polyethylene (LDPE) would be a useful packaging material. However, one serious problem was the difficulty in obtaining good adhesion to the polymer. This led to much research and development to obtain effective pretreatments for LDPE. By about 1950, a number of pretreatments had been developed, including the use of chlorine + UV radiation, chromic acid etching, **Flame treatment** and **Corona discharge treatment**.

The first method, although effective, was abandoned for safety reasons. The other three have remained the most widely used pretreatments, not only for LDPE but also for high-density polyethylene (HDPE) and polypropylene (PP). Other methods have been found to be effective, but for reasons of cost, safety or convenience, they have not been widely used. The pretreatments include fuming nitric acid, potassium permanganate, ammonium peroxydisulphate, ozone, peroxides, UV radiation, grafting of polar monomers, plasmas (see **Plasma pretreatment**) and the use of solvent vapours. The three most widely used methods and the use of trichloroethylene are now discussed briefly; the latter is included because it involves a different mechanism.

Trichloroethylene vapour

Garnish and Haskins[1] found that exposure of polypropylene to trichloroethylene vapour for 10 s resulted in a sixfold increase in joint strength using an **Epoxide adhesive**. The authors concluded that the improved adhesion was due to the removal of a **Weak boundary layer**. However, the treatment causes the formation of a very porous surface, and an alternative explanation for the improved adhesion is the mechanical keying of the adhesive into the porous surface (see **Mechanical theory**). Garnish and Haskins found that the optimum treatment time was about 10 s and that after 25 s the adhesion level was similar to that of the untreated polymer. This reduction is probably due to weakening of the surface region of the polypropylene (see **Weak boundary layers**).

Chromic acid etching

Chromic acid is used to pretreat irregularly shaped objects and is also used with polypropylene prior to metal plating. Several formulations have been recommended, including the following:

Potassium dichromate	7 parts by weight
Water	12
Conc. sulphuric acid	150

Temperatures up to 70 °C are sometimes recommended, but very mild treatments are also effective. For example, 5 s immersion at 20 °C in chromic acid containing 1% of the usual concentration of potassium dichromate gave large improvements in the adhesion of an epoxide to LDPE, HDPE and PP.[2]

X-ray photoelectron spectroscopy (XPS) has been used to study the chemical changes caused by chromic acid treatments.[2] Even very mild treatment conditions result in substantial oxidation. Polypropylene reaches a maximum degree of oxidation after a mild treatment and the depth of modification is limited to a few nanometres because chain scission is occurring simultaneously with oxidation. With LDPE and HDPE, the degree and depth of oxidation increase with the severity of the treatment, i.e. temperature and time.

Flame treatment

Flame treatment is used to improve the adhesion to relatively thick polyolefin objects. For many years the method has been used to improve the adhesion of printing ink to polyethylene bottles. More recently, flame treatment has been used to improve the paint adhesion (see **Paint constitution and adhesion**) to polypropylene car bumpers.

The object to be treated is passed over one or more burners, each of which possesses a large number of closely spaced jets. The burners are fed with an air–hydrocarbon gas mixture whose proportions are carefully controlled. Ayres and Shofner[3] found that with methane the optimum treatment time for an unspecified polyolefin was 0.02 s. A study of the flame treatment of LDPE[4] showed very high levels of oxidation, although the oxidized layer was only 4–9 nm thick.

Corona discharge treatment

The **Corona discharge treatment** has been widely used since about 1950 to treat LDPE, especially to enhance **Printing ink adhesion**. The method has also proved to be very successful with HDPE and PP.

A schematic representation of the treatment is given in Fig. 109. Film is passed, typically at $1-2$ m s^{-1} over an earthed metal electrode covered with an insulator. The distance between the electrode and the film is usually $1-2$ mm. A high-frequency (10–20 kHz) generator and step-up

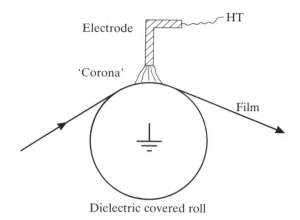

Fig. 109. Schematic representation of the corona discharge treatment

transformer produce a high voltage which causes the electrical breakdown of the air, with the formation of atoms, ions and electrons, which in turn leads to the formation of ozone, etc.

Surface analysis has shown that the **Corona discharge treatment** leads to the introduction into the polyolefin surface of various groups, including carbonyl, carboxyl, hydroxyl and ether.[5]

It is known that additives such as slip agents and antioxidants can adversely affect the treatment if this is done out of line, i.e. if the additives migrate before the treatment is carried out, then inferior adhesion may result. Another problem is that the effectiveness of the treatment can deteriorate over a period of a few weeks.

See also the more general article **Pretreatment of polymers**.

References

1. E W Garnish, C G Haskins in *Aspects of Adhesion 5*, ed. D J Alner, University of London Press, 1969, p. 259.
2. D Briggs, D M Brewis, M B Konieczko, *J. Mater. Sci.*, **11**, 1270 (1976).
3. R L Ayres, D L Shofner, *SPE J.*, **28** (12), 51 (1972).
4. D Briggs, D M Brewis, M B Konieczko, *J. Mater. Sci.*, **14**, 1344 (1979).
5. D Briggs, C R Kendall, *Int. J. Adhesion and Adhesives*, **2** (1), 13 (1982).

Pretreatment of steel

J F WATTS

The pretreatment of steel or any other metallic substrate can be taken, in the broadest sense, to refer to any method of surface preparation that prepares the substrate for the subsequent application of an organic coating or adhesive. In this category one would find the physical cleaning processes of **Degreasing** and **Abrasion treatment**, but by the term 'pretreatment' some chemical modification of the surface is often implied. Such a chemical modification may simply be the dissolution of the native oxide, referred to as pickling, alternatively pretreatment of the steel may involve the application of a conversion coating. The most widely employed conversion coating is a phosphate pretreatment although other methods such as chromating are sometimes employed.

Pickling

The removal of mill scale or other surface oxide layer by acid pickling is a well-established method of surface preparation of steels. The exact composition of the acid bath used varies widely, but in general they are based on hot sulphuric or cold hydrochloric acids. As such reagents will inevitably attack the underlying metal, giving rise to a rough undulating surface profile, inhibitors will usually be added. Care must of course be taken to remove all residues of acid by thorough washing and drying after treatment. The composition of the pickling solution must be carefully optimized as it will have a marked effect on the adhesion of any organic coating or adhesive subsequently applied.[1] Stainless steels require rather more sophisticated pickling solutions; a widely used example is a sulphuric acid/dichromate etch which leads to chromium enrichment in the oxide film and a subsequent improvement in adhesion.[2]

In the case of rusty steel surfaces or those with inorganic salt contamination (such as those resulting from storage of steel structures in a marine environment) pretreatment is readily effected with a hot dilute solution of phosphoric acid. As well as removing the oxide and etching the steel substrate it deposits a thin layer of iron phosphate on the steel. This not only provides improved corrosion resistant properties but also leads to improved adhesion of organic coatings.[3] The type of treatment is a very simple example of a conversion coating, that is one that is deposited on the surface by the chemical interaction between substrate and treatment bath. However, most commercially available conversion coating systems are more complex than the simple phosphoric acid wash and are designed to deposit complex mixed phosphates or chromates.

Conversion coatings

The most widely used conversion coating is zinc phosphate which is used on low-carbon steels in addition to zinc, aluminium, cadmium and tin. In this process a steel substrate, for example, is treated in a

Fig. 110. Phosphate morphology produced by a commercial process, (a) light-duty cleaner, (b) medium-duty cleaner and (c) heavy duty cleaner. Magnification × 1000. (Courtesy of Brent Europe Ltd)

solution of zinc phosphate and phosphoric acid (together with processing aids such as oxidizing agents) to produce a characteristic acicular intermeshed deposit on the metal surface.[4] The deposit is traditionally a mixture of hopeite $[Zn_3(PO_4)_2.4H_2O]$ 90%, phosphophyllite $[Zn_2Fe(PO_4).4H_2O]$ 3% and zinc phosphate dihydrate $[Zn_3(PO_4)_2.2H_2O]$ 7%. Although the various proportions of each phase are determined to a large extent by process chemistry, the concentration of phosphophyllite seems to be particularly important in terms of subsequent performance. The morphology of the deposit is shown in Fig. 110, but is critically dependent on surface-cleaning procedures, temperature of application and the nature of additions to the bath. The acicular morphology promotes good adhesion to paint films because of the large surface area generated; however, it has little mechanical strength and is quite unsuitable for adhesive bonding. It also shows poor performance in situations of **Cathodic disbondment** where the cathodically generated alkali leads to premature failure of the phosphate coating at the point where the crystals join the basal plane. For applications that require a high resistance to cathode disbondment conversion coatings of the chromate family (e.g. Accomet PC, Brent Chemicals International Group) have produced very impressive results.

ASTM D 2651[5] gives recipes for pretreatment of steels.

References

1. J M Sykes in *Surface Analysis and Pretreatment of Plastics and Metals*, ed. D M Brewis, Applied Science Publishers, London, 1982, pp. 153–74.
2. T Smith, R Hank, *Treatment of AM 355 Steel for Adhesive Bonding*, AD-AO 74113, 1979.
3. J F Watts, *J. Mater. Sci.*, **19**, 3459 (1984).

4. W Machu in *Interface Conversion for Polymer Coatings*, eds P Weiss, G D Cheever, American Elsevier, New York, 1968, pp. 128–45.
5. ASTM D 2651 in *Annual Book of ASTM Standards*, American Society for Testing and Materials, Philadelphia, published annually.

Pretreatment of titanium

D M BREWIS

Bonded titanium alloys are used in **Aerospace applications**, chemical and general engineering. Where hot humid conditions are experienced, it is essential that the correct pretreatment is used or serious loss of joint strength may occur (see **Durability**). A detailed account of pretreatments for titanium is given by Mahoon.[1] In recent years, various new pretreatments have been developed. For example, an alkaline peroxide etch gives very durable adhesive bonds.[1,2] However, this treatment is unsuitable for continuous large-scale operation due to the instability of the peroxide at the temperature used, namely 65 °C, and the emission of noxious fumes.[2] Excellent durability (Fig. 111) has also been achieved with a sodium hydroxide anodizing treatment[3] (10 V in 5 M NaOH at

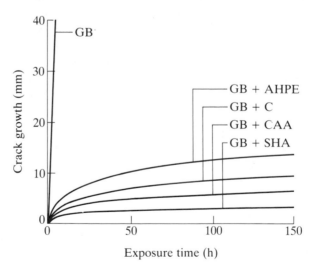

Fig. 111.   Wedge test crack growth data for Ti-6% Al-4% V alloy joints exposed to 50 °C/96% RH. (GB) grit blast; (AHPE) 65 °C alkaline hydrogen peroxide etch; (C) catalytic AHPE; (CAA) chromic acid, fluoride anodize; (SHA) sodium hydroxide anodize

room temperature) and a chromic acid-fluoride **Anodizing** treatment.[4] It has been argued[2] that the former method is superior on environmental grounds.

The reasons for the variation of durability with surface pretreatments are not fully understood. However, the high stability of the oxide layer is likely to be important (see **Pretreatments of aluminium**). Work by the Martin Marietta Company[5] has shown a good correlation between durability and the microroughness produced by a pretreatment[5] (see **Microfibrous surfaces**).

Further information, including sources of pretreatment recipes, can be found in the general article on **Pretreatment of metals**.

References

1. A Mahoon in *Durability of Structural Adhesives*, ed. A J Kinloch, Applied Science Publishers, London, UK, 1983, Ch. 6.
2. P Poole in *Industrial Adhesion Problems*, eds D M Brewis, D Briggs, Orbital Press, Oxford, 1985, Ch. 9.
3. A C Kennedy, R Kohler, P Poole, *Int. J. Adhesion and Adhesives*, **3** (2), 133 (1983).
4. Y Moji, J A Marceau, US Patent 3 959 091 (1976).
5. B M Ditchek, K R Breen, T S Sun, J D Venables, *National SAMPE Symposium*, **25**, 13 (1980).

Primary bonding at the interface

J F WATTS

In considering the adhesion of an organic material to an inorganic substrate (or vice versa) the vast majority of systems rely on the formation of secondary bonds between the materials. Indeed until a decade or so ago it was universally accepted that if the **Adsorption theory** was relevant to a particular system then the formation of an interfacial bond would be the result of **Dispersion forces** and **Polar forces**. In recent years it has become clear that interfacial reactions may involve the formation of very specific bonds with their own bond angles and bond distances. One very important member of this class are **Acid–base interactions** (e.g. hydrogen bonds, see **Hydrogen bonding**) the others are primary bonds. By primary bonds we mean those which may be exclusively responsible for the bonding in a solid and which may lead to compound formation at an interface. The three types of primary forces that occur are ionic bonding (with bond energies in the range $590-1050\ kJ\ mol^{-1}$), covalent bonding

$(63-710 \text{ kJ mol}^{-1})$, and metallic bonding $(113-347 \text{ kJ mol}^{-1})$. Clearly in any investigation of the polymer–metal interface we shall only be concerned with the first two categories.

There are relatively few well-documented examples of interfacial primary bonding in the literature, but it is possible to find examples in the areas of organic coatings on steel, metallized plastics and adhesion promoters.

Organic coatings

Probably the most widely investigated organic coatings system is polybutadiene,[1] and there is now unequivocable experimental evidence from several analytical methods that a chemical reaction occurs between the polymer and the oxidized metal substrate. An investigation by **X-ray photoelectron spectroscopy**[2] showed that at the interface the iron(III) oxide is reduced to iron(II) by the polymer; this is not surprising bearing in mind that the polybutadiene cures by an oxidative mechanism and thus acts as a reducing agent for the iron oxide. This reaction leads to the formation of an iron carboxylate compound at the interface which in turn is responsible for the development of a discrete interphase as illustrated in Fig. 112. The formation of such an interphase has important ramifications in terms of the mechanical properties of the polymer–metal couple. Instead of an abrupt change in stress distribution, as one sees in the case of a two-dimensional interface, there is now a gradual change across the width of the interfacial zone. In this way the load-bearing properties of an interface can be improved by the development of an interphase.

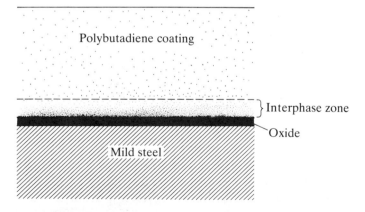

Fig. 112. Schematic representation of the interaction between polybutadiene and mild steel. The interphase zone contains iron carboxylate

Metallized plastics

Plastics can be metallized by a variety of techniques including metallization from solution and vacuum deposition. In the case of vacuum metallization of polyimides with titanium it is known that a two-stage process is involved.[3] In the early stages of deposition (i.e. low coverage) there is a strong interaction between the metal and the polymer by way of electron transfer from the titanium to the carbonyl oxygen of the polymer backbone. As the coverage of metal increases so Ti–C bond formation becomes involved in the reaction, a feature readily identified by **X-ray photoelectron spectroscopy** and **Auger electron spectroscopy**. At sub-monolayer coverage the titanium is already being deposited in the metallic state indicating that the metal–carbide complex is present as a discontinuous interfacial layer. A similar model is thought to apply for aluminium metallization.

Adhesion promoters

Adhesion promoters are large organic molecules designed to improve adhesion between organic and inorganic materials. There is evidence to suggest that specific groups present in the molecule interact with one or other of the two phases to be joined. Silanes are widely used for both steel and glass substrates. XPS studies show that the silicon-containing head adsorbs on to the oxide surface by a chemisorption process and **Secondary ion mass spectrometry** has identified the formation of an Fe–O–Si bond on steel substrates.[4] On glass fibres primary bond formation has been identified by infra-red spectroscopy.[5] See also discussion in **Primers for adhesive bonding**.

Thus when investigating the nature and mechanism of adhesion between an adhesive, coating or polymer matrix and the substrate it is important to consider the possibility of primary bond formation in addition to the interactions that may occur as a result of **Dispersion** and **Polar forces**. In addition to the **Adsorption theory** adhesion interactions can sometimes be described by the **Diffusion, Electrostatic** or **Mechanical theories of adhesion**.

References

1. H Leidheiser, *J. Adhesion Sci. Technol.*, **1**, 79–98 (1987).
2. J F Watts, J E Castle, *J. Mater. Sci.*, **18**, 2987–3003 (1983).
3. F S Ohuchi, S C Freilich, *J. Vac. Sci. Technol.*, **A4**, 1039–45 (1986).
4. M Gettings, A J Kinloch, *J. Mater. Sci.*, **12**, 2511–18 (1977).
5. S S Shyu, S M Chen, *J Chinese Inst. Chem. Eng.*, **16**, 311 (1986).

Primers for adhesive bonding

D M BREWIS

Introduction

Primers usually consist of polymers dissolved in organic solvents. They are applied to the substrate by spraying, dipping or brushing. After the solvent has evaporated a thin coating should be firmly adhered to the surface. Primers may be used in addition to, or as an alternative to, a **Pretreatment**.

Primers for metals

Mechanical and chemical pretreatments of metals result in the formation of surfaces with high surface energy. These surfaces will rapidly become contaminated by the adsorption of organic molecules and water from the atmosphere (see **Engineering surfaces of metals**). Such contamination may seriously affect joint performance, especially in relation to **Durability** under hot moist conditions. It is therefore standard practice in **Aerospace applications** to prime aluminium within a few hours of a pretreatment. Bonding may then be carried out weeks later without loss of performance.

The incorporation of corrosion inhibitors such as strontium chromate into primers can greatly increase the **Durability** of aluminium joints. For example, Scardino and Marceau[1] found that a corrosion-inhibiting primer considerably reduced the rate of crack growth in salt spray tests both with chromic acid etching (see **FPL etch**) and with **Anodizing**. Bethune[2] found that a corrosion inhibiting primer greatly increased the time to failure of stressed joints exposed at 78 °C and 100% RH.

Adhesives used in the aerospace industry often have high viscosities at the temperature of application. An advantage of primers is that they can be formulated to have low viscosities and will therefore achieve better contact with the substrate (see **Wetting kinetics**). The results in Fig. 113 demonstrate that primers, even without a corrosion inhibitor, can provide considerably improved durability.[3] Aluminium single-lap joints were bonded with a nitrile–phenolic adhesive and exposed to 100% RH and 50 °C for up to 10 000 h. Joints which had been primed with a nitrile–phenolic system showed much better resistance to the hot humid conditions.

Primers often consist essentially of the adhesive to be used diluted with a suitable solvent. However, chemically dissimilar materials may be used and, of these, the silanes are of particular importance. Silanes which were originally developed as **Coupling agents** to increase the interaction between glass fibres, and thermosetting resins have been found to improve the

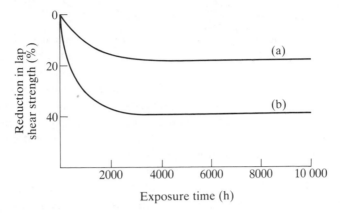

Fig. 113. Strengths of aluminium single lap joints exposed to 100% RH and 50 °C, (a) with primer and (b) without primer[3]

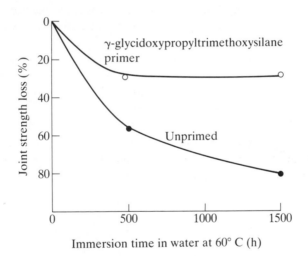

Fig. 114. Effect of γ-glycidoxypropyltrimethoxy silane primer on the durability of grit-blasted mild steel joints bonded with an epoxide[4]

durability of aluminium and steel joints. For example, Gettings and Kinloch[4] showed that using γ-glycidoxypropyltrimethoxysilane considerably improved the durability of grit-blasted mild steel joints (Fig. 114). On the basis of static **Secondary ion mass spectrometry** (SIMS) results, the improved durability was attributed to the formation of **Primary bonding at the interface** between the silane and the iron oxide.

Primers for polymers

Polymers are sometimes primed prior to bonding, but much less frequently than metals. As noted above, there are three clear advantages in using primers with metals, namely: (1) serious contamination can be avoided; (2) corrosion can be reduced and (3) better wetting can be achieved. However, polymers have relatively low surface energies and contamination from the atmosphere is not normally a problem. Clearly the bonding of polymers will not benefit from the presence of corrosion inhibitors. Except in the case of thermosets pretreated by mechanical methods, polymer surfaces to be bonded will normally be smoother than those of metals, as these will usually have been mechanically or chemically pretreated. Hence the benefits of using a primer with a low viscosity are likely to be small in the case of polymers.

Despite not benefiting from the same factors as metals, primers can still provide great improvements in the bonding of polymers. For example, primers consisting of chlorinated polyolefins in a hydrocarbon solvent can improve joint strengths considerably between untreated polypropylene or high-density polyethylene and an **Epoxide adhesive**.[5] Likewise the use of isocyanate primers results in large increases in joint strengths of styrene–butadiene elastomers bonded with **Polyurethane adhesives**.[5] The reasons for the improvements are unclear, but improved wetting due to the low surface energies of primer solutions and the presence of a thin, tough layer between adhesives and substrate, may be important.

Further related information may be found under **Accelerated ageing, Pretreatment of aluminium, Pretreatment of metals, Pretreatment of steel, Pretreatment of polymers, Roughness of surfaces, Stress in joints, Surface analysis** and **Wetting and spreading**.

References

1. W Scardino, J A Marceau, *J. Appl. Polym. Sci., Appl. Polym. Symp.*, **32**, 511 (1977).
2. A W Bethune, *SAMPE J.*, **11**, 4 (1975).
3. J Comyn, D M Brewis, S T Tredwell, *J. Adhesion*, **21** (1), 59 (1987).
4. M Gettings, A J Kinloch, *J. Mater. Sci.*, **12**, 2511 (1977).
5. D M Brewis, *Progr. in Rubbers and Plast. Technol.*, **1** (4), 1 (1985).

Primers for sealants

G B LOWE

There is substantial disagreement on the use of primers for **Sealants**. While there are a number of systems that are truly primerless, such as insulating glass sealants and silicone sealants for glass bonding (see **Sealants in double glazing**), there are a number of applications that require primers to ensure adequate surface preparation. It is usually considered advisable to prime the surface in general construction applications as the site quality may not be totally reliable. For the most part construction surfaces are divided into porous and non-porous. The former covers concrete, brickwork, blockwork, etc. while the latter covers glass, metals, some plastics, etc.

Apart from determining surface type, the service conditions have to be considered when selecting primers. For example, different primers would be required for long-term water immersion compared to above-ground systems. While there are a wide variety of primers they generally fall into clearly defined groups.

Primers for non-porous surfaces

Glass primers are generally dilute solutions of silane or silicone resins. Concentrations vary between 0.5 and 5% by weight depending on type and use. Silanes used are of the reactive side group type such as γ-glycidoxy-propyl-trimethoxy-silane and amino-propyl-triethoxy-silane. The alkoxy group is generally short to allow rapid hydrolysis, the active group selected will often depend on the polymer base of the sealant (see **Coupling agents**).

Metal primers are generally similar to those used for glass, but at the higher concentration level. As a consequence many manufacturers have primers at the 1–2% concentration suitable for both glass and metal. Suitable silanes are described by Comyn[1] in the context of their evaluation as coupling agents.

Primers for porous surfaces

Primers for porous surfaces are generally film-forming resin solutions. They perform a dual function of preparing the surface to receive the sealant and to bind the friable surface to prevent stress damage. This is because most concrete and stone surfaces have a weak layer which can delaminate under load.

Normal porous surface primers are based on solutions of chlorinated rubbers or isocyanate prepolymers. Concentrations are generally 20–30% by weight. A number of porous primers will use a mixture of both chlorinated rubber and urethane prepolymer, and where higher exposure is anticipated these may well incorporate cross-linking catalysts. The porous nature of the surfaces works against the use of dilute silane solutions as the primer is lost due to penetration into the matrix.

In order to prepare the surface it is usual to wire brush or mechanically abrade the surface. Where long-term water immersion is envisaged cured primers are preferred. These are normally modified epoxy systems with aliphatic amine curing agents.

Performance of primers has been studied by Aubrey and Beech[2] and indirectly by a long-term water immersion study.[3] Further information may be found in articles on **Selection of joint sealants, Sealant joint design** and **Paint primers** and **Primers for adhesive bonding**.

References

1. J Comyn, Silane coupling agents, in *Structural Adhesives*, ed. A J Kinloch, Elsevier, London, 1986, Ch. 8.
2. J Beech, D Aubrey, The influence of primers on the performance of building sealant joints, *Proc. Adhesives, Sealants and Encapsulants Conference*, London, Nov. 1986.
3. Construction Industries Research and Information Association, Report on Sealants for Potable Water, Construction Industries Research and Information Association, London, 1989.

Printing ink adhesion

G G BATTERSBY

The general principles governing adhesion (see **Theories of adhesion, Pretreatment of polymers**) apply to printing inks, but there are particular problems associated with printing water-based inks on impervious substrates[1] and with intercoat adhesion.[2]

Assessment of adhesion

Adhesion is frequently assessed by the tape test where pressure-sensitive tape (e.g. Sellotape, Scotch tape) is applied to the ink surface of the dried print and then stripped off.[3] The quality of adhesion is usually judged from visual inspection of the resulting surfaces.

Operators of the test know that the results are affected by the way in which the test is performed, particularly by the type of tape used and the pressure applied, speed and angle of peeling and length of time the tape is in contact with the ink surface of the print before removal.

The results for a specific ink are also affected by such process variables as the film weight of ink (thin films adhere more strongly than thick films), the age of the print (adhesion usually reaches a performance plateau 24 h after printing), the drying temperature and degree of drying achieved (prints show better ink adhesion after drying at higher temperature), and treatment level of the substrate (inks have better adhesion to freshly treated polymeric substrates with treatment level of $38-42 \text{ mJ m}^{-2}$, see **Critical surface tension**). See also **Surface characterization by contact angles – polymers**.

Water-based inks

The problem with printing water-based inks on an impervious substrate is not one of adhesion but of wetting. There are numerous articles on the influence of **Contact angle** on wettability and adhesion.[4] Some general conclusions are drawn here regarding surface tension studies on water-based inks. It is difficult to relate the static surface tension of an ink to its printability or to correlate the critical surface tension of a dried ink film with its adhesion. The treatment level of the substrate is more critical with water-based inks than solvent-based inks. Poor adhesion is likely on substrates with a critical surface tension less than 38 nM m^{-1} ($1 \text{ mN m}^{-1} = 1 \text{ mJ m}^{-2}$), but substrates with a high surface tension, $>44 \text{ mN m}^{-1}$, give irreproducible results, due perhaps to boundary-layer effects.

Fatty amide slip agents (see **Release**) on the substrate surface significantly affect the printability of water-based inks. Perhaps water-based inks containing isopropanol offer an advantage in their ability to penetrate or dissolve this fatty amide layer. Treating the film in line may also help to 'burn off' this surface and improve wetting.

Screen inks

Multilayer printing of screen inks presents a particular problem where adhesion to the substrate is already difficult (e.g. on acrylic sheet, untreated polyester, etc.). Usually the inks are applied at relatively high film weight (e.g. $15-30 \text{ gm}^{-2}$ compared with $2-6 \text{ gm}^{-2}$ for flexography) and for more demanding conditions of use than with print for flexible packaging. Inks cured by UV light differ chemically from the typical solvent-based inks used in screen printing.[5] Adhesion is influenced by shrinkage of the ink layers, which change in volume during the UV curing/drying processes.

References

1. D M Brewis, D Briggs (eds), *Industrial Adhesion Problems*, Orbital Press, Oxford, 1985, Ch. 8.
2. R H Leach (ed.), *Printing Ink Manual*, 4th edn, Van Nostrand Reinhold, New York, 1988.

3. G V Calder in *Adhesion Aspects of Polymeric Coatings*, ed. K L Mittal, Plenum Press, New York, 1983, p. 569.
4. W Hansen in *Advances in Printing Science and Technology*, Vol. 14, ed. W H Banks, Maxwell House, Elmswood, New York, 1977, p. 288.
5. R Holman, *UV and EB Curing Formulations for Printing Ink Coatings and Paints*, Selective Industrial Training Association – Technology, London, 1984.

R

Radiation cured adhesives

M R HADDON and T J SMITH

The term 'radiation curable' is applied to adhesives which use UV light or electron beams for curing. These forms of radiation have the correct energy to initiate polymerization of low-molecular-weight, unsaturated resins. The type of cross-linking achieved with electron beams and UV radiation is very similar, but the way in which the curing is brought about differs. The polymerization reaction is not directly initiated by UV light and a photo-initiator is required to interact with the UV radiation and produce the initiating species. Electron beams are associated with higher energy, and the electron beam itself has sufficient energy to initiate polymerization.

Electron beams are more penetrating than UV radiation and therefore can be used to bond some opaque substrates which are not suitable for UV-curable adhesives. The absence of photoactive fragments in electron beam curable adhesives results in greater stability of both the cured and the uncured adhesive. However, these advantages are offset by the high capital costs of electron beam curing equipment, especially when compared to UV lamps which are both cheap and readily available. This has restricted the use of electron beam curing to large installations. Ultraviolet curing is used in a far wider range of applications and therefore is considered in more detail.

Many potential UV curable systems exist, but only two have found significant commercial use in adhesives, namely free radical polymerization of (meth)acrylates (see **Acrylic adhesives**) and cationic polymerization of epoxies (see **Epoxide adhesives**). The chemistry of these systems is outlined in the following paragraphs. For further information the reader is referred to the standard texts.[1-3]

The essential ingredients of a free radical formulation are an acrylate terminated prepolymer and a photo-initiator. The photo-initiator absorbs the UV radiation and is promoted to an excited state. Subsequently the excited photo-initiator falls apart to form radicals which initiate the polymerization.

A wide range of prepolymers can be acrylated including epoxies, urethanes, polyesters, polyethers and rubbers. Those most commonly used in adhesive formulations are epoxy and urethane acrylates. Epoxy acrylates have properties similar to those of the parent epoxy resin with excellent adhesion, chemical resistance and toughness. Urethane acrylates are noted for their flexibility and tear resistance. They also offer high reactivity and good adhesion.

Acrylated prepolymers usually have high viscosities and require a diluent. This can be an acrylate monomer which itself takes part in the curing reaction. The earliest monomers were highly irritating and sensitizing, but more recently low-toxicity monomers have been developed without loss of performance. The use of such reactive diluents results in changes to the viscosity, flexibility and cure speed of a formulation.

Additives are used to improve the performance of an adhesive and these might include adhesion promoters, fillers, light stabilizers, antioxidants and plasticizers. Oxygen scavengers may be required as oxygen inhibits the curing of acrylates. The inhibition occurs in two ways – quenching of the excited photo-initiator and scavenging of free radicals. Scavenging produces relatively stable peroxides which both slow down the overall rate of cure and have a detrimental effect on the properties of the adhesive. Other methods by which oxygen inhibition can be overcome are nitrogen blanketing, the use of high-intensity lamps and by varying the initiator type and concentration.

A typical cationic UV adhesive formulation contains an epoxy resin, a cure-accelerating resin, a diluent (which may or may not be reactive) and a photo-initiator. The initiation step results in the formation of a positively charged centre through which an addition polymerization occurs. There is no inherent termination which may allow a significant post cure.

The advantages of a cationic formulation over that of a free radical are the lack of oxygen sensitivity, less shrinkage on curing and improved adhesion. The disadvantages are that the photo-initiators are sensitive to moisture and basic materials and that the acidic species can promote corrosion. The result is that the vast majority of current UV formulations are acrylate based and cured by a free radical mechanism.

Adhesives curable by UV account for only a small volume of the total structural adhesives currently sold. However, growth rates are estimated at 15–20% per annum. These high rates reflect the growing awareness of the advantages which UV curing can offer over conventional methods of cure. These include the significant productivity benefits which arise from the low running costs of UV lamps and the very rapid cure of the adhesives.

Adhesives curable by UV contain no solvents which are damaging to the environment. They are suitable for heat-sensitive substrates and are single component.

The disadvantages are that one transparent substrate is normally required, they suffer from oxygen inhibition and only a limited depth of cure can be achieved. The disadvantages have been tackled by the development of dual-cure adhesives. In these systems two independent curing mechanisms are incorporated into a single formulation.

The adhesive is cured to a first chemically stable state by UV irradiation and subsequently advanced to full cure by a second means, for example, thermal cure.

Ultraviolet-curable adhesives are currently being used in a large number of industrial applications. Some of the most significant markets include plastics and glass bonding, as well as applications in electronics. These applications demand high-performance adhesives which can bond to difficult substrates. They must have excellent mechanical strengths in both shear and peel, with good clarity and non-yellowing properties. These properties can all be obtained with current technology, and although certain performance limitations do exist radiation-curable adhesives can offer unmatched benefits in appropriate applications.

References

1. S P Pappas, *UV Curing, Science and Technology*, Vol. 1, Technology Marketing Corporation, Norwalk, Conn., 1978.
2. S P Pappas, *UV Curing, Science and Technology*, Vol. 2, Technology Marketing Corporation, Norwalk, Conn., 1985.
3. R Holman, *UV and EB Curing Formulation for Printing Inks, Coatings and Paints*, Selective Industrial Training Association, London, 1984.

Reaction setting adhesives

D M BREWIS

Reaction setting or reactive adhesives are those which harden by means of chemical reactions. These chemical reactions usually involve either **Addition polymerization** or **Condensation polymerization**. Depending on the functionalities of the molecules involved, either thermoplastics or thermosets are formed, the latter being much more common. Some of the reactants present **Health and safety** hazards.

Reaction setting adhesives usually consist of two individually stable parts which react when mixed, in the cold or when heated, to form a polymer. The products are usually rigid, but some systems produce flexible products, as in the case of some polyurethanes (see **Polyurethane adhesives**). One part of a system is a monomer or a prepolymer and the other is termed a 'hardener' if the reaction proceeds stoichiometrically, or a catalyst or initiator if its function is to start a chain reaction. A common example of a stoichiometric system is the reaction between an epoxide resin and a polyfunctional amine. An example of a catalyst for acrylic systems is benzoyl peroxide plus a tertiary amine.

Two-part adhesives are based on a variety of systems, including: acrylics, amino resins, di-isocyanates (to form polyurethanes), epoxides (see **Epoxide adhesives**), phenolic resins (see **Phenolic adhesives**), silicones, anhydrides–aromatic amines (to form polyimides, see **Polyimide adhesives**).

One-part adhesives have the advantage that formulation is carried out by the adhesives manufacturer and no mixing by the user is required. However, because of problems of shelf life, only a limited range of one-part systems exists. These may harden by a variety of mechanisms, including the presence of moisture (**Cyanoacrylate adhesives**, RTV silicones and some **Polyurethane adhesives**), absence of air (**Anaerobic adhesives**), **Radiation cured adhesives** (UV or electron beam) or by the action of heat. Clearly the last type must be stored at a temperature low enough to provide a reasonable shelf life. The prepolymer or resin is mixed with the hardener and stored at 0 °C or lower until required. Polymerization is subsequently brought about by heating to temperatures usually in the range 120–200 °C.

The formulations of reactive adhesives can be complex. For example, an **Acrylic adhesive** might contain monomer, an elastomer, an adhesion promoter, a cross-linking agent, an initiator and a stabilizer. Details of formulations can be found elsewhere.[1]

Individual reaction setting adhesives do have disadvantages. For example, phenolic systems require high pressure during bonding. In general, reactive adhesives are more toxic than emulsion and hot melt adhesives and they are often much more expensive than other types of adhesive. However, reactive adhesives offer a number of advantages, which are essential in many applications.

Reaction setting adhesives are usually designed to produce cross-linked structures which provide good resistance to temperature and solvents when compared with thermoplastic adhesives. A special type of reactive adhesive, namely the rubber-toughened acrylics and epoxides[2] are able to combine high peel strengths with high shear strengths.

The range of applications is very diverse, some examples being given in Table 32.

Information on particular types of adhesive may be found in the

Table 32. A selection of applications of reaction setting adhesives

Application	Adhesive
Primary and secondary aircraft structures	Toughened epoxides Phenolic/poly(vinyl formal)
Assembly of steel car bodies	Single-part epoxides
Assembly of truck doors from aluminium panels	Toughened acrylics
Bonding fibre-reinforced plastics	Polyurethanes
Mounting microchips on a printed circuit board	UV curing acrylics
Assembly of computers	Cyanoacrylates
Household repairs (DIY)	Two-part epoxides

following entries: **Acrylic adhesives, Anaerobic adhesives, Cyanoacrylate adhesives, Epoxide adhesives, Hot melt adhesives, Phenolic adhesives, Polyurethane adhesives, Selection of adhesives, Silicones, Toughened adhesives**.

References

1. S R Hartshorn (ed.), *Structural Adhesives*, Plenum Press, New York, 1986.
2. D J Stamper in *Synthetic Adhesives and Sealants*, ed. W C Wake, John Wiley, Chichester, 1987, p. 59.

Release

D E PACKHAM

In most practical instances of adhesion the aim is to produce a high-strength bond. There are examples, such as the moulding of plastics and rubber, where low, but controlled adhesion, is required. Problems of mould release are of long standing, but have become more acute with the increase in automation of moulding processes. It is essential that the piece reliably remains in the same part of the mould when it opens at the end of the moulding cycle, so that it can be removed by a robotic arm.

In general, polymers have low values of **Surface energy** and metals have high ones. During moulding there will be a consequent tendency for the polymer to wet the mould and on solidifying or cross-linking to adhere (see **Adsorption theory**).

Low-mould adhesion is sometimes obtained by treating the mould surface so as to lower its surface energy or deliberately to provide a weak layer. Both silicone and fluorocarbon polymers are widely used for this purpose. An enormous number of mould release agents are described in the patent literature.

Low adhesion between polymer and mould alloy is sometimes a consequence of the formation of **Weak boundary layers** by the polymer on solidifying. Slip additives, such as stearates may be added to the polymer for this purpose. In other examples components 'naturally' within the polymer form such a layer (see **Compatibility, Surface nature of polymers**). During the moulding of nitrile–butadiene rubber a complex interfacial layer including emulsifiers and coagulant residues controls the release properties.[1]

Measurement of the low levels of adhesion involved under conditions resembling commercial moulding produces some unusual problems. Briscoe had adapted the **Blister test** to this purpose.[2] Turner has used the Turner, Moore and Smith (TMS) rheometer, a cone and plate development of the Mooney rheometer.[3]

References

1. M Lotfipour, D E Packham, D M Turner, *Surf. and Interface Anal.*, **17**, 516 (1991).
2. B J Briscoe, S S Panesar, *J. Phys. D*, **19**, 841 (1986).
3. R K Champaneria, B Harris, M. Lotfipour, D E Packham, D M Turner, *Plast. and Rubber Process. and Appl.*, **8**, 185 (1987).

Repair methods

K B ARMSTRONG

Introduction

The high cost of replacement parts made from adhesively bonded metal, metal/honeycomb, composite/honeycomb or monolithic composite laminates ensures that the interest in repairing them is very high indeed. In the case of aircraft, in particular, the cost of downtime can exceed the cost of a spare part, and if the required item is not available then repair becomes very urgent.

Temporary repairs may be resorted to, but they mean that the work has to be done again and the second repair is usually larger than the first. This is especially true if the temporary repair is a riveted or bolted one. If the facilities are available a permanent repair does not usually take much longer than a temporary repair, and a good permanent repair should always be made the first time if possible.

Repairs to a structure such as an aircraft are particularly critical. Procedures have to be rigorously evaluated, and specifications carefully followed. Repairs need to take account of aerodynamic considerations, mechanical clearance, lightning protection and drainage requirements, and in the case of control surfaces weight and balance is also important.

Some areas are more critically loaded than others: each aircraft has a structural repair manual (SRM) in which established repair procedures for different areas are set out, with the maximum size and location of permitted repairs indicated.[1,2]

Surface preparation

Good surface preparation is important in this context (see **Pretreatment of metals prior to bonding**), as in most examples of adhesive bonding, as it influences the durability of the bond, especially in wet areas (see **Durability – fundamentals**).

With aluminium alloys phosphoric or chromic acid anodizing are the best methods (see **Pretreatment of aluminium, Anodizing**). Portable anodizing equipment is available: one type uses a gauze saturated with gelled electrolytes.[3] As an alternative, blasting with alumina grit (see **Abrasion treatment**) followed by a silane coating has been found to be quite effective. Simpler methods such as use of abrasive papers give very poor durability.

Choice of adhesive

In general the same adhesive and bonding conditions of temperature and pressure should be employed for repair as were originally used. Elevated temperatures can be achieved using lamps or heater blankets, and pressure can be applied with vacuum bags. Using the vacuum-bag techniques, described below, the atmosphere can be exploited to apply pressure to the skin in the repair of a **Honeycomb structure**. However, it should be noted that it can be dangerous to use an adhesive curing at the original cure temperature, as the remaining adhesive may soften and the internal pressure blow the skin from the sound cells adjacent to the repair. A repair adhesive curing 25–50 °C lower should be selected.

Types of repair

Metal skins with bonded metal doublers If the doubler has disbonded, and no corrosion is present, repair may be effected by riveting. If skins are too thin for countersunk rivets then repair by adhesive bonding may be necessary.

Metal-skinned honeycomb panels The type of repair will vary according to the damage. A dent in the skin, if the honeycomb is metal and it is not disbonded, may be repaired by treating the surface of the dent and some distance around, filling the dent with epoxy potting compound and bonding on an external plate to restore the buckling stiffness of the skin.

Puncture of the skin may be treated in the same way, except that the damage should be cut out to a round or elliptical hole and new honeycomb of the correct type potted in before fitting a repair plate. Many aircraft manufacturers allow very thin aluminium alloy skins to be repaired with fibreglass overlays and they provide a table relating skin thickness to the number of fibreglass layers, or rather the thickness of fibreglass required.

A common problem with honeycomb panels is disbond between the skin and the honeycomb, especially at the edges. This often leads to corrosion. Usually both skin and honeycomb have to be replaced over the whole disbonded area. This may be done using two-part epoxy paste adhesives curing between room temperature and 80 °C or using a film adhesive curing at 120 °C or 180 °C. If high temperature (i.e. 120 °C and above) repairs of any size are required, then tooling may be necessary to maintain the shape of the part.

Employing a vacuum-bag technique, atmospheric pressure can be used to hold the skin to the honeycomb during cure. A 'sandwich' of the new skin, a film adhesive and polyester non-woven monofilament fabric (3M-AF 3306 is satisfactory) is placed on the honeycomb. The vacuum bag is sealed around the damaged area and evacuated, the non-woven fabric providing a conduit for the air pumped out.[2] Elevated pressure may also be required to prevent honeycomb/skin disbond adjacent to the repair area. Complete rebuilding can be done in this way.

Temporary repairs on lightly loaded parts are sometimes made by drilling a pattern of holes through the panel at about 50 or 75 mm pitch in all directions, fitting spacers and then bonding and bolting plates on both sides. This may meet an immediate need to keep an aircraft flying, in which case it is justified, but it does mean a more extensive bonded repair at some later date.

Composite parts Solid composite parts may be repaired by making carefully tapered scarf joints. Hot-bonded repairs are preferred provided that the surface is first adequately dried (Brown[4] and Armstrong[5]). Parts made from structural fibres i.e. carbon fibre, Kevlar or boron fibres and some glass-fibre parts require that each repair layer be of the same type as the original, or a direct equivalent, and laid up in the correct orientation.

Composite skinned honeycomb panels These may use either Nomex nylon paper honeycomb core, aluminium honeycomb or sometimes PVC (polyurethane or acrylic foam core). Dents are not acceptable in composite skins as they indicate fibre damage. Nomex core may split under a dent so any such damage needs to be cut out and repaired. Nomex honeycomb can absorb considerable amounts of moisture. Wet honeycomb should be thoroughly dried or replaced. Moisture meters are available for use with Fibreglass, Kevlar or Nomex, but these meters will not work with carbon fibre because it is electrically conductive. No instrument is known that can indicate when carbon-fibre composites are dry enough to repair.

Bolted repairs

Bolted metal plates are sometimes used for temporary repairs to composite-skinned honeycomb panels but a result is the need for more extensive permanent repairs.

Permanent bolted repairs Some aircraft such as the British Aerospace Harrier GR5 or its American counterpart the McDonnell-Douglas AV8B, use thick monolithic composite skin and stringer structures. In such cases lighter, smaller and quicker repairs can be made using specially designed titanium plates bolted in place. Composites and bonded metal structures are often associated with thin-skinned honeycomb parts. However, either metal or composite skins of 3 mm thickness or above can be repaired with lighter and simpler bolted joints. Adhesive bonding is preferred for materials usually significantly thinner than 3 mm.

Composite repairs to metal parts Alan Baker of the Aeronautical Research Laboratories, Melbourne, Victoria, Australia, has pioneered this type of work, which has been very successful. Repairs around a fuel drain hole in the lower wing of the Mirage fighter and repairs to stringers in the wing of the Hercules transport have been particularly successful.[6] Carbon-fibre patches on the wing leading edge panels of Concorde have doubled the life of some of these items.

References

1. K B Armstrong, The design of bonded structure repairs, *Int. J. Adhesion and Adhesives*, **3**, 37 (1983).
2. K B Armstrong, *The Repair of Adhesively Bonded Aircraft Structures Using Vacuum Pressure*, Society for the Advancement of Material and Process Engineering Series 24, Book 2, Azusa, Calif. 91702, 1979, pp. 1140–87.
3. R E Horton et al., Bonded structure repairs, in *Adhesives and Sealants*, ed. H F Brinson, Engineered Materials Handbook, Vol. 3, ASM, Ohio 44073, 1990, pp. 799–847.
4. H Brown (ed.), *Composite Repairs*, SAMPE Monograph No. 1, SAMPE, Covina, Calif. 91722, 1985.

5. K B Armstrong in *Bonding and Repair of Composites*, Anon., Butterworths, London, 1989, pp. 93–9.
6. A A Baker, R Jones (eds), *Bonded Repair of Aircraft Structures*, Martinus Nijhoff Publishers, Dordrecht, 1988.

Rheological theory

D E PACKHAM

The **Theories of adhesion** discussed in the articles **Adsorption theory, Diffusion theory, Electrostatic theory** and **Mechanical theory** aim to describe the forces which cause the adhesive and substrate to adhere; the rheological theory is concerned with explanation of the values obtained for adhesion measured by destructive **Tests of adhesion**.

In its extreme form, as put forward by J J Bikerman,[1] the theory maintains that the measured adhesion simply reflects the rheology of the joint, especially of the adhesive, and that it can give no information about the forces acting at the interface.

Bikerman maintained that the mode of failure of an adhesive bond was always cohesive, and that adhesive failure (interfacial failure, see **Locus of failure**) could never occur. When adhesive failure appeared to occur, more careful examination would reveal a thin layer of adhesive remaining on the substrate. When failure occurred close to the interface it was often in a **Weak boundary layer**.

Bikerman's theoretical arguments need not concern us in detail: they are set out in his book.[1] Briefly, he argued as follows:

1. That forces across an interface were the geometric mean of those in the two phases joined (cf. Eqn 3 in **Dispersion forces**, Eqn 4 in **Polar forces** and Eqn 7 in **Surface energy components**) and therefore were necessarily stronger than the cohesive forces in the weaker phase where failure would occur;
2. That even if a crack happened to initiate at the interface the probability of its propagating along the interface for more than a few atomic spacings was infinitesimally small.

Bikerman's theoretical arguments were often the subject of heated debate. A particularly telling critique by Good,[2] produced some interesting discussions.[3]

For a particularly telling response see Good.[2] Few would hold them in their absolute form today. Despite this, Bikerman's influence on the development of the understanding of the phenomenon of adhesion was on the whole benign. Much more careful examination of failure mode became more common and with it a better understanding of reasons for failure of adhesive bonds. Similarly, there is now broad agreement that the measured adhesion reflects the rheology of the joint, especially of the adhesive. However most would argue that it also reflects the interfacial forces, although in a subtle and often obscure way. The most important advances in the subject in the last 20 years have been associated with a better understanding of the relationship between measured adhesion, interfacial forces and joint rheology.

A contemporary form of the rheological theory which would command wide support would be as follows. 'The measured adhesion is strongly influenced by the rheology of the joint, but also depends upon the interfacial forces. The interaction between these two factors may be obscure, but in some types of test it is already possible to isolate and separately gauge the influence of each factor.'

Some of these points are developed in the article **Adhesion – fundamental and practice**.

References

1. J J Bikerman, *The Science of Adhesive Joints*, 1st edn, Academic Press, New York, 1961.
2. R J Good in *Adhesion Measurement of Thin Films, Thick Films and Bulk Coatings*, ASTM STP No. 640, ed. K L Mittal, American Society for Testing and Materials, Philadelphia, 1978, p. 18.
3. R J Good, *ibidem*, pp. 27–9, 38–40.

Roughness of surfaces

D E PACKHAM

It is common to talk loosely of surfaces as being 'rough' or 'smooth'. What is being referred to is better described by the broader term 'surface texture'. Surface texture is bound to be relevant to a phenomenon such as adhesion.

Apparently smooth surfaces will often show features of roughness when examined by **Optical microscopy** or **Scanning electron microscopy**. Adhesion depends on molecular contact between adhesive and substrate (see **Wetting and spreading**), so it is important to recognize that on the molecular scale all practical surfaces are rough: the only question is 'how rough?'

Fig. 115.   Roughness and waviness. Schematic representation of roughness
superimposed upon waviness

Roughness, waviness and form

Many surfaces of practical importance are complex: they may display
features of roughness on widely differing scales. It is common to distinguish
between the 'roughness', 'waviness' and 'form' of a surface.[1] 'Waviness'
is applied to undulations on which the finer roughness is imposed
(Fig. 115). 'Form' refers to the general shape of the surface on a still larger
scale. It might refer to some curvature, perhaps irregular. The borderlines
between features described by these three terms are necessarily arbitrary,
but it is desirable to keep in mind whether a particular method for
characterizing surface texture is sensitive to all, or only to some, of these
features.

Characterization of surface texture

This major topic[1-3] can only be briefly discussed here.

Qualitative methods The usefulness of optical and electron microscopy has
already been mentioned. By examination at various magnifications they
give an overall qualitative impression of the surface. For relatively smooth
surfaces optical interference methods may be used. Among other methods,
the surface profilometer is important as it is a popular method which can
readily provide routine results.

The surface profilometer consists of a fine stylus which is drawn across
the surface and is supposed to follow the contours. The vertical motion
of the stylus is converted to an electrical signal which may be processed
to present the results in several different ways.

A useful form of output of a profilometer is a surface profile graph.
Magnifications appropriate to the context may be chosen, but, in order
to display the relatively fine features of roughness, the vertical
magnification is always many times greater than the horizontal. The profile
graph then is a distorted representation of the shape of the irregularities:
gentle undulations are made to look like sharp Alpine peaks.

A single graph gives an indication of roughness along a single line in
the surface. Instruments are available which make a series of traverses

over an area, and produce a map showing the roughness as contours or as an isometric view.

Quantitative parameters A very large number of parameters giving a quantitative representation of surface roughness have been defined.[1,3] Depending on the complexity of the instrument, the profilometer output may be processed to give the value of a number of these. Some parameters characterize the vertical features of roughness, others the horizontal. The only one to be discussed here is the commonly used 'rough average' R_a. The roughness average (previously called the 'centre-line average') represents the average vertical displacement from the mean line of the surface. It is defined as

$$R_a = \frac{1}{L} \int_0^L |z| \, dx \qquad [1]$$

where L is the sampling length, x the horizontal coordinate and z the vertical coordinate.

Roughness average may be the only indication of surface roughness given: it is necessary to be aware of its limitations. Surfaces may have the same roughness average, but an entirely different distribution of peaks and valleys. Here R_a characterizes only the vertical aspects of roughness, so surfaces with the same R_a value may differ widely in the horizontal distribution of peaks. For these reasons surfaces with the same R_a may differ widely in surface texture.

The extent to which R_a is sensitive to the waviness of the surface depends on the relation between the waviness and the sampling length. Most profilometers enable a range of 'cut-off wavelengths' to be selected, essentially altering the sample length. Thus the same surface may give different values of R_a if they are measured at different cut-off wavelengths.

Roughness and adhesion

The relationship between roughness and adhesion is a complex and somewhat problematic one.[4] The presence of roughness will alter the apparent contact angle (see the Wenzel equation in **Contact angle**). Depending on the nature of the roughness and the surface tension and viscosity of the adhesive, a rough surface may not be completely wetted by an adhesive that would wet a corresponding smooth one. This incomplete wetting may be the result of either thermodynamic or of kinetic factors. Thermodynamic considerations show that it may not be possible for an adhesive to penetrate far into a closed pore, especially if it has an 'ink bottle' (re-entrant) shape.[4] Further, the adhesive may set before it has time to reach an equilibrium configuration (see **Wetting kinetics**). Either way this incomplete wetting is likely to lead to poorer adhesion.

Many **Pretreatments of metals** increase the roughness of the surface. This is clearly a result of **Abrasion treatment** (see figures in that article), but also occurs during many types of chemical etching. The enhanced adhesion after such pretreatments is unlikely in general to be a consequence of the rougher surface. Some examples where the roughness as such does appear to have a beneficial effect on adhesion are discussed under **Mechanical theory, Friction–adhesion aspects** and **Rubber adhesion**; see also **Microfibrous surfaces** and **Anodizing**.

References

1. H Dagnall, *Exploring Surface Texture*, Rank Taylor Hobson, Leicester, 1980.
2. British Standards Institution, *Method for the Assessment of Surface Texture*, BS 1134, Part 1, *Methods and Instrumentation* (1988), Part 2, *Guidance and General Information* (1990).
3. T R Thomas (ed.), *Rough Surfaces*, Longman, Harlow, 1982.
4. D E Packham in *Adhesion Aspects of Polymeric Coatings*, ed. K L Mittal, Plenum Press, New York, 1983, p. 19.

Rubber adhesion

A D ROBERTS

The latex from certain plants, in particular *Hevea brasiliensis*, can be coagulated with acid to yield a solid mass of rubber characterized by its elasticity, ability to flow and tackiness to the touch. Its industrial use dates from the beginning of the nineteenth century (varnish, waterproofing). A great drawback was its softening when hot, its brittleness when cold and its unwanted tackiness in contact with other materials. A 'cure' was found by Goodyear in 1839 when he discovered that heating raw rubber with sulphur mixed into it yielded a product stable to temperature extremes and almost non-tacky. This is 'vulcanized' rubber. In any discussion of rubber adhesion a distinction must be drawn between 'uncured' and vulcanized rubber.

Uncured rubber

When this is pressed against a hard substrate, it can easily flow into intimate contact resulting in firm adhesion. Separation is made difficult because it flows. If the substrate is roughened then even more work must be expended to pull rubber plugs out of crevices.[1] Thus the peel energy (work done per unit area to separate surfaces – see article on **Peel tests**) is considerably greater than the thermodynamic surface energy due to work unavoidably dissipated in the bulk. Raw natural rubber gives as much as a tenfold increase in peel adhesion on roughened substrate surfaces compared with a smooth substrate (see **Roughness of surfaces**).

The way in which roughening improves the bond given by soft elastomer adhesives (on tapes) can be understood from the flow mechanisms. In factory-processing operations such as rubber mixing, strong adhesion between the machinery surfaces and the raw rubber is required so that a high shear deformation can be imposed upon the rubber, and here the influence of surface roughness can be profound.

Tack

Raw rubber bonds to itself rather well, a property much appreciated in tyre construction. Considerable efforts have been made to explain this in terms of interdiffusion of rubber molecules across the interface.[2] **Tack** is a key material property in a number of rubber product manufacturing operations.

If a smooth-surface rubber track is pressed on to a steel ball and then raised, the ball so adhered will generally detach itself in time under the action of gravity. This phenomenon is recognized as 'tackiness', and tackmeters have been devised to give a measure of it. An analysis has been made[1] which relates this type of phenomenon to the adhesion behaviour described above. The relation of interface tack to **Surface energy** and **Viscoelasticity** is of practical significance.

Rubber to metal bonding

For this operation it is generally advisable to roughen the metal surface in order to achieve good bonding. Certain proprietary bonding agents may be applied to the adherends before hot bonding is attempted in order to promote 'chemical' bonds. An example where chemical bonds between metal and raw rubber are considered important is in tyre cord adhesion. Complex chemistry between the brass plating on the steel cords and the unvulcanized rubber mixture leads to very strong bonding[3] (see the article **Rubber to metal bonding**).

Vulcanized rubber

The adhesion of fully cured rubber to various substrates has been the subject of considerable research in recent years. When optically smooth rubber touches a glass plate, its high elasticity enables short-range surface forces to draw the bodies into intimate contact over a considerable area.

A force must be exerted to pull the surfaces apart. The adhesion depends principally upon the factors of interfacial surface energy and rubber hysteresis. The reversible work of adhesion Γ_0 (see **Contact angles and interfacial tension**) which has been related by Dupré to the individual free surface energy of the contact bodies ($\Gamma_0 = \gamma_1 + \gamma_2 - \gamma_{12}$) arises from several kinds of interactions that may be physical and/or chemical in nature. Physical thinking tends to be in terms of van der Waals attractive forces (see **Dispersion forces**). Some experiments suggest the adhesion of polymeric particles and films owes much to electrostatic attraction (see **Electrostatic theory**), but this does not appear to be significant in the case of the self-generated charge associated with rubber contacts. Some results imply the importance of **Acid–base interactions**.

For a sphere-on-flat geometry, the static contact adhesion energy, Γ, can be calculated[4] from measurements of the contact area according to the general expression

$$\Gamma = \frac{1}{6\pi K a^3}\left(\frac{Ka^3}{R} - W\right)^2$$

where a is the observed radius of the circle of contact between touching surfaces of reduced elastic modulus

$$K = 4/(3\pi[k_1 + k_2])$$

where $k = (1 - v^2)/\pi E$. Each surface has a Young's modulus E and a Poisson's ratio v. The applied load is W and the reduced radius of curvature is

$$R = R_1 R_2/(R_1 + R_2)$$

R_1 and R_2 being the radii of curvature of the two bodies in contact. One may note that the expression for the adhesion energy, denoted Γ_0, corresponds to what is termed in fracture mechanics as the strain energy release rate.

The validity of this surface energy expression has been tested. Generally, the best measure of Γ at 'equilibrium' (no hysteresis or rate dependence) is by 'touch-on' of two rubber surfaces at or near, zero applied load. Experiments yield values of about $\gamma = 30$ mJ m^{-2} for each surface, which is reasonable for rubber. If a negative force is applied, then surfaces will just separate, assuming no viscoelastic effects, when that force is $-3\pi R\Gamma/2$.

In rapid peeling the energies may be 100 times greater than the equilibrium value, because at the peel edge significant viscoelastic losses occur. The instantaneous peel energy Γ_p appears to be a function of both the equilibrium energy and an hysteresis factor H the magnitude of which varies with rubber type, temperature and peel rate (except for speeds at which inertial effects arise). It was suggested[5] that $\Gamma_p = \Gamma_0 f(H)$ though the range of applicability of this simple expression has not been fully established. Despite this, it serves to emphasize the dependence of the adhesion on these two terms. Here Γ_p is measured as a function of peel rate in **Peel tests**; its value is given by that of the instantaneous strain energy release rate.

If the influence of the network molecular structure is considered, then it needs to be remembered that the flexible rubber chains adhered to a substrate become stretched by the detaching force. This stores elastic energy. When the critical force for detachment of a particular network chain is reached, the chain will retract freely and the energy stored in it between the two cross-linking points is lost from the system. It has been proposed that

$$\Gamma_p = \Gamma_0 g f(V, T)$$

where the product $\Gamma_0 g$ represents the limiting value of Γ_p in the absence of viscoelastic losses. The parameter g is related to the length of the stretched chains[1] and $f(V, T)$ is the macroscopic dissipation factor that varies with peel rate V and temperature T.

References

1. A D Roberts (ed.), *Natural Rubber Science and Technology*, Oxford University Press, Oxford, 1988.
2. R P Wool, *Rubber Chem. Technol.*, **57**, 307 (1984).
3. W J van Ooij, *Rubber Chem. Technol.*, **52**, 605 (1979).
4. K L Johnson, K Kendall, A D Roberts, *Proc. Roy. Soc. Lond.*, **A325**, 301 (1971).
5. A N Gent, J Schultz, *J. Adhesion* **3**, 261 (1972).

Rubber-based adhesives

J PRITCHARD

Definition

Rubber-based adhesives are adhesives having one or more rubbery materials as the major constituent, thereby establishing the characteristic properties in terms of performance and application characteristics (see **Adhesive classification**). They can be used in a variety of forms including solvent based, emulsion or latex based, hot melt, permanently tacky or pressure-sensitive coatings or preformed masses and two-part curing systems. This article will outline the properties and uses of adhesives made from the more important rubbery materials in use today. Hazards are mentioned under **Health and safety**. See also **Solvent-based adhesives**.

Natural rubber

Compounding and manufacture Natural rubber can be used as an adhesive as it occurs naturally in latex. Antioxidants, filler and resins can be added as required usually by ball mills or stirrers, extra soaps or wetting agents and stabilizers are usually required.

If it is to be used in solvent, the natural rubber is first masticated or broken down on a rubber mill. This reduces the molecular weight of the polymer making subsequent dissolution in hydrocarbon solvent easier and the resulting solution smoother and less stringy. To improve adhesion a variety of resins can be added to natural rubber adhesives. These include natural rosin, rosin salts and esters, coumarone-indene and various phenolic resins.

Uses Latex adhesives are used extensively with porous substrates. In the footwear industry (see **Footwear applications of adhesives**) they are used for sticking socking material inside shoes. Carpets can be joined with latex adhesives and indeed some carpets are manufactured by binding the fibres together with a latex adhesive. Solvent-based adhesives and reclaim adhesives find wide use as general-purpose adhesives sticking well to natural rubber, wood and metal. They are used extensively for bonding proofed fabrics.

Polychloroprene

Compounding and manufacture Polychloroprene, like natural rubber, can be used either in latex or solution form. The polymer, being synthetic, can be produced having different properties. Hence the molecular weight, degree of crystallinity, branching and gel structure can be controlled making the compounding easier and the choice of properties greater compared to natural rubber. The solvents used are usually a combination of aliphatic aromatic and oxygenated.

Compounding the latex is by addition of preformed dispersions in water – usually made with a ball mill or high-speed stirrer. For solvent adhesives, various compounding ingredients can be added on the mill, but it is now possible to buy grades of polymer that will dissolve easily by simple stirring. The main compounding ingredient is a phenol-formaldehyde resin, present in up to 40–50 parts per hundred (pph) rubber. Magnesium oxide is also added up to 6–8 pph, but as it reacts with some resins, in these cases more is required. Antioxidants are necessary to protect long-term ageing characteristics.

Single-pack adhesives have good all-round properties, but the strength diminishes rapidly above 70–80 °C. To combat this, two-pack adhesives are possible – the second part being usually an isocyanate cross-linking agent. This effectively cross-links the rubber at room temperature in a matter of hours to give non-softening bonds.

Uses Polychloroprene adhesives are used extensively throughout industry as general-purpose adhesives sticking to many different substrates. As contact adhesives they have been mentioned in another article. Latex adhesives are used with fabrics and porous substrates. They have better oil resistance than natural rubber and can be used for adhering to lightly plasticized polyvinyl chloride.

Polychloroprene latex can now be compounded into satisfactory contact adhesives thereby removing the problems associated with the use of solvents. Their main drawback is the dependency of the drying time on ambient conditions.

Butyl rubber and polyisobutylene

Polyisobutylene is an amorphous rubber depending mainly on chain entanglement for its cohesive strength. Because of its structure the polymer chains are very flexible which leads to a permanently tacky condition; it is this property that gives rise to the main use for polyisobutylene which is a base for pressure-sensitive or permanently tacky adhesives. Butyl rubber is polyisobutylene with a few per cent of double bonds in the molecule which act as sites for cross-linking. These can be utilized to bring about improved creep resistance, especially at above ambient temperatures. Polyisobutylene has little adhesion in its own right due to lack of polarity within the molecule.

Manufacture and compounding Being a synthetic rubber, polyisobutylene can be produced in a variety of molecule weights which vary the properties of the material from low-molecular-weight viscous liquids to high-molecular-weight extremely tough rubbery solids. For low-viscosity solutions the rubber can be dissolved in hydrocarbon solvents with stirring provided there is a comparatively small amount of rubber. For high-viscosity solutions and doughs, heavy equipment of the Banbury type is necessary.

For permanently tacky adhesives, the polyisobutylene is usually present in a blend of high and low molecular weights, while tackifiers, resins and plasticizers can be added.

Uses Polyisobutylene is mainly used for **Pressure-sensitive adhesives** for labels and tapes.

Styrene–butadiene rubber

A random copolymer of styrene and butadiene, i.e. where the styrene and butadiene groups are spaced by accident along the polymer chain, gives rise to a comparatively cheap general-purpose rubber which can be compounded with the usual additives to produce adhesives.

Of much more interest is the development over the last 20 years of block copolymers in which styrene and butadiene are polymerized to form an ordered chain consisting of polystyrene at both ends and polybutadiene in the middle. This gives properties similar to a cross-linked polymer since the styrene forms a separate phase thus locking the ends of each polymer chain. In practice this leads to low-viscosity solutions giving high cohesive-strength adhesives.

Manufacture and compounding Styrene–butadiene is easily dissolved using conventional equipment. Random copolymers will dissolve mainly in aliphatic hydrocarbons while block polymers will dissolve in blends of aliphatic and aromatic hydrocarbons. Resins, fillers and antioxidants are normally added. In the case of block copolymers, resins are chosen to be compatible with either or both of the blocks, depending on the properties required.

Uses Random polymers are used for general-purpose adhesive applications. Block copolymers are used for sprayable adhesives and for **Pressure-sensitive adhesives** with enhanced heat resistance.

References

1. G S Whitby, *Synthetic Rubber*, John Wiley, New York, 1954.
2. J Shields, *Adhesives Handbook*, 3rd edn, Butterworths, London, 1984.
3. R J Coresa, *Block & Graft Copolymerization*, Vol I, John Wiley, New York, 1973.

Rubber fillers

J A LINDSAY

Fillers are particulate materials that are added to raw polymers to modify the polymer characteristics (see **Filler polymers**): they are widely used in polymer technology, and play a particularly significant role in modifying the properties of rubber.[1] In this article various effects of fillers are discussed with some emphasis on the formulation of rubber for **Rubber to metal bonding**.

Many of the properties of a rubber can be significantly altered by incorporation of a filler; these include: cure rate, tensile properties (strength, modulus, ductility), tear strength, hysteresis and resistance to permanent set, abrasion resistance, swelling behaviour in various fluids, density and, of course, cost.

Practical experience has shown that all fillers in general use can be used in rubber compounds for bonding, during the moulding process, to almost any substrate. Use of fillers is not restricted to the rubber compound. The bonding agents employed throughout the industry also contain fillers. Certain grades of carbon black are claimed to improve adhesive strength and are set out in a patent awarded to a leading maker of bonding agents.

The earlier bonding agents were basically highly filled natural rubber solutions containing iron oxide and similar materials, together with a generous helping of sulphur. Current leading bonding agents are generally based on solutions of halogenated polymers, carbon black, basic lead–acid acceptors and dinitroso-benzene.

Types of filler

Two classes of filler may be distinguished, reinforcing and non-reinforcing. Reinforcing fillers again fall into two types, white and black. The latter are invariably carbon black, but white fillers include silica, silicates and calcium carbonate. Black fillers are the more important. Grades range in particle size from 10 to 500 nm and are generally classified according to size and surface area by an ASTM designation as described in ASTM D 1765. At an equivalent volume loading, reduction of particle size increases reinforcement, leading to higher tensile strength and modulus, increased abrasion and fatigue resistance and higher loss angle. The processing is also affected, mixing cycles being longer and the cure rate higher. Carbon blacks of smaller particle size are also more expensive.

The level at which any of these properties may be at its optimum varies from grade to grade. It is therefore necessary to investigate rubber formulations very thoroughly before use. This is best achieved through factorial design of experiments, results of which can then yield an optimized solution through the use of regression analysis.

The nature of carbon black is extremely complex, and different grades[2] differ according to their particle size, surface character (including surface area) and structure.[3,4] The mechanisms of reinforcement are not fully understood, but it is likely that different levels of adhesion between filler and polymer play an important part.[4] The **Surface energy** and structure will vary, not only between grades, but between different parts

Table 33. A typical experimental rubber formulation

Natural rubber (SMRCV)	100
Zinc oxide	4
Stearic acid	1
6PPD (antiozonant)	2
Microcrystalline wax	2
Carbon black	0–100
Process oil (paraffin)	0–10
Sulphenamide (CBS)	0.5–1.5
Booster (DPG)	0–0.3
Sulphur	1–3

[6PPD: *N*-(1,3,dimethylbutyl)-*N*-phenyl-*p*-phenylenediamine.
CBS: cyclohexyl-2-benzothiazole sulphenamide. DPG: diphenyl guanidine]

of the same filler. Diverse chemical groups have been detected on the surfaces of carbon fillers: alcohol, phenol, quinone, lactone, carboxyl as well as free radicals. Thus, depending on the chemistry of the filler surface and of the polymer, different levels of adhesion associated with van der Waals forces or chemical bonds (see **Adsorption theory, Dispersion forces, Polar forces**) may occur leading to different effects on the properties of the filled elastomer.[4] Most carbon black is produced by the furnace method of burning oil in a controlled oxygen environment. The price therefore follows that of oil.

For the automotive industry (see **Automotive applications**) the manufacture of engine mountings and anti-vibration mountings generally requires control over stiffness (or modulus), fatigue life and loss angle. If these are correct then it is usually possible to engineer bonding processes to produce components which satisfy the specifications. A general formulation is shown in Table 33 as a basis for experiments.

White reinforcement This is obtained through the use of silica. Grades are defined by size surface area and method of manufacture, although there is no universally accepted grading system such as exists for carbon black. A comparison of the effects of silica and carbon black as reinforcing fillers is given in Table 34. In general, unless one needs a non-black rubber, less trouble is experienced with black reinforcement.

Non- or semi-reinforcing fillers These are usually added to reduce cost. In natural rubber or polychloroprene they may be used alone, but with non-crystallizing polymers such as butadiene–acrylonitrile or styrene–butadiene copolymers, they can only be used in conjunction with a reinforcing filler. Their effect is to reduce tensile strength and elongation, tear resistance and resistance to set. The effect on modulus varies

Table 34. Comparison of reinforcement by
silicon and carbon black

Silicon better for	Carbon better for
Tear resistance	Processing (mixing)
Non-staining	Compression set resistance
Fire resistance	Mould flow
	Resilience
High modulus	Moisture absorption

Table 35. Some non-reinforcing and semi-reinforcing filters

Filler	Reinforcing capability
Coal dust	None
Ground whiting (calcium carbonate)	None
Precipitated whiting (stearate coating)	Slight
Calcium silicate	Slight
Clays	Semi-reinforcing
Aluminium silicate	Semi-reinforcing

according to choice of filler, but it is always much weaker than that of a reinforcing filler.

Examples of semi-reinforcing and non-reinforcing fillers are given in Table 35. The ranking of such fillers is imprecise because it depends upon the use which may be made of them. Certain clays perform well in insulation compounds and precipitated whitings are more than adequate to extend rubbers for components which are not subjected to high levels of stress.

Selection criteria for rubber bonding

Fillers must be selected first on their ability to meet end-use requirements. These may be set out formally through specifications such as seen in SAE J200 or ASTM D 2000. Such specifications define tensile strength, hardness, age resistance, compression set, etc. The level of these requirements will be a good guide to filler types and grades. Tensile strengths higher than 14 MPa will certainly require a reinforcing filler at a size of 50 nm or finer (N660) and for strength above 20 MPa a particle size of 30 nm or less is recommended (N330).[2] If dynamic properties are important, coarser fillers will give high resilience at equivalent levels of filler than will the finer grades.

Of equal importance to the properties of the moulded component is the influence of the filler during the mixing and moulding stages. Ease of incorporation in the rubber is important if a homogeneous mixture is to be produced. The more fine the filler, the more difficult it is to disperse.

Flow properties during moulding influence the final quality of the component. High viscosity leads to poor flow precuring of the rubber and hence to weld lines, risk of early fatigue failure and poor-quality bonds. The finer the filler the higher the viscosity for any given volume of filler. The moisture content of any filler needs careful control. Adverse levels of moisture ($>0.5\%$) will affect rubber cure rate, final modulus and bond strength. If these considerations are made prior to the formulation of the compound, a high quality of rubber bonding can be achieved.

References

1. B B Boonstra in *Rubber Technology and Manufacture*, ed. C M Blow Newnes–Butterworths, London, 1975, pp. 227–61.
2. *ASTM Book of Standards*, ASTM-D 1765, annually.
3. F W Barlow, *Rubber Compounding*, Marcel Dekker, New York, 1988, Chs 9, 10.
4. W Hofmann, *Rubber Technology Handbook*, Hanser, Munich, 1989, pp. 277 ff.

Rubber to metal bonding – applications

G J LAKE

Adhesive bonding of rubber to metal, or to other materials, occurs in many applications (Fig. 116). This may simply be for fixing purposes, as is often the case with springs and mountings, but may be a necessary part of the design. Examples in the latter category include bonding to fabric or cords (metal or polymer) in tyres, hose and many other articles where increased stiffness and strength are required; bonding to internal steel plates in bearings of various sorts (Fig. 116) can increase the shape factor and hence the compression stiffness without affecting the low shear stiffness of the rubber.[1] In many, though not all such applications, the integrity of the bond is vital to satisfactory performance and life. Bonding prevents

Fig. 116. Various articles in which rubber is bonded to metal or other materials: (a) bridge bearing; (b) cylindrical bearings; (c) conical bearing; (d) spherical segment bearing; (e) cross-section of radial tyre. (Photographs (b), (c) and (d) courtesy of Lord Corporation)[4]

(a)

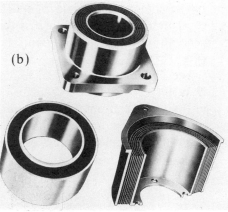

(b)

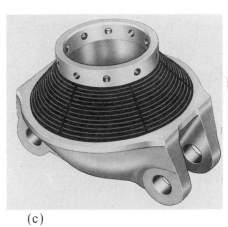

(c)

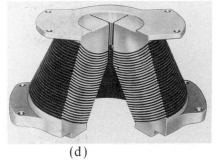

(d)

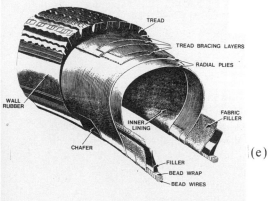

(e)

slip at load-bearing surfaces thus ensuring reliable load–deflection characteristics. However, in some applications where compression is involved, friction alone may suffice to prevent slip. In addition to some bearings, coaxial bushes are an example where one or both metal surfaces may be unbonded. A reason for this is that hydrostatic tensile stress may arise if a bonded bush is made in a single-stage vulcanization process owing to the difference in the thermal contraction of the rubber and metal parts on cooling. Avoidance of such stresses is very important for bonded units as otherwise internal cracking may occur at very low stress levels.[1,2]

The large change in modulus occurring across a boundary between rubber and a more rigid material may, in itself, give rise to stress concentrations. Avoidance of sharp corners or other features causing stress concentration is also important, as when designing with metal alone[1] (Fig. 117).

The environment in which an article is used may influence bond **Durability**. Atmospheric ozone can cause time-dependent crack growth in vulcanized elastomers; in addition it can induce failure at a bond with certain bonding agents. Although only slightly soluble, water can permeate elastomers by an osmotic mechanism induced by salt-like impurities. As a result the uptake in salt water is generally less than that in pure water. Rubber to metal bond failure has been found to occur in

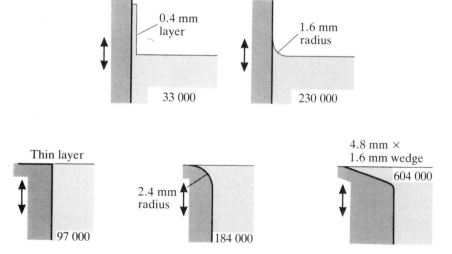

Fig. 117. Effect of various internal or external shapes on bond durability: the number of cycles to the development of appreciable cracks are shown for a natural rubber vulcanizate subjected to shear strains of ±100% in the direction indicated (the rubber is on the right in all cases[1])

a time-dependent manner under salt water in the presence of electro-chemical activity but much more slowly, if at all, in its absence[3] (see also **Cathodic disbondment**). In the absence of imposed electrochemical activity, effects are likely to depend particularly on the metal used and its corrosion resistance. Provision of a bonded rubber cover layer over all metal surfaces subject to immersion is likely to enhance bond durability.

Means of effecting good bonds between rubber and metal are discussed in articles on **Rubber to metal bonding – basic techniques, Rubber to metal bonding – pretreatments** and **Rubber to metal bonding – testing**.

References

1. P B Lindley, *Engineering Design with Natural Rubber*, Malaysian Rubber Producers' Research Association, Hertford, 1974, revised 1992.
2. G J Lake, A G Thomas, Strength of rubber, in *Handbook of Engineering Design of Rubber Components*, ed. A N Gent, American Chemical Society, Washington, DC, 1992.
3. A Stevenson, On the durability of rubber/metal bonds in sea water, *Int. J. Adhesion and Adhesives*, **5**, 81–91 (1985).
4. P T Herbst, Lord Corporation, *Natural Rubber as a Bearing Material for Rotary Wing Aircraft Application*, NRPRA 3rd Rubber in Engineering Conference 1973.

Rubber to metal bonding – basic techniques

P M LEWIS

In many elastomeric products the rubber is bonded to a metal component either for location purposes or in order to meet certain design requirements such as a given combination of compression and shear stiffness (see **Rubber to metal bonding – applications**). High rubber to metal bond strengths can be obtained by a number of techniques, of which the most common are described below. The choice of method will depend on a number of factors including service requirements of the product, ease of assembly of components and the types of rubber and metal being bonded. All methods require careful surface preparation, see **Rubber to metal bonding – pretreatments**.

In-vulcanization bonding

While in-vulcanization bonding[1] suffers from the disadvantages associated with having to incorporate and heat metal parts in moulds, it has been satisfactorily used by the rubber industry for many years. It is particularly well suited to bonding of complex surfaces where the flow of unvulcanized rubber ensures intimate contact with the bonding surface.

There are three principal methods of bonding the metal component to the rubber during vulcanization, given under the headings below.

Proprietary chemical bonding agents These are usually mixtures of an undisclosed composition, consisting of reactive ingredients suspended or dissolved in organic solvents. The solids content includes resins, elastomers and fillers. Some water-based adhesives are also available. Cyclized, chlorinated and carboxylated rubbers, rubber hydrochloride, poly-isocyanates and phenolic resins have all featured in bonding agents. They offer a convenient and versatile means of bonding a wide range of rubbers and metals and constitute the major method of in-vulcanization bonding for non-tyre applications. For most rubbers it is usual to use a two-coat system. In some cases the primer may be omitted without affecting the initial bond strength, but the resistance to under-bond corrosion may be diminished. One-coat systems have been developed for some combinations of rubber and metal.

The bonding agent should be applied to the prepared metals as soon as possible, in order to prevent corrosion or contamination. Application can be carried out by painting, dipping or spraying. Whichever method is used, a minimum thickness of bonding agent, typically 25 μm, must be obtained.

The bonding agent may be diluted with a suitable solvent, preferably the one used in the original material. Use of a solvent with a lower boiling-point than that used in the bonding agent may lead to premature drying during spraying or the formation of an incoherent film or bonding agent. The use of a higher-boiling solvent may lead to incomplete drying and also lead to poor bond strength.

Coated metals should be stored in such a way that risk of contamination or mechanical damage is minimal. A minimum period of storage should be specified to ensure complete drying of the bonding agent; typically 30 min at room temperature should be adequate.

Ebonite bonding[2] The adhesive layer in this case is unvulcanized ebonite (or hard rubber, USA) applied to the metal either as a solution or as a thin sheet). Ebonite contains 30–50 parts sulphur per hundred parts of rubber (phr), whereas soft rubber seldom has more than 3 parts phr of this vulcanizing agent. The method dates back to the mid-nineteenth century and is most closely identified with natural rubber.

During vulcanization sulphur can migrate from the ebonite layer to the rubber and may result in the formation of an interlayer of highly cross-linked rubber which has poor physical properties. Formation of such a layer may be avoided if a second adhesive coat, consisting of the rubber mix without sulphur, is applied. Since ebonite softens at higher temperatures (e.g. 80 °C) the use of ebonite bonding is restricted to service temperatures near to ambient.

In applications where tack of the adhesive is required, for example, tank linings or rubber-covered rollers, ebonite may be more suitable than proprietary bonding systems.

Brass plating[3] Sulphur-vulcanized rubber will form a good bond to brass during vulcanization provided:

1. The brass contains between 60 and 70% copper.
2. The rubber mix contains sufficient sulphur. Maximum bond strength may require 4–6 phr of sulphur, although good bonds to rubbers containing lower levels can be obtained by using an intermediate layer of a mix containing a high sulphur level.
3. The rate of vulcanization of the rubber is adjusted to give optimum bond strength.

If the brass can be strongly bonded to a metal, brass plating can be used as a basis for rubber to metal bonding, and under the appropriate conditions can give excellent results. The bond is not unduly sensitive to influences such as humidity, low mix viscosity and low vulcanizing pressures. The main drawbacks are connected with the plating process, which requires considerable investment and expertise. Brass plating has now been largely superseded by the use of proprietary bonding agents, except in the tyre industry where bonding to brass-coated steel cords is still undertaken.[3]

Moulding For most reliable results transfer or injection moulding is preferred to compression moulding, as this ensures that a fresh rubber surface is presented to the coated metal, and the higher mix temperature improves contact with the bonding surface.

Mould design should allow for rapid filling of the cavities, but removal of bonding agent by rubber flowing over the metal surface must be avoided. Sprues and flash lines should be located away from the bonding areas, and runners in moulds should be designed to minimize the possibility of scorch. When high-temperature vulcanization is carried out the number of cavities in the mould should be restricted to enable rapid loadings of metal parts.

As bond strength decreases with increasing temperature, the mould should be designed to allow easy removal of components so that little or no strain is put upon the bond. Easy removal also reduces the need for mould release agents, an over-enthusiastic use of which can cause rejects.

Safe temperature ranges for bonding are specified by manufacturers, and are generally restricted to the range 120–180 °C, although special systems are available for high-temperature vulcanization. One factor limiting the applicability of many bonding agents is their effective lifetime at elevated temperatures. At temperatures of 170 °C and above, bond strength may be drastically reduced if coated metals are heated for more than a certain time before coming into contact with the rubber. This phenomenon of pre-bake becomes troublesome if mould loading and filling times exceed the safe time limits. The problem can be minimized by loading metals into multi-cavity moulds using jigs or restricting the number of cavities in the moulds. Other precautions include reducing the mix viscosity and using wider runners and/or sprues.

The level and consistency of bond strength are influenced by the nature of the rubber;[1] factors include the mix viscosity, level of filler, scorch safety and, in the case of sulphur-vulcanized rubbers, the concentration of elemental sulphur in the vulcanizing system.

Post-vulcanization bonding[4]

Vulcanized rubber can be bonded to metals by a number of methods using proprietary bonding agents. The rubber must be freed from blooms, oils or dust by washing with solvent or detergent.

Post-vulcanization (PV) bonding offers the advantage of not requiring the incorporation of metal parts into moulds, resulting in an increase in moulding productivity. On the other hand, it introduces an extra stage in the production of an article which must be carried out by either the rubber manufacturer or the end user. As PV bonding can be carried out at lower temperatures than in-vulcanization bonding, in-built stresses in the vicinity of the bond resulting from rubber shrinkage will be lower and thus the method can be attractive for bushes.

In hot PV bonding, the bonding agents react with the rubber and no further preparation of the rubber is required. The bonding agents used are often suitable for in-vulcanization bonding, but the reverse is not true.

Rubbers can be bonded to metals at room temperature with a wide variety of bonding agents, including solvent-based cements.[5] If, however, high bond strengths are required **Cyanoacrylate** or **Epoxide adhesives** should be used. Cyanoacrylates do not require treatment of the rubber other than cleaning, but epoxides can only be used successfully if the rubber surface is modified prior to bonding. The modification can be carried out by chlorinating the surface with acidified hypochlorite solution or less commonly by dipping the rubber in sulphuric acid. Other halogen-donating materials have been patented for this treatment and are widely used in the footwear industry (see **Footwear applications of adhesives**).

For further information see the articles **Rubber to metal bonding – testing** and **Solvent-based adhesives**.

References

1. E Cutts in *Developments in Adhesives – 2*, ed. A J Kinloch, Applied Science Publishers, London, 1983, Ch. 10.
2. S Buchan, *Rubber to Metal Bonding*, Crosby Lockwood, London, 1959.
3. W J van Ooij, *Rubber Chem. Technol.*, **52**, 437 (1979).
4. M A Weih, C E Silverling, E H Sexsmith, *Rubber World*, **194** (5), 29 (1986).
5. P B Lindley, *NR Technol.*, **5**, 52 (1974).

Rubber to metal bonding – pretreatments

P M LEWIS

Many engineering applications of rubber depend upon the polymers being bonded to metal (see **Rubber to metal bonding – applications**). Successful bonding of rubber to metal requires both the rubber and the metal to make intimate contact with the applied bonding agent. To minimize risk of bond failure, surface contamination of the three components of a rubber to metal bond must be avoided.[1,2] Scrupulous cleaning of the metal should be followed as soon as possible by application of bonding agent and vulcanization or assembly. Handling of coated metals should be minimized and, when used, gloves should be changed frequently.

The following methods are used to clean and prepare metal surfaces prior to application of bonding agent.

Degreasing

Metal parts should be degreased with an organic solvent, for example trichloroethylene, or a detergent wash. Degreasing with solvent vapour is often found most effective because it minimizes contamination, but proprietary aqueous treatments may be used satisfactorily and may be preferred on grounds of health and safety. If a chlorinated solvent is used, the pH should be monitored to ensure that the solvent remains neutral; otherwise under-bond corrosion may be encountered. The vapour degreasing time should be sufficient to ensure that all parts are well washed with condensate, and the efficiency of the operation should be regularly checked by wetting a part with water. If **Degreasing** has been achieved a coherent water film will remain on the surface.

Mechanical cleaning

This is best carried out by grit-blasting. Steel grits (40–80 mesh) are suitable for ferrous metals, but sand or, preferably, aluminium oxide may also be used and must be so for non-ferrous metals. Compressed air used for blasting or spraying should be free from oils and water. It is often recommended that grit-blasting be followed by a second degreasing cycle (see **Abrasion treatment**).

Chemical cleaning

A number of proprietary chemical cleaning treatments may be used as alternatives to degreasing with organic solvents and grit-blasting. These treatments normally consist of an acid or alkaline cleaning bath and a phosphating bath, with washing after each stage. These are followed by a very dilute phosphoric/chromic acid rinse. Chemical cleaning treatments can be used satisfactorily if care is taken to maintain concentrations, temperatures and dwell times to within the supplier's recommendations. Baths should be changed regularly to avoid the build-up of contaminants. Phosphate treatment is suitable for mild steel, but not always for other metals. For example, treatment with hydrochloric or chromic acid is recommended for stainless steel and with hydrofluoric acid for titanium.

Contamination of rubber surfaces by release agents (in particular silicone fluids), dusting agents such as zinc stearate, or blooms should be avoided. For in-vulcanization bonding contamination of the rubber surfaces can be minimized by using transfer or injection moulding. Compression moulding can be used satisfactorily provided blanks are freshly prepared or, in some cases, the surfaces are freshened by a solvent wipe.

For further information see the articles **Rubber to metal bonding – basic techniques, Rubber to metal bonding – testing, Degreasing** and **Pretreatment of metals prior to bonding**.

References

1. J D Hutchinson, *Elastomerics*, **110** (4), 35 (1978).
2. J Shields, *Adhesives Handbook*, 3rd edn, Butterworths, London, 1984.

Rubber to metal bonding – testing

J A LINDSAY

The general principles of testing adhesion and a wide range of types of test can be applied to the rubber to metal bonds: relevant articles include

Blister test, Fracture mechanics test specimens, Non-destructive testing, Peel tests, Rubber adhesion, Shear tests, Tensile tests, Tests of adhesion and **Wedge test**. This particular article is concerned with those aspects of practical concern in the rubber processing industry.

Bond testing can be broadly classified into two categories, laboratory tests and product tests. In each category, tests may be either qualitative or quantitative depending upon circumstances. Qualitative tests are sometimes useful for rapid screening of a wide range of conditions or materials. For optimization and specification purposes quantitative tests need to be employed where possible.[1]

Peel tests

The simplest type of test to perform is the peel test. The test piece is a flat strip made from the substrate to which is bonded over two-thirds of its length a layer of the rubber, preferably 4–6 mm thick. The test piece is conveniently dimensioned 150 × 25 mm. The thickness of the substrate is usually of the order of 2–3 mm.

A qualitative indicator of the bond may be made simply by applying a 180° peel force to the unbonded part of the rubber by hand. It is important to examine a bonded length of 20–30 mm back from the leading bonded edge. This is achieved by applying, carefully, cross-cuts to the rubber, across the bonded front, immediately adjacent to the substrate, while maintaining the peeling force.

With a satisfactory bond there must be no separation of the rubber from the substrate under such peel forces. Weak bonds, similar to that obtained with an adhesive cellophane tape, is an immediate indication that the rubber bond is totally inadequate.

It is now satisfactory to make quantitative measurements of bond strength. To this end, the peel test piece may be mounted in a straining frame. The angle at which the peeling force is applied will considerably effect the result (see **Peel tests**). Experiment has found that an angle of 30° between the substrate and the direction of peel will give the most useful results. For a further description of the method the reader is referred to ASTM D 429, Method B.

Advantages Test pieces are relatively cheap and are easily prepared.

Disadvantages The method often produces results which are more a measure of tear resistance of the rubber. This can be overcome by incorporation of a backing cloth to the rubber, but this is an added complication.

Tests in tension

Some tests specify the use of two circular plates which may attach directly to the straining frame. Rubber is bonded between the plates which are then separated in a mode perpendicular to the plane of the plates (see **Tensile tests**).

Very high loads should be expected in this type of test. On failure there should be no sign of bonding agent–rubber or bonding agent–substrate separation. The reader is referred to ASTM D 429, Method A.

Disadvantages Analysis of stress in this type of test piece shows that stresses are not specifically concentrated at the bond interface. Mould is more expensive than for a peel test. It does not reveal much about the bond.

Advantages. It is an easy test to perform.

Cone test piece

In this test a 25 mm diameter cylinder of rubber is moulded around two cone-shaped test pieces such that the tips of the test pieces are separated by 10 mm. This test is fully described in ASTM D 429, Method C. It has the distinct advantage that due to the shape of the test piece, a high strain is directed to the rubber–substrate interface when the ends of the cylinder are separated by the straining frame. Results are sensitive to the tip radius of the cone which must be carefully controlled.

Arrowhead test piece

The arrowhead test has been suggested by Aubrey et al.[2] as incorporating some of the advantages of the cone while being easier to prepare. The test piece consists of a flat arrowhead, 25 mm wide and tapering to a point with a 60° included angle, using steel strip 6 mm thick. This is bonded, using the rubber and bonding agent under test, to a 25 mm square of a similar piece of metal. Like the cone test piece, failure is concentrated at the bond interface when the test and pieces are separated by a straining frame. Unlike the cone test piece there is no sensitivity to tip radius, and test pieces can be prepared from strip metal by conventional shearing or punching techniques.

Obtaining consistency in test results

Typical procedures for rubber to metal bonding involve a large number of steps (see **Rubber to metal bonding – basic techniques** and **Rubber to metal bonding – pretreatments**). Consequently a lack of repeatability associated with inconsistencies in materials and procedures can easily occur. It is important to check critical properties of bonding agents and of the rubber compound. Substrate preparation and application of primer layers must be standardized with careful attention paid to the time intervals between the different stages involved. The cure of the rubber and thence the properties of the bond will be affected by the times and temperatures employed in the moulding sequence. For example bonds made when the moulds are warming up will have different properties from those produced when thermal equilibrium is established: failure to control the 'open time' or the mould will also lead to inconsistencies.

Testing the service life

The usual way in which the life of differently made bonds can be compared is to prepare types of test pieces and then place them in a jig which will allow a strain to be placed upon the rubber and the bond edge. The test pieces are then subjected to contact with a hostile environment such as hot water or salt fog (see ASTM B 117). The resultant failures generally depend upon the type of surface preparation used and may be used to monitor these procedures.

Component testing

A range of interesting engineering components depend on rubber to metal bonding in order to function (see **Rubber to metal bonding – applications**). Components themselves may be tested to obtain a measure of how well the bond process has been carried out. The method is obviously very much dependent upon the component type and shape. Whatever the method, observation of failure in the rubber rather than cement–rubber or cement–metal separation is the desired result.

References

1. E Cutts in *Developments in Adhesives – 2*, ed. A J Kinloch, Applied Science Publishers, London, 1981, Ch. 10.
2. D W Aubrey, E Southern, A D Harman, *Rubber–Metal Bond Testing*, London School of Polymer Technology, n.d.

S

Scanning electron microscopy

B C COPE

Electron microscopy exploits the general microscopy relationship that resolution is improved by reducing the wavelength of the illuminating radiation. Whereas the resolution routinely obtained by **Optical microscopy** is typically around 1 μm transmission, **Electron microscopy** improves this to 1 nm. The scanning electron microscope (SEM) sacrifices some resolution for the ability to handle large, rough specimens and is routinely capable of resolving to about 10 nm. All three techniques may improve their performance under controlled conditions and with suitable specimens.

The SEM operates, in its normal secondary electron mode, by rastering an electron beam over the surface of a conducting specimen mounted in an evacuated column. The secondary electrons emitted from the surface regions of the sample are collected by a scintillator and the resulting signal used to control the generation of an image on a cathode ray tube (CRT) display screen by modulation of the beam brightness in sychronization with the raster. Contrast results from surface topography and an image with a strong impression of spatial depth is displayed and may be photographed. Figure 118 shows the surface of zinc specially prepared for adhesive bonding by electrodeposition (see **Microfibrous surfaces**). Although the magnification ($\times$ 750) is within the capability of an optical microscope, the depth of field can only be obtained with the SEM. Figures in **Abrasion treatment** show contrasting micrographs of surfaces differently grit-blasted.

Other imaging modes than secondary electron are available on the instrument, but only backscattered electrons, which give a topographical image with some compositional contrast, are likely to be of very much interest in solving adhesion problems.

Fig. 118. Scanning electron micrograph of zinc dendrites electrodeposited on to a zinc substrate as a pretreatment for adhesive bonding (P J Hine, S El Muddarris, D E Packham, unpublished micrograph, see *J. Adhesion*, **17**, 207 (1984))

The salient advantage of the SEM over the reflectance optical microscope perhaps lies less in the much greater magnification available and the superior resolution than in the vast depth of field which permits the study of comparatively rough surfaces.

Samples for the SEM may typically be a centimetre or so in diameter and about 1 cm deep, smaller than can be accommodated by the reflectance optical microscope but much larger than the small, ultrathin sections required for the transmission electron microscope. It is sometimes possible to work at low acceleration voltages with non-conducting specimens, but the secondary electron mode normally demands that the specimen has a conducting surface. Metals thus present no problems, but organic

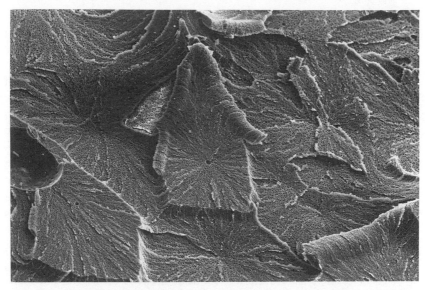

———— 200 μm

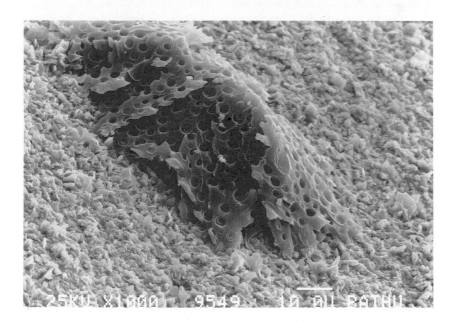

Fig. 119. Failure surfaces of bonds between zinc and toughened epoxy resin: (top) region of cohesive failure within the resin; (bottom) region of apparently mixed failure (P J Hine, S El Muddarris, D E Packham, unpublished micrographs, see *J. Adhesion*, **17**, 207 (1984))

specimens usually need to be coated with a thin layer of metal applied by sputtering or vacuum evaporation. Careless preparation can lead to the masking of fine structure and the production of artefacts. Other changes in the specimen can occur as a result of the removal of volatiles under vacuum in the coater or the microscope or as a consequence of 'beam damage', degradation of various kinds resulting from the concentrated input of energy from the electron beam.

Scanning electron microscopy is essentially a technique for the observation of surfaces and is thus especially applicable to the examination of adherends and failure surfaces. Observations of the results of surface preparation techniques for metals and polymers, investigations of failure modes (see **Locus of failure**), and examinations of finishes on fibres are typical of work reported. Figure 119 shows fracture surfaces for bonds between zinc and rubber toughened epoxy resin (see **Toughened adhesives**). In Fig. 119(a) is a region of cohesive failure within the resin: the sites of bubbles which may have initiated the fracture can be seen. In Fig. 119(b) a piece of resin is seen adhering to what appears to be the bare zinc substrate. There is less place for SEM work in the study of intact joints and the technique is not suited for observing sections.

Scanning electron microscopes range from single bench-top instruments fitted only for secondary electron imaging to large, versatile and extremely expensive machines capable of several imaging modes and fitted for **Electron probe microanalysis**. The former may cost less than research grade optical microscopes, but the latter are a substantial multiple of the cost of these.

Select references

P J Goodhew, F J Humphreys, *Electron Microscopy and Analysis*, 2nd edn, Taylor and Francis, 1988.

B C Cope in *Surface Analysis and Pretreatment of Plastics and Metals*, ed. D M Brewis, Applied Science Publishers, London, 1982.

Scratch test

D E PACKHAM

Conventional **Tests of adhesion** are not well suited to measurement of thin films. The scratch test has been used to characterize the adhesion of thin films, such as plasma polymerized polymers or evaporated metals, on hard substrates.

The test uses a vertically loaded stylus of a hard material such as diamond or tungsten carbide with a tip radius of the order of tens of micrometres. A micro-hardness tester may be used to make the test.[1] The stylus is drawn across the film a number of times, the vertical load being increased for each traverse. The critical load at which the film is just removed from the substrate is usually taken as a measure of the adhesion.[2,3] Oroshnik and Croll have proposed a somewhat different criterion as characterizing the adhesion.[4]

Benjamin and Weaver argued that the critical load was determined only by the properties of the interface, the radius of the stylus and the hardness of the substrate.[2] On this basis the scratch test would give a measure of 'fundamental' adhesion (see **Adhesion – fundamental and practical**). However, others have reported dependency on the thickness and mechanical properties of the thin film[3] and even upon which of several supposedly similar styluses was used.[4]

A simple form of the scratch test is used for assessing paint adhesion[5] (see **Paint constitution and adhesion**). The scratch test, as described in this article, is perhaps best regarded as a useful comparative method of characterizing thin film adhesion.[1] The test variables used should be carefully characterized and caution exercized in comparing results obtained in different laboratories.

References

1. A J Pedraza, M J Godbole, D H Lowndes, J R Thompson, *J. Mater. Sci.*, **24**, 115 (1989).
2. P Benjamin, C Weaver, *Proc. Roy. Soc.*, **A254**, 163 (1960).
3. J Ahn, K L Mittal, R M MacQueen in *Adhesion Measurement of Thin Films, Thick Films and Bulk Coatings*, ASTM STP 640, ed. K L Mittal, American Society for Testing and Materials, Philadelphia, 1978, p. 134.
4. J Oroshnik, W K Croll in *Adhesion Measurement of Thin Films, Thick Films and Bulk Coatings*, ASTM STP 640, ed. K L Mittal, American Society for Testing and Materials, Philadelphia, 1978, p. 158.
5. T R Bullett, J L Prosser in *Industrial Adhesion Problems*, ed. D M Brewis, D Briggs, Orbital Press, Oxford, 1985, p. 197.

Sealants

G B LOWE

Sealants may be described as thick bed adhesives some with elastic properties to allow movement. In general the low movement systems are referred to as mastics, whereas the elastic and elasto-plastic type systems are referred to as sealants. More detailed descriptions are discussed by Damusis.[1]

The mastic-type sealants include hand-applied systems based on linseed oil (putty), gun-applied oil-based systems, using one or more drying oils and gun-applied 'butyl' systems based generally on liquid polyisobutylenes (PIB) with oils and solvents. These function primarily as gap fillers to prevent the ingress of dust, rain and wind in substantially static conditions.[2,3]

Intermediate-type sealants include solvent-based and emulsion-type acrylic systems, higher-quality butyl/PIB blends, neoprene and styrene–butadiene rubber (SBR)-based compounds. These have a degree of movement tolerance and will accept low levels of thermally induced contraction/expansion across the sealant. These intermediate type systems 'cure' by solvent evaporation and auto-oxidation or a mixture of the two processes.

The elastomeric sealant types are chemically curing systems based on a limited variety of polymers. The best known of these being polysulphide (Thiokol), silicone and polyurethane and are available as both single-part and two-part compositions.

The single-part systems generally require atmospheric moisture to effect cure either by inducing oxidation (single-part polysulphides cured with calcium peroxide) or condensation of a reactive end group as in the polyurethanes and silicones.

Speciality sealants based on modifications of the above find wide usage not only in construction but also in civil engineering, **Aerospace**, shipbuilding and **Automotive applications**. Some of these applications make use of specific properties of the polymer systems. Such examples are high-fuel-resistance polysulphides for sealing aircraft fuel tanks and adhesion to glass for all glass façade buildings using silicone sealants.

Common usage of sealants in glazing applications (see **Sealants in double glazing**) can be found in BS 6262 and some of the more widespread systems are covered by British Standard Specifications (5212, 5215, 4254, 5089) and are tested to the general performance requirements as outlined in BS 3712.

Further information may be found in ref. 4 and in articles on **Selection of joint sealants, Primers for sealants** and **Sealant joint design**.

References

1. A. Damusis (ed.), *Sealants*, Van Nostrand Reinhold, New York, 1967.
2. W R Sharman, J I Fry, R W Whitney, Six years natural weathering of sealants, *Durability of Bldg Mater.*, **2**, 79–90 (1983).
3. D Aubrey, J Beech, The influence of moisture on building joint sealants, *Bldg and Environ.*, **24**, 179–90 (1989).
4. H F Bronson (ed.) *Adhesives and Sealants*, Engineered Materials Handbook, Vol. 3, ASM, Ohio, 1990.

Sealants in double glazing

G B LOWE

Insulated glass or double glazing units consist of two panes of glass separated by a spacer medium and all held together by an edge **Sealant** (Fig. 120). These sealants seek to exploit the specific properties of low-moisture vapour transmission, low inert gas transmission and high **Durability**.

The most common system used is based on polysulphide (Thiokol) polymers cured with manganese dioxide and uses a low-volatility plasticizer, usually a high phthalate. These systems retain the benefit of oil resistance when used in areas of timber windows (putty and solvent from the preservative (see **Selection of joint sealants**).

Other systems that have been introduced include polyurethanes based on a hydroxy terminated polybutadiene polymer, hot applied butyl/polyisobutylene (PIB) systems, being a mixture of polybut-1-ene, polyisobut-1-ene and butyl rubber containing isoprene, and silicone sealants. Silicones do not have a sufficiently low moisture vapour transmission rate and must be used with a PIB primary seal (Fig. 120(b)). The primary seal is optional with polysulphides and some polyurethanes.

The **Durability** of insulated glass units is normally assessed by measuring the rate of change of the internal dewpoint or by quantitative evaluation

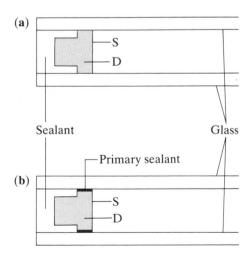

Fig. 120. Most common insulated glass unit construction, (a) without primary sealant and (b) with primary sealant (S = spacer, usually aluminium; D = desiccant, either zeolite or silica gel)

of the water absorbed by the desiccant. This latter property called the penetration index is given by

$$I = \frac{T_f - T_i}{T_c - T_i}$$

where T_f, T_c and T_i are the weight loss in ignition of the desiccant on the tested unit T_f, the reference unit (T_i) and the capacity of the desiccant (T_c).

Because these systems are used in a different way from their construction sealant counterparts, extensive studies on long-term adhesion have been carried out to evaluate adhesion promoters capable of giving long (10–12 years) service life. The most successful have been the alkyl silanes specifically γ-glycidoxy-propyl-trimethoxy-silane or mercapto-propyl-trimethoxy-silane, although workers have found amino substituted silanes to be preferable (see **Coupling agents, Primers for sealants**).

The gas-retention qualities of the sealants have become of increased importance as it has been found that use of higher-molecular-weight inert gases influences the transmission of sound and heat. Gases such as argon, sulphur hexafluoride, krypton and carbon dioxide have been suggested, although only the first two have been widely used. These gases show greater absorption than air of sound at high frequency and of IR radiation. Their greater viscosity also reduces heat loss by convection between the panes.

Rates of gas loss from a unit have been established by many workers:[1]

	Argon (% per annum)
Polysulphide	0.5
Polyurethane	1.0
Silicone	10–30

The water vapour transmission rate (WVTR) through the sealant may be measured using such apparatus as is shown in Fig. 121. Because of differences in details, such as film thickness and humidity differential, results from different workers[2] are not comparable. However, the following figures give an indication of relative transmission rates in g m^{-2} day^{-1} for various sealant materials.

PIB/butyl	0.5–1.0
Polysulphides	1.0–2.0
Polyurethanes	1.5–3.0
Silicones	3.0–10.0

The WVTR and gas transmission must be considered along with other physical properties such as T_g[3] for a better understanding of the sealant's potential performance.

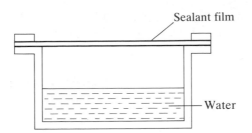

Fig. 121. Determination of moisture vapour transmission. The apparatus is stored at constant temperature and humidity. Weight loss is taken as a measure of water vapour transmission rate

Ageing of insulating glass units have been studied extensively.[4,5] These tests have resulted in a wide variety of test specifications from Europe as well as North America. The most notable of these are BS 5713 (UK), DIN 1286 (Germany), UEA T_c (European Agrément), ASTM E773/E774 (USA). These specifications can be met by all the systems mentioned earlier. However, polysulphides still account for 78% of all insulated glass units made in Europe.

References

1. B Streeter, Gas filled insulating glass sealants, *SIGMA Meeting*, New Orleans, Oct. 1989.
2. J A Box, L Aruscavage, Comparing sealants requires understanding several terms, *Glass Dig.* **67**(12), 52–4 (Dec. 1988).
3. G B Lowe, *Dynamic Mechanical Thermal Analysis of Sealants for Insulated Glass*, Morton International, Coventry, 1988.
4. S. Brolin, Penetration of water vapour into insulated glass units, *The Swedish National Test Institute Symposium*, Borås, Sweden, 1989.
5. M Schulman, M Scherrer, Adhesion to aluminium (in I. G. units), *Glass Dig.* **59**(12), 52–3 (Dec. 1980).

Sealant joint design

J C BEECH

A preliminary article indicates the scope and types of **Sealants**. In designing a sealant joint, it is essential to select a suitable sealant, with adequate capability to withstand the expected movement at the joint.

A generalized procedure for the design of a building joint employing a sealant is shown schematically in Fig. 122. In practice the designer will be subject to constraints which may require modification of this procedure. The design procedure must take account of the influence of temperature change as a primary cause of movement at the joint. Guidance is given in British Standard BS 6213[1] and in Building Research Establishment (BRE) Digests 227 and 228.[2]

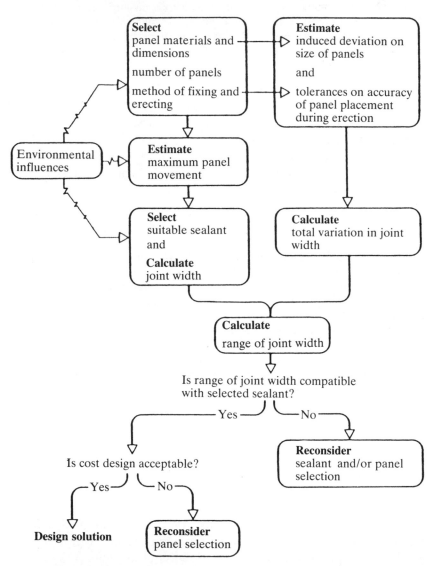

Fig. 122.   Procedure for the design of a building joint employing a sealant

An important decision for the designer is the type of joint to use. Joints at present used for buildings of pre-cast wall panel construction are of two main types: those relying exclusively on the seal provided by the sealants and/or gaskets, known as filled joints; and those referred to as open-drained joints, in which the geometry of the joint, in conjunction with an airtight seal, is used as a two-stage defence against the weather.[3] These latter are usually very reliable, although additional capital costs will usually be incurred.

The two main types of filled joint are butt joints and lap joints (Fig. 123); each has advantages and disadvantages. The lap joint gives the sealant some protection against the weather, and a further advantage of this type of joint over the butt joint is that the sealant will be less prone to failure for a given amount of joint movement. This is because it is predominantly shear forces which are created during lap joint movement, and sealants more readily resist these forces than the tensile forces associated with butt joint movements. However, access to the joint to insert joint fillers and to ensure effective application of the sealant is more difficult than for the butt joint. Also inaccuracies in component dimensions or their placing on site can mean that it is impossible to apply sufficient sealant in the joint to ensure satisfactory performance. The degradative effects of surface heat and UV light may be reduced by using a design in which the sealant is recessed within the joint. This could be simply achieved by designing a butt joint in which the sealant is applied to the back of the joint. However, in open-drained joints, the sealant isthoroughly protected from the weather by the sides of the components.

The choice of joint design should also take account of the fact that the lifetime of the chosen sealant may well be less than that of the building and that at some future date the sealant will have to be replaced. If such

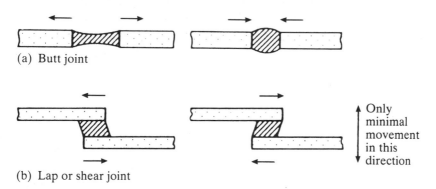

(a) Butt joint

(b) Lap or shear joint

Only minimal movement in this direction

Fig. 123. Sealant-filled joints: arrows indicate direction of movement of jointed component

an operation is likely, a design of joint must be used which allows defective sealant to be removed and the joint surfaces cleaned and prepared before reapplication of fresh sealant. Because this procedure is fairly straight-forward in the butt joint, this is the solution most frequently adopted by designers. In the discussion of some other aspects of joint design which follows, it is assumed that the butt joint is the option chosen.

Component tolerances and joint widths

When calculating the design width of joints in a building, it is essential to take account of the tolerances on the relevant dimensions of the components and the accuracy in placing them which is likely to be achieved on site.[4,5] For these reasons there may in practice be considerable variation in the widths of a number of ostensibly identical joints. The joint width for the design must ensure that in no joint will the sealant be subjected to a level of compression or extension which, when expressed as a proportion of the achieved joint width, exceeds its movement capability. (See the **Selection of joint sealants**, where Table 36 gives maximum movement permitted for each major type of sealant.) However, considerable variation in formulation and properties may occur between different brands of the same chemical type which may affect this value. The maximum tolerable joint movement quoted by the manufacturer should be used in joint design calculations.

A simple formula for the calculation of joint width has been derived.[1] The minimum joint width achieved must be sufficient to allow the sealant to be effectively gunned into the joint. The minimum width quoted for most sealants is 5 mm. The maximum joint width achieved must not exceed that quoted as the maximum service value by the manufacturer. If it does, the sealant is likely to slump out of the joint, most probably in the period before the material fully cures, though longer-term failures of this kind may be experienced by non-setting types of sealant. Some elastomeric types of sealant may be applied to wide joints in two or more stages, the material finally being 'tooled' in the normal way.

Sealant geometry

After application the sealant is forced into the joint and into good contact with the side and back surfaces by means of a wetted spatula or similar tool. This procedure is called 'tooling'. It also ensures the correct front profile of the sealant surface, which should be slightly concave for elastomeric types, since this reduces stresses at the surface induced by joint movement.

The choice of the correct depth–width ratio for each type of sealant minimizes local stresses which could promote premature failure of the sealant. The recommended depth:width ratios for different types of sealant[1] are: plastic 3:1 to 1:1, phaso-elastic 2:1 to 1:1, elasto-plastic 1:1 to 1:2, elastic 1:2.

Notwithstanding these recommendations, it is important to achieve a minimum depth of sealant in a joint to ensure adequate adhesion to the joint sides. For porous surfaces this depth should be not less than 10 mm, while for non-porous materials the minimum depth is 6 mm.

The back profile of the sealant and its correct depth are controlled by the back-up material which is placed in the joint before sealant application. A number of readily-compressible materials, such as closed-cell foamed polyethylene, are used for this purpose. It is essential that the sealant does not adhere to the back of the joint and is free to extend with joint movement. Where appropriate, therefore, a bond breaker consisting of a thin film of plastics material is inserted between the back-up material and the sealant.

Cure of sealants and joint movement

Elastomeric sealants are only capable of fulfilling their functions in a building joint once the chemical reactions resulting in their cure are complete. The time taken to attain effective cure varies considerably between different types; it is longest for those sealants which rely on the diffusion of atmospheric moisture into the bulk of the material to complete the cure process. The rate of cure will obviously also depend on environmental conditions, principally air temperature and humidity, at the time of applying the sealant and for some time afterwards. One part polysulphides usually take up to 3 weeks to cure completely, and this period can be considerably longer if the air remains dry and cold. Clearly, these relatively slow-curing sealants are vulnerable to damage if subjected to deformation during the period immediately after application to the joint. This may be caused either by cyclic movements at the joints or by irreversible joint movements, such as drying shrinkage of components or settlement of the structure.

Further information may be found in the articles **Primers for sealants** and **Sealants for double glazing**.

References

1. British Standards Institution, Guide to the Selection of Construction Sealants, BS 6213: 1982, London, BSI, 1982.
2. Building Research Establishment, Estimation of Thermal and Moisture Movements and Stresses: Parts 1 and 2. BRE Digests 227 and 228, 1979.

3. Building Research Establishment, Principles of Joint Design, BRE Digest 137, 1977.

4. British Standards Institution, Code of Practice for Design of Joints and Jointing in Building Construction, BS 6093: 1981.

5. British Standards Institution, Tolerances for Building, BS 6954: 1988.

Secondary ion mass spectrometry (SIMS)

D BRIGGS

Three different experiments for surface and interface analysis are possible by SIMS: static, dynamic and imaging SIMS.

The sample, in ultra-high vacuum is bombarded with a beam of positively charged ions (e.g. Ar^+, O_2^+, Ga^+, Cs^+). Material is sputtered from the surface, mostly as neutral species but with a few per cent of charged species (positive and negative secondary ions). These are collected by a mass spectrometer and mass analysed. Charging of insulators is controlled by simultaneous 'flooding' with relatively low-energy electrons.

In static SIMS the primary ion current density is very low (~ 1 nA cm^{-2}) so that the spectrum is representative of undamaged surface. The aim is to optimize the collection of large fragments (cluster ions) which provide information on surface molecular structure, corresponding to only ~ 10 Å sampling depth. This technique is particularly valuable for identifying surface contaminants (e.g. Fig. 124).

In dynamic SIMS the primary ion current density is very much higher (usually > 10 μA cm^{-2}) so that the surface is rapidly eroded. The intensity of elemental ions is followed as a function of time (i.e. eroded depth) to provide composition depth profiles. This technique is particularly valuable for studying buried interfaces and thin film structures.

In imaging SIMS the two-dimensional distribution of species on the surface is studied either by scanning a focused ion beam over the surface and mapping the intensity of useful secondary ions (ion probe) or by using a defocused ion beam and imaging mass selected species by ion–optical means (ion microscope). Imaging can be performed either in the static or dynamic mode with spatial resolutions down to 1000 Å.

The advantages of SIMS are: (1) surface sensitivity (~ 2 monolayers); (2) detection limits (as low as parts per billion (ppb) in dynamic SIMS); (3) molecular specificity and high degree of structural information from both organic and inorganic materials (in static SIMS). The main disadvantage is the inherent lack of quantitation.

An example of an application is given in **Primes for adhesive bonding**.

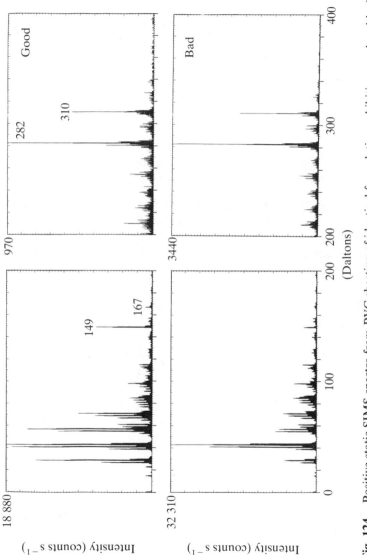

Fig. 124. Positive static SIMS spectra from PVC sheeting of identical formulation exhibiting good and bad adhesion to printing ink. In each case the surface composition is dominated by contaminants: dialkylphthalate (peaks at 149, 167 daltons) and ethylene-bis-stearamide (peaks at 282, 310 daltons). The latter is at much higher concentration in the 'bad' case and causes the loss of adhesion

Select references

D Briggs, A Brown, J C Vickerman, *Handbook of Static Secondary Ion Mass Spectrometry* (*SIMS*), John Wiley, Chichester, 1989.
A Benninghoven, R G Rudenauer, H W Werner, *Secondary Ion Mass Spectrometry*, John Wiley, New York, 1987.

Selection of adhesives

D M BREWIS

Introduction

With the exception of the pressure-sensitive type (see **Pressure-sensitive adhesion**), adhesives change from a fluid to a rigid state during bonding. This hardening process may be brought about by one of four mechanisms.

1. **Solvent-based adhesives** harden by evaporation of solvent, although with porous substrates, hardening will be accelerated by capillary action.
2. **Emulsion adhesives** harden by evaporation of the dispersion medium which is usually water; with porous substrates, hardening is again aided by capillary action.
3. **Hot melt adhesives** are thermoplastics which are heated above their softening-point; hardening occurs when the molten polymer is cooled between the two substrates.
4. **Reaction setting adhesives** or reactive adhesives harden when a monomer or resin undergoes a polymerization reaction.

There are, therefore, four main types from which to select an adhesive. The choice of an adhesive depends upon several factors which are now considered.

Setting time

The setting time for solvent-based adhesives depends largely on the boiling-point of the solvent used and can therefore be varied over a very wide range. Mixed solvents are often used to give the required evaporation rate.

With emulsion adhesives, drying can be accelerated by using heaters or by spraying the adhesives on to the substrates.

The setting times for reactive adhesives vary from a few seconds with **Cyanoacrylate adhesives** to several hours for a typical DIY epoxide cured at room temperature. For a given reactive system, the setting time can be varied by a large factor by changing the temperature and/or catalyst.

Melt adhesives can give the fastest setting rates of all and this is critical where very high production rates are required, e.g. edge veneering of blockboard.

Hazards

Adhesives may be associated with toxicity and/or inflammability and pollution hazards (see also **Health and safety**).

Harmful effects can occur due to inhalation of vapour or skin absorption. Solvent-based adhesives are probably the greatest potential problem as far as toxicity is concerned because their mode of operation involves the evaporation of large quantities of solvents. Some of the components of reaction setting adhesives are highly toxic and very efficient ventilation is necessary. Molten polymers have very low vapour pressures. However, additives in hot melt adhesives and decomposition products may produce toxic vapours, especially if reservoirs of melt adhesives are used. Emulsion adhesives where the dispersion medium is water represent the safest group of adhesives from the toxicity viewpoint.

Fire hazards are at a minimum with emulsion adhesives and at a maximum with many solvent-based adhesives because large quantities of organic compounds are involved. Chlorinated solvents are not inflammable, but have a relatively high toxicity.

Because of toxicity and inflammability problems, solvent-based adhesives are being replaced by other types of adhesive.

Initial performance

The high mechanical strengths associated with many cured reaction setting adhesives are often reflected in high lap shear strengths which may be up to 70 MPa with some metals. **Structural adhesives** are usually of the reactive type, with **Epoxide adhesives** being especially important.

Previously, these cured adhesives had been associated with poor peel, cleavage and impact strengths (see **Tests of adhesion**). However, with the incorporation of a rubber phase into the adhesive, the fracture toughness can be much increased and high peel strengths can be obtained while retaining high shear strengths (see **Toughened adhesives**). Toughened epoxides and **Toughened acrylics** are discussed elsewhere in this book.

High peel strengths can be obtained with adhesives possessing moderate cohesive strengths. For example, polyurethanes and nitrile rubbers give high peel strengths with a wide range of flexible substrates.

Durability

Adhesive joints may be subjected to a variety of adverse service conditions, including elevated temperature, organic solvents, water and stress (see **Durability**). Solvent-based, emulsion and melt adhesives are normally based on thermoplastics with fairly low softening temperatures. If a loaded joint is subjected to elevated temperature, failure may occur due to **Creep** unless the adhesive is cross-linked. Likewise, attack by organic solvents can be minimized by cross-linking. Solvent-based, emulsion and hot melt systems are available which cross-link after the initial bonding has been carried out. These systems provide improved in-service performance.

The more common environmental problem with metal joints is attack by water (see **Durability of coatings in water**). In addition to alloy composition, pretreatment and primer, durability is affected by the adhesive used (see **Pretreatment of metals, Pretreatment of polymers, Primers for adhesive bonding**). Sell[1] has ranked adhesives in the order of durability they provide with aluminium substrates:

nitrile-**Phenolics** > high-temperature curing epoxides
> medium-temperature curing epoxides > two-part room
temperature curing epoxides > two-part **Polyurethanes**

Cost

In evaluating the total cost of applying an adhesive, the following must be considered:

1. The cost of the adhesive based on the area of substrate covered;
2. Labour costs;
3. The cost of special equipment such as hot melt applicators, ovens for reactive adhesives and ventilation equipment for solvent-based adhesives;
4. Energy costs.

Overall assessment

No adhesives will be ideal with respect to all the above factors and it is therefore necessary to decide which are the most important properties required and then to make the appropriate compromise. In particular cases there will be other factors to consider, for example if it is necessary to bond oily steel then an **Acrylic adhesive** might be the best choice.

Further details regarding the advantages and limitations of different adhesives are given in articles already cited and in the articles listed below. Other sources include refs 1–4 and works discussed in **Literature on adhesion** and Appendix 3. See also **Adhesive classification, Anaerobic adhesives, Cyanoacrylate adhesives, High-temperature adhesives, Rubber-based adhesives**.

References

1. W D Sell, Some analytical techniques for durability testing of structural adhesives, *Proc. 19th Nat. SAMPE Symp.*, April 1974
2. J Shields, *Adhesives Handbook*, 3rd edn, Butterworths, London, 1984.
3. W C Wake, *Adhesion and the Formulation of Adhesives*, 2nd edn, Applied Science Publishers, London, 1982.
4. I Skeist (ed.), *Handbook of Adhesives*, 3rd edn, Van Nostrand Reinhold, New York, 1990.

Selection of joint sealants

J C BEECH

The primary function of a joint sealant is to prevent ingress of water or particulate matter to the building or structure. It must withstand the effects of changing environmental conditions throughout its effective service life, including the effect of joint movements. To do this it is required to maintain adhesion to the adjoining joint surfaces. Despite this **Sealants** are not adhesives, having entirely different functions, and they must be assessed by different performance criteria.[1,2]

A large variety of different sealants products is available, ranging from plastomeric oil-based mastics which are little more than joint fillers, to high-performance elastomeric sealants which are capable of withstanding large joint movements of up to 60% of the nominal joint width.

Selection of sealants

This is made primarily on the basis of the intended joint movement (see **Sealant joint design**), though other criteria are important such as compatibility with joint surface materials.[3,4] To achieve a satisfactory bond to certain surfaces under the prevailing site conditions, it may be essential to employ an appropriate primer (see **Primers for sealants**). Table 36 summarizes the types of sealant available commercially and indicates their performance characteristics.

The cheapest types of sealant, often referred to as 'mastics' are plastic in character, that is, they show little or no recovery when subjected to deformation. They have limited durability and will withstand only slight joint movement; they are most suitable for pointing around door and window frames. After application a tough skin is formed by oxidation of the surface, protecting the material inside which remains plastic. Progressive hardening through the bulk of the sealant leads eventually to cracking and to failure, though over-painting can considerably prolong its life.

Table 36. The main types of commercial joint sealants

Sealant type	Maximum tolerated movement (as % of joint width)	Character (after cure)	Typical uses	Comments	Expected service life (years)
Bituminous and rubber/bitumen	10	Plastic	In contact with bituminous materials	Poor durability in external movement joints	5
Oleo-resinous ('oil-based mastics')	10	Plastic	Pointing around window and door frames	Regular maintenance necessary	
Butyl rubber	10	Plastic	Pointing, bedding	Properties vary with formulation	Up to 10
Acrylic (solvent)	15	Plasto-elastic	Pointing, e.g. around timber frames treated with exterior wood stains, etc.	Good adhesion. May need warming before application	
Acrylic (emulsion)	10	Plasto-elastic	Internal pointing	Low health hazard. Durability uncertain	Up to 15
1-part polysulphide	25	Elasto-plastic	Movement joints in heavy structures	Slow curing: vulnerable to damage by movement until fully cured	
1-part polyurethane	30	Elastic	Movement joints with light (e.g. metal) components		
2-part polysulphide	30	Elasto-plastic	Both fast-moving joints in lightweight structures and slow moving joints in large heavy structures	Mixed on site so must be used within 'application life'. Maintenance costs low	Up to 20
2-part polyurethane	30	Elastic	Joints between plastics and metal components	High initial cost. Careful surface preparation essential	
1-part silicone (low modulus)	50	Elastic			
1-part silicone (high modulus)	20	Elastic	Sanitary ware. Fast-moving joints	Unsatisfactory on porous surfaces	

Notes: (1) Sealant types are listed in order of increasing initial cost; the list does not claim to be comprehensive. (2) Gun-grades are listed: for joints in horizontal surfaces the following self-levelling types are available: (a) hot-poured bituminous and rubber/bitumen; (b) cold-poured polysulphide and polyurethane.

The rest of the sealants in Table 36 are said to be elastomeric; they will all withstand some degree of joint movement and undergo change to a more or less elastic state after application. Elastomeric sealants which are predominantly plastic in character but which show limited deformation are termed plasto-elastic. The butyl-rubber and acrylic sealants are typical of this type, and are limited in their uses to joints in which moderate movement occurs and to pointing around window and door frames. Most of these change to a plasto-elastic state by loss of volatile constituents which can cause shrinkage, and this must be borne in mind when designing joints.

When the amplitude of joint movement is likely to be 20 per cent or more of the joint width, it is essential to use one of the elasto-plastic or elastic types of elastomeric sealant.[2] These also have greater inherent **Durability** than other types of sealant; the silicones, polysulphides and polyurethanes can be expected to perform for at least 20 years when used correctly in appropriate joints. All these sealants 'cure', that is, they are converted from a paste-like consistency to a rubbery state after application to the joint. This curing can be the result of a catalysed chemical reaction or by reaction with atmospheric moisture.

The sealants termed 'elasto-plastic' exhibit predominantly elastic behaviour but are also subject to some stress relaxation under deformation. These sealants perform well in slow-moving joints between large, heavy components of materials such as precast concrete. Stresses tend gradually to decay when the sealant is extended for long periods, so reducing the internal forces which may cause cohesive failure of the sealant or pull it from the joint faces. Those sealants which are elastic virtually recover completely after deformation within the expected range of joint movement. They are most appropriate for use in joints between lightweight components of high thermal conductivity, such as metals and plastics, where it may be expected that the rate of movement will be fairly high.

The sealants discussed so far are primarily supplied as 'gun-grades' for application to joints on vertical building surfaces. A number of types (e.g. polysulphides) are also supplied as self-levelling or pourable grades for use in horizontal surfaces; these sealants can be supplied with special properties, such as abrasion resistance for joints in traffic-bearing surfaces. Bitumen and bitumen/rubber-based sealants are also supplied for these purposes; both these types can withstand only limited movement.

Environmental factors

The location of the building and of the individual joints, together with the aspect of the face of the building, will largely determine the environmental factors affecting the performance of a sealant through its lifetime in the joint (see **Durability – fundamentals**). Dark coloured building surfaces exposed to sun in temperate climates may attain surface temperatures in excess of 70 °C in summer, while in the tropics temperatures above 100 °C will often occur; allowance must therefore be made for the sealant in the adjacent joints to be subjected occasionally to temperatures of this order. High temperatures will tend to accelerate certain chemical and physical changes in the sealant which, in conjunction with the cyclic movement to which it is subjected, cause gradual deterioration of the material. This degradation will be exacerbated by direct exposure to UV light. For non-curing and solvent-based sealants, this deterioration is manifest as hardening of the surface by oxidation or in shrinkage through loss of volatiles. Elastomeric sealants are prone to some surface degradation in these conditions; this may result in discoloration, chalking, crazing and, in extreme cases, surface cracking which leads eventually to cohesive failures. The formulation and colour of these sealants will influence the kind of degradation observed in service.

Direct exposure to moisture may also have adverse effects on the performance of some sealants, either by its effects on the sealant or on the joint substrates.[5] Absorption of water may cause softening of the sealant material, and some acrylic-based sealants are particularly affected in this way. If absorption is followed by prolonged near-freezing temperatures, the sealant may become stiffer, leading to an increase in internal stresses as the joint opens. These stresses tend to promote failures of adhesion at the interface of the sealant and the joint surfaces. The repeated absorption of moisture by components of porous materials, such as concrete, may cause a gradual migration of soluble salts towards the sealant interface which could lead to adhesion failure. This effect can be reduced by the application of appropriate primers to the joint faces before application of the sealant. It is essential to follow the advice of the sealant manufacturer in this respect.

Sealants in double glazing are the subject of a separate article.

References

1. J C Beech, The performance concept and jointing products, *Build. Res. and Pract.*, **8** (3), 158–69; **8** (4), 212–21 (1980).
2. J R Panek, J P Cook, *Construction Sealants and Adhesives*, John Wiley, New York, 1984, Chs 2, 5.
3. British Standards Institution, Guide to Selection of Construction Sealants, BS 6213: 1981, BSI, London, 1981.
4. *Manual of Good Practice in Sealant Application*, British Adhesives and Sealants Association and Construction Industry Research and Information Association, London, 1990.

5. J C Beech, C Mansfield, The water resistance of sealants for construction, *Building Sealants: Properties and Performance*, ASTM STP 1069, ed. F O'Connor, American Society for Testing and Materials, Philadelphia, 1990.

Shear tests

A D CROCOMBE

Introduction

Shear tests are generally carried out for one of two purposes. The first is to serve as a performance indicator for various adhesives subjected to the most common type of adhesive loading, namely shearing. The second is to determine actual material characteristics of the adhesive itself to be used in subsequent joint design, see **Engineering design with adhesives**. In this article a number of shear tests will be considered and the purpose of each discussed. Further information on these tests may be found in **Tests of adhesion** and also in other sources.[1,2] Many of the tests have been formalized in international standards and where possible examples have been cited from the ASTM, which is possibly the most comprehensive. The reader may find reference to other standards in **Standards on adhesion and adhesives**, Appendices 1 and 2 or elsewhere in this book.

All shear tests are variants of either an overlap or a torsional joint and these are illustrated schematically in Fig. 125. The shear in the first is induced by transfer of the tensile/compressive load from one adherend to the other, while in the second it is caused by transfer of torsional loads from adherend to adherend. The tests have been grouped according to the manner in which the shearing has been induced.

(a)

(b)

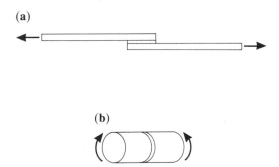

Fig. 125. Schematic diagram of (a) overlap and (b) torsional joint

Joints in shear by tension

Classification	Purpose
D 1002	Static strength
D 3528	Static strength
D 3165	Static strength
D 2295	Static strength, high temperatures
D 2557	Static strength, low temperatures
D 2294	Creep properties
D 3166	Fatigue properties
D 2919	Joint durability
D 3983	Shear modulus

The first of these tests (D 1002) is the standard single lap joint (single lap joint) test as portrayed in Fig. 125(a). Another article, **Stress in joints**, illustrates that, although this is a simple configuration, rotation of the overlap causes stresses that are complex, including shear and direct stresses both of which are highly non-uniform. The result of this is that the test does not measure any true shear properties. It is, however, valuable because it is simple and is typical of many bonded joint configurations. The double lap joint (D 3528) is essentially two single lap joints, back to back. This eliminates the gross joint rotation, and generally the strength of such configurations are more than twice the equivalent single lap joint strength. However, the adhesive stresses still include non-uniform direct and shear components and thus this suffers the same limitations as the single lap joint. The third test (D 3165) is a variant of the first but produced from two sheets bonded to form a laminate. This does not offer a significant improvement over the original single lap joint except that it might be easier to produce. All but one of the other tests in the table above use the single lap joint to measure joint performance under other types of load and give joint and not adhesive properties. The remaining test worthy of separate comment is the thick lap shear test (D 3983). This is an attempt to obtain true shear data by removing the non-uniformity in the adhesive stresses by significantly increasing the thickness of the adherends in a single lap joint. There are, however, a number of limitations which include difficulties in measuring the adhesive shear displacement accurately and also the continued presence of direct stresses in the adhesive. A number of improvements on this have been suggested. These include the development of a much more sophisticated extensometer to measure adhesive shear strains[3] and by the use of even more rigid adherends.[4]

Shear-strength tests are discussed in **Pressure-sensitive adhesives – adhesion properties**.

Joints in shear by compression

Classification	Purpose
D 2293	Creep properties
D 4027	Static shear strength and failure criteria
D 4562	Pin and bush shear strength

When shearing the joint using eccentric compressive loads as in the first two of these tests, it is necessary to prevent rotation of the specimen and this is usually done by physical constraint of the joint. This does not eliminate the direct stresses but rather reverses their sign and thus the same limitations apply. In the first of these tests a fairly standard single lap joint is loaded in compression within a simple constraining jig. The second text fixes the joint to massive carriers and uses a four-bar link principle to ensure that the adherends remain parallel. The specimen can also be subjected to tension or compression and the response of the adhesive to a known combined state of loading can be obtained. The final test is normally used to assess the performance of thread-locking adhesives, etc. Once again the stresses are not uniform and thus only the load required to shear the pin and bush assembly is recorded.

Shear by torsion

Classification	Purpose
E 229	Shear properties of the adhesive

This is generally known as the **Napkin ring test** and is similar to the configuration shown in Fig. 125(b). Essentially a thin-walled cylindrical tube is made up by bonding annular discs, typically made of aluminium. This tube is then subjected to a torque and the adhesive is subjected to almost pure shear which is essentially constant if the radial dimension of the adhesive layer is small. As with the thick adherend test (D 3983) careful extensometry is required to measure the small shear displacements that occur. A variant of this is the bulk torsion test where a thin-walled cylinder is made completely from cured adhesive (either by casting or machining) and this is tested as an ordinary torsion specimen. As the shear displacement now occurs along the complete specimen length, rather than only the bondline thicknesses, the displacements are much larger and easier to measure. Clearly, it is an easier test but specimen manufacture is more difficult; indeed it is not applicable to film adhesives.

Stresses in joints contains a discussion of shear stresses; see also **Creep**.

References

1. R D Adams, W C Wake, *Structural Adhesive Joints in Engineering*, Elsevier Applied Science Publishers, London, 1984.
2. A J Kinloch, *Adhesion and Adhesives*, Chapman and Hall, London, 1987.
3. R B Kreiger, Stress analysis concepts for adhesive bonding of aircraft primary structure in *Structural Adhesives in Engineering*, Mechanical Engineering Publications, London, 1986.
4. K M Leichi, T Freda, T Hayashi, Determining the constitutive and fracture properties of structural adhesives, *Proc. 5th Adhesion Science Review*, Virginia Polytechnic, 1987.

Solvent-based adhesives

J PRITCHARD

What are they?

For an adhesive to work it must start out in a fluid state to enable adequate wetting of the substrates to take place, and then the fluid state must be replaced by a solid state to enable the adhesive bond to gain strength.

Solvent-based adhesives accomplish this by having suitable solid materials dissolved in appropriate liquids. The solution is the liquid phase which carries out the wetting of the substrates, then, by evaporation, the solvent is removed leaving a solid adhesive film.

In practice there is a large range of materials that can be classified as solvent-based adhesives; the solvent can be either water or any of a number of organic liquids, the choice of the latter depending on what is to be dissolved, availability, cost and health and environmental considerations.

Aqueous solvent-based adhesives

This class of adhesives contains some of the most traditional materials, for example animal and fish glue and inorganic materials such as soluble silicates. Modern technology has brought other water-soluble materials into use including modified cellulose, soluble vinyl polymers; for example **Polyvinyl alcohol adhesives**, and a host of resins based on phenol, urea and formaldehyde (see **Phenolic resins**).

Generally these adhesives are used where at least one of the substrates is porous or permeable to moisture vapour. The two substrates to be joined are laminated while the adhesive is still wet, and the water escapes either by initial absorption into one of the substrates or by evaporation due to permeability. See also **Emulsion and dispersion adhesives**.

Uses of aqueous-based adhesives Because of the need to remove water, most applications for aqueous adhesives are with materials able to allow transmission of water vapour; hence the major uses are with paper, wood and fabric (see **Adhesives in the textile industry**), either as binders or laminating adhesives. A list of applications would include the following: remoistenable gummed tape, tube winding, box manufacture, plywood manufacture, reconstituted wood products, abrasives, woodworking.

Organic solvent-based adhesives

The progress of technology over the last 100 years has made available a number of organic liquids in large quantities, together with a range of synthetic solids that can be dissolved in them. Various combinations of these materials form a substantial part of the adhesives industry.

The method of use is similar to the aqueous products in that the solvent has to evaporate. If a wet laminating process is adopted then the solvent vapour must be able to escape through one of the substrates. A special and most important type of organic solvent-based adhesive is the so-called 'contact' adhesive. This makes use of the fact that certain elastomeric or 'rubbery' solids have the property of autohesion, i.e. they can stick readily to themselves, especially if compounded with resins and containing small amounts of solvents. This means that substrates may be coated with a contact adhesive, the adhesive can be allowed to dry till most of the solvent has evaporated, when the two adhesive surfaces can be brought together. Bonding takes place immediately and without much pressure. This property has meant that contact adhesives find wide use in industry for sticking many different materials together, easily and without the need for clamping.

Manufacture and compounding The majority of organic solvent-based adhesives are based on rubbery polymers, the main ones being natural rubber, polychloroprene, butadiene–acrylonitrile, styrene–butadiene and polyisobutylene. Traditionally the rubber was placed in a heavy-duty mixer and solvent was added slowly till a smooth solution was formed. In some cases the rubber was milled beforehand to reduce viscosity and produce smoother solutions. Nowadays it is possible to obtain some grades of material that only require stirring in comparatively simple churns.

Depending on the rubber, the solvent can either be petroleum hydrocarbon of differing boiling ranges, aromatic hydrocarbon – usually toluene – or oxygenated solvents such as ketones or esters. If non-flammability is required then chlorinated solvents are used. See also **Rubber-based adhesives**.

Various resins are added to improve adhesion to different substrates. Antioxidants will be present to improve ageing characteristics while fillers are often added to improve performance.

Applications of organic solvent-based adhesives Organic solvent-based adhesives find wide application in general industry. They can be used in the footwear industry (see **Footwear applications of adhesives**) for sticking soles to uppers and a variety of other operations. The automotive industry (see **Automotive applications**) uses them for a variety of jobs including stick-on trim. Foam seat manufacturers use them in brushable and sprayable form. Paper conversion employs solvent-based permanently tacky adhesives in the manufacture of labels and sticky tapes. Wherever a variety of different materials has to be stuck together quickly and easily, then organic solvent-based adhesives will often be the answer.

Advantages and disadvantages of solvent-based adhesives

Advantages They are very easy to use. They can be taken directly from the can and applied with brush or scraper or applied easily by large spreading machines. They can be spray applied. Viscosities can be altered between wide limits to suit the application. Heat does not have to be applied. The standard one-can products have an extensive shelf-life (12 months minimum is quite normal) and do not require expensive metering/mixing equipment. They can be compounded to stick to a wide variety of materials and can stick many dissimilar materials together: extensive surface preparation is usually unnecessary. They are easy to use and can withstand some abuse. The presence of rubbery elastomer gives tenacious impact-resistant bonds and the 'contact' variety can make bonds immediately that do not need clamping.

Disadvantages Water-based adhesives can take a long time to dry in cool, damp conditions. Organic solvents often constitute a risk to **Health and safety**, unless used properly.

References

1. R L Patrick, *Treatise on Adhesion and Adhesives*, Vol. II, Marcel Dekker, New York, 1969.
2. J Delmonte, *The Technology of Adhesives*, Van Nostrand Reinhold, 1947.
3. I Skeist, *Handbook of Adhesives*, Van Nostrand, New York, 1977.
4. R Houwink, G Saloman, *Adhesion and Adhesives*, Elsevier, Amsterdam, 1967.

Solvent welding

M A GIRARDI

Introduction

Solvent welding (also known as solvent cementing) over the years has become one of the ways of mass-producing strong and reliable joints for thermoplastic materials.[1-3] These materials are best suited for solvent welding, because of the importance of polymer chain interdiffusion, where there should be no major structural restrictions on this movement (see **Diffusion theory**). For this reason solvent bonding is substantially confined to the amorphous and some partly crystalline types.

Moreover, in non-crystalline polymers the possibility is reduced of local morphological differences arising in the material of the bond region of the completed joint to introduce areas of potential weakness through the creation of internal stresses and/or localized phase boundaries.

Some of the amorphous and partly crystalline polymers with which solvent welding has been used, are shown in Table 37.[1]

Procedure

Prior to making a joint, the fit and alignment of the mating parts should be checked and masked to ensure ready assembly in the correct position. Once a cemented joint has been made it cannot be dismantled. A solvent may be chosen from tabulations of solvents (e.g. Table 37) or by matching polymer and solvent solubility parameters (see **Compatibility**).

The two surfaces to be joined must be pre-cleaned with a suitable reagent (which has no solvent actions on the polymer). A suitable solvent or mixture of solvents, which may contain some of the polymer already dissolved, is then applied. The joint is left 'open' for a short time for any excess solvent to disappear from the surface, partly by penetration inwards and partly by evaporation.

The treated surfaces are brought together and held under pressure (clamped), to ensure intimate surface contact between the components. Depending on the material, design and geometry of the joint, anything from contact pressure to several meganewtons per square metre ($MN\ m^{-2}$) may be employed. Some figures for weld strength are given in Table 38; these show that, with suitable bonding conditions, a strength equivalent to the cohesive strength of the polymer may be obtained.

Since the solvent is only used to soften the surface, its rate of evaporation influences both the time needed to prepare the bond (set-up time) and the time needed for the bond to set (cure time). If too much solvent evaporates prior to clamping, the bond can become weak. If solvent is

Table 37. Polymers suitable for solvent welding[1]

Polymer		Solvents employed	
Name	Normal morphology	Type	Examples
Acrylic (polymers and copolymers)	Glassy	Chlorinated hydrocarbons dichloroethane	Dichloromethane
PVC and some vinyl chloride copolymers	Glassy	Homopolymer: selected ketones	Tetrahydrofuran
		Copolymers: some ketones and esters	Tetrahydrofuran, dioxane, butal acetate
Polystrene and styrene copolymers	Glassy	Aromatic hydrocarbons (especially for the homopolymer), chlorinated hydrocarbons, ketones (especially for the copolymers)	Toluene, dichloroethane, methylethyl ketone
Modified polyphenylene oxide	Glassy	Chlorinated hydrocarbons dichloroethane	Dichloromethane
Polysulphone	Glassy	Chlorinated hydrocarbons dichloroethane	Dichloromethane
Polycarbonate	Glassy (can be partially crystallized by heating or solvent treatment)	Chlorinated hydrocarbons dichloroethane	Dichloromethane
Cellulose esters (especially nitrate, acetate and acetate/butyrate)	Some may be partly crystalline	Selected ketones, esters and chlorinated hydrocarbons	Methylethyl ketone, acetone, butyl acetate, dichloromethane
Polyamides	Partly crystalline	Phenolic compounds, some bivalent metal salt solutions	Phenol, resorcinol, cresols (aqueous or alcoholic solutions), calcium chloride (alcoholic solution)

Table 38. Shear strength of solvent-promoted joints[1]

The joint		After treatment (at a clamping pressure of ~ 1 MN m^{-2})	Shear strength (MN m^{-2})
Polymer	Solvent		
1. Polycarbonate	1,2-dichloroethane	85 °C for 6 h then 120 °C for 5 h	> 38.43
2. Polycarbonate	1,2-dichloroethane	Kept at room temp. 25 h	16.77
3. Polycarbonate	Chloroform	As 1	29.77
4. Polycarbonate	1,2-dichloroethane	55 °C for 4 h 80 °C for 14 h 120 °C for 5 h	35.8
5. Polysulphone	1,2-dichloroethane	55 °C for 4 h 80 °C for 14 h 120° for 5 h	38.5
6. Polycarbonate	None ('solid polymer specimens')	—	40.90
7. Polysulphone	None ('solid polymer specimens')	—	39.8

trapped in the bond area, evaporation can take days, and the part can be damaged. An increase in strength may be achieved by post-heating the joint.

Many forms of joint may be required during the fabrication of components, e.g. tee joints, 90° bends, lap and butt joints of several types. Each should be assessed individually to ensure that it is the appropriate joint for the purpose.

Safety

Organic solvents are highly toxic and volatile, and it is therefore important that precautions advised by the manufacturers should be followed. This essentially includes good ventilation of the work area, and avoidance of excessive skin contact. Volatile liquids and cements should always be kept in cool storage and containers kept sealed (see **Health and safety**).

Advantages and disadvantages

Solvent welding is a simple, clean and comparatively inexpensive way of joining plastic parts. It can also lend itself to joining large components compared to other joining methods, which are restricted to the size of components that can be welded. In addition, other adhesive bonds, although usually effective, introduce a new material into the assembly (and thus the threat of incompatibility), but solvent welding barely leaves a visible joint.

Solvent cements – solvents with pre-dissolved polymer – have a limited storage life; deterioration shows as a marked increase in viscosity or as the formation of a gel. Such solvent cements are then difficult to spread and make very weak joints. A cement which has started to gel may be thinned by adding small amounts of original solvent and mixing well, but there is a risk that it may be over-thinned and/or will not be properly mixed; if so a poor cement will result. To reduce or slow down the deterioration of a solvent after unsealing the container, the container should be effectively resealed after use and even between the making of joints if practicable.

References

1. W V Titow in *Adhesion–2*, ed. K W Allen, Applied Science Publishers, London, 1978, p. 181.
2. W V Titow, R. J. Loneragan, J H T Johns, B R Currell, *Plast. Polym.*, **41**, 149 (1973).
3. A H Lemdrock, *Adhesives Technology Handbook*, Noyes Publications, Park Ridge, N.J., 1985.

Standards for adhesives and adhesion

G R DURTNAL

Standards play an important part in a technologically related topic such as adhesion and adhesives. They are published by both national and international bodies. Important sources of standards include the British Standards Institution (BSI), the American Society for Testing and Materials (ASTM) and International Organization for Standardization (ISO). An indication of the wide scope of relevant standards can be judged from Appendix 1 which lists standards in the field, and from Appendix 2 where a range of test methods for adhesive joints are compared. This list will doubtless continue to grow, as requirements are established for new test methods to cover novel adhesives, or perhaps well-established materials in novel applications. In addition, the increasing importance of international and European standardization will add methods new to the UK list.

The reason for all the effort expended on these standards lies with the increasing importance of quality assurance. To assure the quality of a material implies the need to test it, to obtain consistent and reproducible results from such testing requires closely controlled test methods. Hence the use of standards as documents to be quoted by the purchaser and the supplier of the material. Even today, when disagreements or discrepancies arise relating to the testing of an adhesive to a recognized standard, the reason very often can be found in some uncontrolled variable between two test sites, which had resulted in a large effect upon the recorded value of the result.

It is therefore of paramount importance to lay down strict conditions of test in standards. This has a negative side, however, in that it encourages a test to be as simple as possible, which decreases its relationship with the conditions experienced by the material in service. Accordingly, great caution must be used when looking at the quantitative results from standard tests, for example, the bond strength attained by a structural adhesive in a simple lap-shear test (see **Shear tests**) at room temperature may bear no relationship to the level it achieves in service, where the substrate, nature of the applied stresses and environment will be totally different (see **Durability**).

Adhesives standards

In the UK there exist relatively few British Standards for specific adhesives materials, the main ones being BS 1203 and 1204 for wood adhesives and BS 6209 for solvent cement. The tendency has been to prepare test methods for adhesives as British Standards in the BS 5350 series (British Standards Methods of Test for Adhesives) which can then be quoted in specifications for adhesive materials issued by suppliers or users of the adhesive. It is interesting to note in passing that the situation regarding adhesive tapes is quite the opposite, with most common varieties of tape being covered by individual British Standards.

With adhesives, therefore, the same test method may be quoted for widely differing materials; the parts of BS 5350 are written in such a way as to take account of this. For instance, BS 5350, Part C5, the simple lap-shear bond strength test, may be used to determine the bond strength of adhesives of widely different cohesive strengths on substrates of widely differing natures.

In very simple terms, tests on adhesives may be regarded as comprising three main types (see **Testing of adhesives**).

1. Tests carried out on an adhesive as a material, usually to give information concerning its application properties. Typical of these tests are viscosity, solids content and density.
2. Tests on adhesives to determine their curing or setting behaviour, e.g. tack-free time, skin-over time, pot life, etc. (see **Tack**).
3. Tests on the cured adhesive/adherend assembly to determine the bond strength attained (see **Tests of adhesion**).

A wide variety of each test exists, as can be seen from Appendix 1.

In addition, other tests not directly related to the quality of the adhesive may be considered. These include, to an increasing extent, tests concerned with health, safety and environmental properties, e.g. flash point and toxicity (**Health and safety**). Mention must also be made of BS 6138, the Glossary of Terms used in the adhesives industry, which is essential reading if one wishes to interpret some of the more abstruse terms used in the trade.

In general the bond strength tests in the BS 5350 series are short-term quality-control tests not suitable for design purposes; indeed the more recent revisions of such methods include a disclaimer to this effect. They have the advantage of simplicity, rapidity and ability to give reproducible results. However, there is an increasing body of opinion in favour of issuing standards for more complex and longer-term techniques, which could be of greater value to designers. Examples of such tests include cyclic loading of **Fatigue** testing, crack propagation (Boeing wedge, etc., see **Wedge tests**) testing and environmental testing. Despite the problems inherent in standardizing such tests, they will undoubtedly grow in importance in the future.

Formulation of standards

The British Standards referred to here are the responsibility of a British Standards Technical Committee. This is composed of representatives of adhesives suppliers and users, along with any interested Government departments. Such members are generally experts in their field, although for particular topics, other experts may be co-opted on to the committee, or consultants may occasionally be used. The standards are rarely, if ever, prepared 'from scratch'; they are normally based upon existing methods, such as other national standards, standards published by Government departments or perhaps those used by large commercial concerns. It is the function of the committee to discuss new standards and to prepare a draft document suitable for public comment; the public in this context means other parties interested or active in the relevant field. After taking such comments into account, the standard is then issued, but is subject to revision at regular intervals in order to keep abreast of technical progress. The representatives on technical committees perform therefore a vital task which is often undervalued as a mundane and routine job.

Standardization on an international level has existed for a considerable time now, mainly through the ISO. Draft standards are progressed in much the same way as for national standards; indeed they are usually based on one or more published national standards. Following comment and discussion, which by the nature of the work is often protracted, the ISO Standard is issued and, if found acceptable, is adopted by the UK, often as a 'dual-numbered' British Standard.

More recently, greater emphasis has been given to work within the European Community under the auspices of CEN (Comité Européen de Normalisation), the organization responsible for standardization in Europe. This is connected with the single European market planned from 1992. It is likely that CEN will adopt ISO Standards wherever possible to avoid duplication of effort; unlike ISO Standards, CEN Standards once published are required to be adopted by the member nations, with national standards that cover the same area being superseded. This could well lead to new and unusual test methods becoming the required technique in the UK. Through BSI, this country is well represented on the relevant CEN Committee and associated working groups, which should ensure that due weight is given to the current British Standard methods.

The issue is unfortunately rather complicated by the parallel existence of aerospace committees working along similar lines, both within BSI and in Europe within the Association Européene des Constructeurs de Matériel Aerospatial (AECMA) organization. This is the European organization responsible for standardization related to aerospace materials, and hence covers many structural applications of adhesives, as well as aircraft sealants. It is to be hoped that the links existing between AECMA, CEN and ISO will eliminate the risk of multiple standards being published to cover the same basic techniques or materials. Undoubtedly, international standardization is the way forward, with work on a national scale showing a rapid decrease.

Conclusion

Further information on test methods can be obtained from the general references below, from the standards listed in Appendices 1 and 2 and from the following specialized articles: **Blister test, Climbing drum peel test, Fracture mechanics, Fracture mechanics test specimens, Peel tests, Scratch test, Shear tests, Tensile tests, Testing of adhesives, Tests of adhesion, Wedge test**.

Select references

BSI Standards Catalogue, BSI, London, annually.
Annual Book of ASTM Standards, ASTM, Philadelphia, annually.
ISO Catalogue, ISO, Geneva, annually.

Statistics

C CHATFIELD

Statistics may be described as the science of collecting, analysing and interpreting data in the best possible way. Note that it is not just concerned with data analysis as some scientists think.

Random variability

All data in the experimental sciences are subject to some 'random' variability due to a variety of causes such as measurement error, natural variability in laboratory specimens, uncontrolled variation in external conditions and so on. With careful control of the measurement process and of laboratory conditions, this variation can sometimes be kept very small. However, in other cases it may be more substantial and then the need for statistics is more apparent. Random variation is often described, perhaps rather misleadingly, as the 'error' component of variation. However, it is essential to realize that there is nothing abnormal about random variation, though the experimenter will naturally wish to reduce it as far as possible.

A major objective in statistics is to 'make decisions in the presence of uncertainty' and come to conclusions about relationships despite the presence of 'error terms'. These brief notes naturally cannot provide more than a brief introduction to the subject, and should be read in conjunction with an introductory text for scientists – see Select references.

Collecting the data

The first step in any study is to clarify the objectives, decide which variables need to be measured and/or controlled and decide how many observations are to be taken, and how. If the data are to be collected by a technician, then clear instructions and close supervision are advisable to ensure that 'good' data are collected. Poor data cannot be rescued by a statistical analysis. The topic of experimental design does not always receive the attention it deserves.

Analysing the data

The first step in analysing the data is to process them into a suitable form for analysis, often by typing them into a computer. At this stage it is advisable to have a look at the data in order to get a 'feel' for them. It will usually be helpful to plot the data in any way that seems appropriate and also to calculate some simple descriptive statistics. Do not rush into a more complicated analysis before this has been done.

The simplest case is when we have, say, n observations on a single variable x. The observations will be denoted $x_1, x_2, \ldots, x_n$. If x is a continuous variable (which can take any value in a specified range), then it is a good idea to plot a histogram of the data to see what the distribution of the variable looks like. Possible values of x are divided into different class intervals and the number of values in each interval (the frequency) is then found. The histogram plots frequency against x. One can then see if the distribution has a 'bell-shaped' symmetric shape (called a normal distribution) or if one tail is longer than the other (called a skew distribution). Any extreme values, called outliers, should also be visible. If the latter are judged to be errors, then they should be excluded from the analysis. If the observed variable is discrete (so that it can for example only be a non-negative integer), then the frequencies can be plotted in the form of a bar-chart.

It is also helpful to calculate a typical value (called a measure of location) which is usually the (arithmetic) mean $\bar{x} = \sum_{i=1}^{n} x_i/n$. Alternatives include the median (the middle value), which is useful if the data have a skewed distribution, and the mode (the most frequently occurring value). It is also a good idea to measure the variability of the data with a measure of spread such as the range (largest observation minus the smallest observation) or the standard deviation,

$$ s = \sqrt{\left[\sum_{i=1}^{n} (x_i - \bar{x})^2/(n-1) \right]} $$

Note that the square of the standard deviation s^2, called the variance, is not in the same units of measurement as the data and so is no use as a descriptive statistic, although it has many other uses in more complicated analyses.

Having obtained some feel for the data, and revised or excluded obvious errors, a proper statistical analysis can now be carried out. There are numerous statistical techniques appropriate for different situations. Here we can only mention the broad aims of a few selected techniques. With the advent of the computer, the details of mathematical formulae are no longer so important. What is important is to be able to choose the right technique and understand the results.

With observations on a single variable, it may be of interest to assess whether the sample mean, $\bar{x}$, is significantly different from a suggested population 'true' value, say μ. A procedure called a t-test essentially looks at the different $(\bar{x} - \mu)$ to see whether it is too large to have arisen by chance. This is the best-known example of a significance (or hypothesis) test.

The p-value which comes from a significance test is the probability of observing a more extreme value than the one obtained given that the null hypothesis (in this case that $\bar{x}$ really is from a population whose true mean is μ) is true. It is not, as many scientists believe, the probability that the null hypothesis is true. A small p-value (usually <0.05) is taken to be evidence against the null hypothesis leading to its rejection.

The two-sample t-test, as the name implies, is concerned with comparing the sample means of two groups of observations. The null hypothesis is that both groups come from the same population and a 'large' difference between the two sample means leads to rejection of the null hypothesis (as is intuitively obvious).

With three or more groups of observations to compare, it is incorrect to compare each pair of groups with a two-sample t-test. Instead a one-way analysis of variance (ANOVA) should be carried out. The ANOVA technique can be generalized to deal with observations from many other types of experimental design. In each case the analysis separates out the variation due to specified components of variation (e.g. the differences between group means in a one-way ANOVA) and the variation due to the residual or error terms. The former components of variation are then compared with the latter component to see if the systematic components are too large to have arisen by chance.

If the scientist is interested in the relationship between two variables, say x and y, and has observed n pairs of observations, say $(x_1, y_1), (x_2, y_2), \ldots, (x_n, y_n)$, then it is a good idea to begin by plotting y against x to give what is called a scatter diagram. The form of the relationship (e.g. a straight line or a quadratic curve) between x and y, may then be evident. If the relationship is linear, it may help to calculate a statistic, called the correlation coefficient, which measures the degree of association between the two variables. A value near zero indicates little or no relationship, while a value close to $+1$ (or to -1) indicates a strong positive (or negative) relationship. The technique, called regression, is concerned with finding a regression curve to describe the relationship between a dependent variable and one or more explanatory variables.

Quality control

Another important area of statistics is statistical quality control. This is concerned with the problems involved in controlling the quality of a manufactured product. Acceptance sampling is concerned with monitoring the quality of manufactured items supplied by the manufacturer to the consumer in batches. The problem is to decide whether the batch should be accepted or rejected on the basis of a sample drawn randomly from the batch. Statistical process control is concerned with detecting changes in the performance of a manufacturing process and with taking appropriate action, where necessary, in order to keep the process on target. In order to do this control charts, such as Shewhart charts or cusum charts, are often used to plot a variable of interest against time. The recent arrival of Taguchi methods from Japan has heightened interest in quality control. These methods emphasize error prevention rather than detection and recommend the use of experimental design at the product design stage in order to get a process running near its optimum setting with minimum variability. The general philosophy of Taguchi methods is excellent although some of the detailed techniques are open to criticism.

Reliability

The reliability of a product is a measure of its quality and has a variety of definitions depending on the particular situation. For example reliability could be defined as the probability that a device will function successfully for a certain period of time. One method of measuring the reliability of a product is to test a batch of items over an extended period of time and note the failure times. This procedure is called life-testing. The resulting data may enable us to estimate the underlying probability distribution of failure times and/or estimate the conditional failure rate function, which describes the conditional probability of failure given that an item has survived until the particular time in question. The topic of reliability was given a major boost by the American space rocket programme where sufficiently high reliabilities could only be obtained by adding back-up items in parallel.

Select references

G E P Box, W G Hunter, J S Hunter, *Statistics for Experimenters*, John Wiley, New York, 1978.
C Chatfield, *Statistics for Technology*, 3rd edn, Chapman and Hall, London, 1983.
G W Snedecor, W G Cochran, *Statistical Methods*, 7th edn, Iowa State University Press, Ames, Iowa, 1980.

Stresses in joints

R D ADAMS

In most adhesive joints the designer should arrange for the loads to be transferred through the adhesive layer by shear, taking care to avoid any

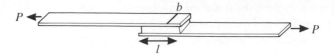

Fig. 126. Lap joint loaded in shear by tensile forces *P*

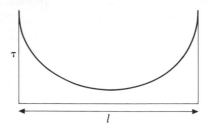

Fig. 127. Volkersen's prediction of the shear stress τ, acting in the joint shown in Fig. 126

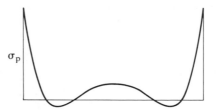

Fig. 128. Goland and Reissner's prediction of the transverse (peel) stresses σ_P, acting in the joint shown Fig. 126

major stress concentrations. It is generally agreed that peel (major transverse tensile loads) should be minimized or avoided.

For a lap joint loaded as in Fig. 126, the average shear stress τ is given by $\tau = P/bl$ where P is the in-plane tensile load, l the length of the joint and b its width. In practice, this gives a very crude approximation to the joint stress. A better solution was given by Volkersen[1] who allowed for the stretching of the adherends under the action of the tensile load, giving rise to differential shears and the bathtub shear stress curve given in Fig. 127 (see also **Shear tests**).

The system of loading induces not only shear stresses, but also tensile (peel) stresses across the adhesive layer thickness. The first analysis of these peel stresses was given by Goland and Reissner[2] and they showed peaks in these at the joint ends, as shown in Fig. 128.

A joint will fail when the stresses and/or strains reach some critical state. In order to predict joint failure, it is necessary to determine accurately the true stress and strain state. This will be a function of the micro and macro geometry, whether there are sharp corners, whether the adhesive is elastic or elasto-plastic, and what is the nature of thermal, cure and hygrothermal loading. The true nature of the stress and strain state can only be ascertained by carrying out a detailed **Finite element analysis**. This requires considerable care in setting up, and even more care in analysing the results. Of particular importance is the input data on the adhesives used, and these are often unreliable. The reader is referred to the book by Adams and Wake[3] for a fuller description of the theories of Volkersen, Goland and Reissner, and the finite element method as applied to bonded joints.

For joints other than overlaps loaded basically in shear, there are few, if any, standard solutions available.

Finally, it must be emphasized that since most adhesives fail in tension and not shear, any proprietary stress or strain analysis which predicts failure using shear parameters for failure is itself doomed to failure. The best method of design is to minimize peel, use $\tau = P/bl$ for all its shortcomings and to build in a factor of safety (ignorance) of 10 to 20.

References

1. O Volkersen, *Luftfahrtforschung*, **15**, 41 (1938).
2. M Goland, E Reissner, *J. Appl. Mech., Trans. ASME*, **66**, A17 (1944)
3. R D Adams, W C Wake, *Structural Adhesive Joints in Engineering*, Elsevier Applied Science Publishers, London, 1984.

Structural adhesives

P CULLEN

Introduction

Despite the success of structural adhesives in **Aerospace applications** over the past several decades,[1,2] growth in their use in the automotive (see **Automotive applications**) and other sheet steel fabrication industries, while positive, has been less spectacular. Several factors have contributed to this, not least of which has been the general unsuitability of many available adhesives to satisfy the much differing requirements of these industries. Thus, auto engineers, when studying the use of structural adhesives in aircraft construction (particularly epoxies, see **Epoxide adhesives**), have found it difficult to capitalize on the undoubted benefits of bonding, given that the adhesives used require rigorous surface pretreatments such as acid etching, use of skilled labour, long clamping and oven-cure cycles. Furthermore, much of this available information on the use of such adhesives obviously concentrates on the bonding of aerospace aluminium and not mild sheet steel coated with the various pressing and protective lubricants used in the fabrication and protection of car body components. However, recently major advances in the formulation and bonding capability of structural adhesives, based on modified polymer technology, have been achieved. Improvements have evolved to address fastening challenges presented by the revolutionary changes in the technology of materials in all industries, especially automotive. Structural thermoplastics have provided the greatest challenge and opportunity because of their performance demand and diversity in composition. For example, today's methacrylate technology represents a quantum leap over the performance of the more traditional acrylic technology of the past. This technology evolved to fill the niche unsuccessfully targeted by early acrylics (see **Acrylic adhesives**) that suffered from shortcomings such as brittleness, limited gap-filling ability and unreliable cure.

Adhesive profile for structural bonding

This challenge is being met as the number of industries now using structural adhesives is large and their activities diversified; their requirements of an adhesive have many common factors. This is not surprising as many construction techniques, such as attachment of stiffeners, brackets and panels, are used in a wide range of structural products from motor cars and commercial vehicles through to electrical equipment, office furniture and agricultural machinery. The most important requirements can be summarized in the following paragraphs.

A high-performance structural adhesive should provide both rigidity for high tensile strength and the toughness and flexibility to resist high peel and impact forces.[3] The latter is important as the minimal joint design that may initially confront any adhesive will undoubtedly mean that bonds will not just be subjected to tensile stress but also to the more difficult to carry cleavage and peel stresses. Numerous types of high-modulus, brittle and glassy adhesives can easily accommodate high-tensile stresses, but show no resistance at all to peel and impact forces. These conflicting requirements can only be met by toughened adhesives, such as **Toughened acrylic adhesives** or **Epoxide adhesives**.

Curing and bonding mechanisms

The condensation cure mechanisms of **Epoxide adhesives** and **Polyurethane adhesives** require stoichiometric or balanced combining ratios and thorough mixing for maximum performance. Unlike methacrylates, epoxies and polyurethanes combine two dissimilar materials to form a polymer by a stepwise process. Therefore, off-ratio mixing or incomplete mixing can alter significantly the structure of the backbone and affect its resulting properties.

On the other hand, the free radical cure mechanism of methacrylates is catalytic in nature (see **Addition polymerization**). As such, the liquid adhesive is a complete mixture of the functional ingredients and needs only the addition or mixing of catalyst components to solidify the adhesive polymer backbone.

Cure of the methacrylates is initiated by the addition of a peroxide to the formulated adhesive which already contains amine. The amine induces decomposition of the peroxide into free radical fragments that trigger room-temperature polymerization of the methacrylate monomer.

Applications

Structural adhesives can be formulated to have low controllable viscosity or to exhibit thixotropy with anti-sag behaviour suitable for bonding joints with large gaps. Both structural acrylic and polyurethane adhesives are generally multi-component. The room-temperature curing of these adhesives makes them particularly useful for structural applications where plastics are being used. For example, the bonding of high-impact polycarbonate/polyester thermoplastic bumper in an automotive application.

Structural epoxides, which can be either mono-component or multi-component are generally selected for strength and/or strength at temperature (hot strength). Typical structural applications for epoxies include numerous joints in the aerospace industry where they are widely used.

Advantages/disadvantages of structural adhesives

Table 39 compares structural acrylics (often referred to as **Toughened acrylic adhesives**) to two other classes of structural adhesives, **Polyurethane adhesives** and **Epoxide adhesives**. Often, the fast room-temperature cure of structural acrylics, together with the non-critical mix ratio, make them ideal for an automated production line, while both the structural epoxies and structural polyurethanes, although slower curing, can often bond a more diverse range of substrates.

Other articles of relevance are **Epoxide adhesives, Polyurethane adhesives, Toughened adhesives** and **Engineering design with adhesives**.

Table 39. Comparative benefits of industrial structural adhesives (structural acrylics, polyurethanes, epoxies)

	Structural acrylics	Structural polyurethanes (with primers)	Structural epoxies	
			Two-part mixed	One part heat cure
Bond strength				
Metals/ferrites	+ +	+	+	+ +
Metals/plastic	+ +	+	0	—
Toughness	+ +	+	+	+
Impact resistance	+	+ +	+	0
Temperature/humidity	+	—	+	+ +
Speed of bonding	+ +	—	0	—
Ease of cure				
Automatic application	+ +	—	—	+ +

Note: + + great benefit, +, significant benefit, 0 no benefit, — significant disadvantage.

References

1. R L Patrick, *Structural Adhesives with Emphasis on Aerospace Applications*, Marcel Dekker, New York, 1976.
2. J C McMillan, Durability test methods for aerospace bonding, in *Developments in Adhesives – 2*, ed. A J Kinloch, Applied Science Publishers, London, 1981.
3. J N DeLollis, *Adhesives for Metals Theory and Technology*, Industrial Press, New York, 1970.
4. R S Charnock, F R Martin, Structure–property relationships of a rubber modified acrylic adhesive, *Proc. International Conference Adhesion and Adhesives*, Plastics and Rubber Institute, Durham, 1980.
5. W A Lees, Use of adhesives in constructing vehicles, *Adhesive Age*, **24**(2), 23–31 (1981).

Surface analysis

D BRIGGS

Surface analysis refers to the characterization of the outermost layers of materials. A series of techniques have been undergoing continuous development since the late 1960s based on ultra-high vacuum (UHV) technology. The most useful of these methods provide information on surface chemical composition and four techniques satisfy this requirement: **X-ray photoelectron spectroscopy** (XPS or ESCA), **Auger electron spectroscopy** (AES), **Secondary ion mass spectrometry** (SIMS) and ion scattering spectroscopy (ISS). The first three are described elsewhere in the book. In ISS the energy of scattered incident ions is measured and the elemental composition of the outermost atomic layer is determined.

Despite this unique capability ISS is very much less used than the other three methods.

By combining surface analysis with ion beam sputtering (etching) a sequential or continuous monitoring of composition as a function of depth below the surface is possible. This is known as depth profiling.

Ultra-high vacuum conditions are important for two reasons. Firstly the material to be analysed is bombarded with particles or photons and particles (electrons or ions) leaving the surface are detected; these particles suffer collisions with residual gas molecules if the pressure is $> 10^{-5}$ torr. Secondly, and more important, surfaces to be analysed can be rapidly contaminated by adsorption of residual gas molecules if the pressure is $> 10^{-9}$ torr. Hence surface analysis instrumentation is constructed routinely to achieve $\sim 10^{-10}$ torr (after bake-out).

As the individual technique descriptions detail, the three techniques of XPS, AES and SIMS are highly complementary in terms of the ability to deal with materials of different types (ranging from metals to polymers), the information available from the spectra, the surface sensitivity, the elemental detection limits and degree of quantitation and the degree of spatial (lateral) resolution available. Hence multi-technique systems are common in which more than one technique can be performed on the same sample, also saving the cost of multiple UHV systems.

Surface analysis techniques have been extensively applied in both fundamental and applied adhesion studies across the whole field of materials science (see references).

Other methods of surface characterization are discussed in **Infra-red spectroscopy of surfaces, Surface characterization by contact angles, Critical surface tension, Electron probe microanalysis** (EPMA).

Some features of surface analysis techniques are compared in Table 40.

Table 40. Comparison of some surface analysis techniques

	Incident radiation	Emitted radiation	Approximate depth analysed
AES	Electrons	Electrons	3 nm
XPS	X-rays	Electrons	3 nm
SIMS	Ions	Ions	2 monolayers
ISS	Ions	Ions	Outer atom layer
EPMA	Electrons	X-rays	1 μm
ATR	IR photons	IR photons	1 μm

Note: ATR – attenuated total internal reflection IR spectroscopy.

References

1. D M Brewis (ed.), *Surface Analysis and Pretreatment of Plastics and Metals*, Applied Science Publishers, London, 1982.
2. D M Brewis, D Briggs (eds), *Industrial Adhesion Problems*, Orbital Press, Oxford, 1985.
3. D Briggs and M P Seah (eds), *Practical Surface Analysis*, Vols 1 and 2., 2nd ed, John Wiley, Chichester, 1990.

Surface characterization by contact angles – metals

M E R SHANAHAN

Metals and metal oxides may be considered to be high-energy solids with **Surface energies** typically (although not always) of the order of several hundred millijoules per square metre or even more for noble metals such as platinum.[1] Characterization of **Surface energy** and its components is a delicate task for several reasons. As far as the **Contact angle** methods are concerned, probably the main cause of imprecision is the difficulty of interpretation of data. Although **Contact angles** may be measured, their meaning is often vague. Is the substrate in question truly a metal or a metal oxide? Since metals tend to be of high **Surface energy**, the equilibrium spreading pressure π_e cannot be ignored. A liquid drop clearly has a surrounding vapour, and this will tend to be adsorbed on to the substrate in order to lower its free energy (a problem rarely encountered with polymers). Some work has been done in UHV conditions,[2] but local adsorption of vapour leading to wetting hysteresis cannot presumably be avoided given the very presence of a liquid. Using classic equations such as that due to Fowkes[3] to evaluate **Dispersion force** interactions can also lead to errors since an assumption of this theory is that the molecular dimensions of the two phases in contact are similar. The **Roughness of surfaces** can also perturb **Contact angles**.

Notwithstanding these drawbacks and the delicate nature of the problem, the methods described in **Surface characterization by contact angles – polymers** may be employed. Clearly the one-liquid method is open to criticism because of potential adsorption problems, and if the substrate is indeed of high energy virtually all liquids will spread anyway, leading to unusable **Contact angle** data. However, the two-liquid method may be attempted and indeed has been used with success in the case of aluminium.[4,5] Combining wetting measurements with corrections necessary to take into account the **Roughness of surfaces**, the surface characteristics of the metal having undergone various treatments have been evaluated. These are given in Table 41, where γ_S^D represents the

Table 41. Surface characteristics of aluminium after various treatments[4,5]

Surface treatment	γ_S^D (mJ m^{-2})	I_{SW}^P (mJ m^{-2})
Hexane extraction	42	38.7
DMF extraction	135	62.5
Phosphatization	150	18
Anodization	125	95
Sealed anodization	41	55.5

dispersion component of the **Surface energy** of the solid and I_{SW}^P its polar interaction with water.

Related articles are **Adsorption theory, Contact angles and interfacial tension, Dispersion forces, Pretreatment of metals, Roughness of surfaces, Surface characterization by contact angles – polymers, Surface energy** and **Wetting and spreading**.

References

1. A J Kinloch in *Adhesion and Adhesives*, Chapman and Hall, London, 1987, Ch. 2.
2. M E Schrader, *J. Phys. Chem.*, **78**, 87 (1974).
3. F M Fowkes, *Ind. Eng. Chem.*, **56**, 40 (1964).
4. A Carré, Ph.D. thesis, Université de Haute Alsace, France, 1980.
5. A Carré, J Schultz, *J. Adhesion*, **15**, 151 (1983).

Surface characterization by contact angles – polymers

M E R SHANAHAN

Whereas the surface tension of a liquid (or the interfacial tension between a liquid and a fluid) may be measured directly by means of techniques such as that of the Wilhelmy plate or capillary rise, the lack of molecular mobility within a solid prevents the deformation necessary for detecting surface forces. As a consequence, indirect methods must be used. The most frequently employed are probably those involving the wetting of the solid by known reference liquids.

Consider a drop of a given liquid, L, resting on the surface of a solid, S, in the presence of a fluid, F (vapour or second liquid immiscible with the first, see **Wetting and spreading**). Assuming the interfacial tension liquid/fluid γ_{LF} to be known, measurement of the equilibrium **Contact**

angle θ leads to the difference between solid/fluid and solid/liquid tensions ($\gamma_{SF} - \gamma_{SL}$), via Young's equation:

$$\gamma_{SF} - \gamma_{SL} = \gamma_{LF} \cos \theta_0 \qquad [1]$$

However, the terms γ_{SF} and γ_{SL} cannot be evaluated separately without further information. In order to go further, certain assumptions must be made. For the case of solid polymers certain simplifications can be made.

Critical surface tension for wetting of Zisman[1]

Pioneering work on the evaluation of solid surface tensions was carried out by Zisman and co-workers. The principle was to measure the **Contact angles** θ_0 of a series of liquids of decreasing surface tension γ on the polymer surface (see **Critical surface tension**). As the value of γ decreases, so does that of θ_0, whereas $\cos \theta_0$ increases. By plotting a graph of $\cos \theta_0$ vs γ, extrapolation can be made to $\cos \theta_0 = 1$ (even if a liquid in this neighbourhood is only hypothetical). The value of γ for $\cos \theta_0 = 1$ is known as the critical surface tension for wetting of the solid γ_c and corresponds to the value of the surface tension of a liquid which will just spread spontaneously. It does not, in fact, correspond exactly to the surface tension for wetting of the solid γ_s, since Young's equation predicts that

$$\gamma_{SF} - \gamma_{SL} = \gamma_c \qquad [2]$$

Nevertheless, the interfacial tension between a polymer and an organic liquid is generally quite low and as a result, to a fair approximation it can be seen that

$$\gamma_c \sim \gamma_{SF} \sim \gamma_s \qquad [3]$$

where the second equivalence is a consequence of assuming negligible spreading pressure, see below. Although simple in principle, the concept of critical surface tension can be very useful. There exists a 'wipe' test in which cotton wool dipped in solutions of different surface tensions are wiped over a polymer surface to observe where the liquid film break occurs (ASTM D-2578-67). The critical surface tension of the solid is situated between consecutive liquids in the series, one spreading and the next merely wetting (see **Wetting and spreading**).

One-liquid method

The original work of Zisman was developed by several workers. Using Young's equation, two problems exist. Firstly, even in a vapour atmosphere the term γ_{SF} (or γ_{SV}) does not correspond necessarily to the surface tension of the solid γ_S. Whereas γ_S corresponds to the surface tension of the clean solid, γ_{SV} refers to the same quantity after any potential adsorption (see **Adsorption theory**) of the vapour of the liquid. The difference, always positive or zero, is called the spreading pressure π_e, with suffix e representing equilibrium:

$$\pi_e = \gamma_S - \gamma_{SV} \qquad [4]$$

The second problem involves interpretation of the interfacial tension γ_{SL} (there is no difficulty concerning $\gamma_{LF} = \gamma$ since the liquid must coexist with its vapour). The interfacial tension must be lower than the sum of the two individual surface tensions (otherwise the interface would not form) by an amount related to them (it is a consequence of the same forces that give rise to the individual surface tension). Good and Girifalco[2] expressed this (see **Good–Girifalco interaction parameter**) as

$$\gamma_{SL} = \gamma_S + \gamma - 2\phi(\gamma_S\gamma)^{1/2} \qquad [5]$$

where ϕ is the interaction parameter, which, for interfaces where only **Dispersion forces** (q.v.) act is close to unity. Here ϕ is often difficult to calculate for interfaces of interest in adhesion. Fowkes[3] proposed (see **Acid–base interactions** and **Surface energy components**):

$$\gamma_{SL} = \gamma_S + \gamma - 2(\gamma_S^D \gamma^D)^{1/2} - I_{SL}^{ND} \qquad [6]$$

where D refers to dispersion components of γ, γ_S (see **Dispersion forces**) and I_{SL}^{ND} represents non-dispersion interactions (**Polar forces, Acid–base interactions**).

Developing the second approach using Eqns 4 and 6 in Young's equation, we obtain

$$\cos\theta_0 = 2(\gamma_S^D)^{1/2}\frac{(\gamma^D)^{1/2}}{\gamma} + \frac{I_{SL}^{ND}}{\gamma} - \frac{\pi_e}{\gamma} - 1 \qquad [7]$$

Although π_e can be important for high-energy solids it is generally negligible for low-energy solids such as polymers.

The experimental procedure is to measure equilibrium **Contact angles** of a series of liquids (showing no chemical interaction with the substrate) and to plot a graph of $\cos\theta_0$ vs $(\gamma^D)^{1/2}/\gamma$. For apolar liquids (1-bromonaphthalene, tricresyl phosphate), I_{SL}^{ND} should be close to zero. A straight line passing through the corresponding data and -1 has a gradient of $2(\gamma_S^D)^{1/2}$. For polar liquids I_{SL}^{ND} may be readily estimated from the excess of the ordinate value above the reference apolar line. Various expressions have been proposed to break I_{SL}^{ND} into its components,[4] but none has yet proved entirely successful. Suffice it to say that a high value of I_{SL}^{ND} for a given liquid indicates a highly polar solid.

Two-liquid method

Following work originally done to elucidate the surface characteristics of a high-energy solid, mica,[5] a two-liquid method has been developed in which the vapour phase is replaced by a second liquid immiscible with the first. In general the initial liquid is water (or formamide) and the surrounding liquid a member of the n-alkane series. Bearing in mind that the non-dispersive interaction between the solid and the hydrocarbon will be negligible and substituting Fowkes's relation 6 for both γ_{SF} and γ_{SL} in expression 1 leads to

$$\gamma_L - \gamma_F + \gamma_{LF} \cos \theta_0 = 2(\gamma_S^D)^{1/2}[(\gamma_L^D)^{1/2} - (\gamma_F)^{1/2}] + I_{SL}^{ND} \qquad [8]$$

(due to the apolar nature of the hydrocarbon, $\gamma_F^D \sim \gamma_F$).

Experimentally, the **Contact angles** of the polar liquid on the solid surface in the presence of a series of n-alkanes are measured and a graph of

$$\gamma_L - \gamma_F + \gamma_{LF} \cos \theta_0 \quad \text{vs} \quad [(\gamma_L^D)^{1/2} - (\gamma_F)^{1/2}]$$

plotted. Again the gradient is equal to $2(\gamma_S^D)^{1/2}$ and the intersection at the origin of the abscissa gives I_{SL}^{ND}. Careful use of this method can eliminate doubt as to the importance of the undetermined π_e term appearing in the one-liquid method since potential adsorption of the vapour of L can be avoided.

An advantage of this method is that with a judicious choice of liquids (water for which $\gamma_L^D = 21.6$ mJ m^{-2} at room temperature and n-octane for which $\gamma_F = 21.3$ mJ m^{-2}), Eqn 8 reduces to

$$\gamma_L - \gamma_F + \gamma_{LF} \cos \theta_0 \sim I_{SL}^{ND} \qquad [9]$$

The non-dispersion interaction between the solid and water may be estimated immediately.

Related articles are **Adsorption theory, Contact angles and interfacial tension, Pretreatment of polymers, Surface energy, Wetting and spreading**.

References

1. H W Fox, W A Zisman, *J. Colloid Sci.*, **5**, 514 (1950).
2. L A Girifalco, R J Good, *J. Phys. Chem.*, **61**, 904 (1957).
3. F M Fowkes in *Treatise on Adhesion and Adhesives 1*, ed. R L Patrick, Marcel Dekker, New York, 1966, Ch. 9.
4. S Wu, *J. Macromol. Sci. – Revs. Macromol. Chem.*, **C10** (1), 1 (1974).
5. J Schultz, K Tsutsumi, J B Donnet, *J. Colloid Interface Sci.*, **59**, 272 (1977).

Surface energy

D E PACKHAM

The basic concept

The curved meniscus on a liquid, the 'head' on a pint of beer and the effectiveness of soap are all phenomena which can be rationalized using the concept of surface energy. The basic idea that there is an 'excess energy in the surface' of a solid or liquid can be accepted by considering the formation of the surface by breaking the bonds across what are to become the two surfaces formed. The energy to break these bonds is the surface energy.

An advantage of this mental picture of 'bond breaking' is that it immediately points to a relationship between the magnitude of surface energy and the bond energy ('strength') of the bonds which have to be broken. Thus the surface energy for a paraffin is about $22 \, \text{mJ m}^{-2}$, for water $72 \, \text{mJ m}^{-2}$ and for mercury $465 \, \text{mJ m}^{-2}$ as the dominant bonds to be broken are **Dispersion forces, hydrogen bonds** and metallic bonds respectively.

A surface cannot exist on its own: it must be part of an interface between two phases, even if one is vacuum. The term 'surface energy' without qualification strictly applies to the substance concerned in contact with a vacuum. For practical purposes the distinction between surface energy against vacuum and against air is usually neglected. The simple mental picture of surface energy's being the energy required to break the bonds to form a surface must be refined. It is better regarded as the algebraic sum of the energy required to break the bonds to form the surface *in vacuo*, and that released when any new bonds are formed on the surface when it is brought in contact with the second phase. Thus the surface energy of a solid in contact with a vapour may well be less than that *in vacuo*, as some components of the vapour are likely to be adsorbed on to the solid surface (this is closely connected with 'spreading pressure').

'Surface tension' is a concept closely allied with surface energy. The meniscus of mercury for example can be explained in terms either of reduction of surface area and therefore of surface energy or of there being a surface tension – a force acting in the surface – trying to contract it. The connection between surface tension and surface energy and the precise meanings of both can be seen from thermodynamic considerations.

Thermodynamic aspects

Consider a soap film held between three fixed sides of a rectangular wire frame and a frictionless wire slider of length *l*. Under the influence of a force F, acting at right angles to the slider, the slider moves a distance dx, and the area of the film is increased by

$$dA = l\,dx \tag{1}$$

The work done by the force is

$$dW = F\,dx$$

$$= F/l \cdot l\,dx = F/l\,dA \tag{2}$$

If the change is done at constant temperature and pressure, the increment of work will equal the increment of Gibbs free energy dG. Moreover, in elementary treatments the surface tension γ is defined as the force per unit length in the surface. Thus Eqn 2 can be written as

$$dW = \gamma\,dA = dG \tag{3}$$

which leads to the thermodynamic definition of surface tension γ as the partial differential of the free energy of the system with respect to area:

$$\gamma = (\partial G / \partial A)_{T,P} \tag{4}$$

Now the surface energy, more strictly the Gibbs surface free energy G^s, is the excess free energy of the system associated with the surface (per unit area), and so is defined as

$$G^s = \frac{G - G^B}{A} \tag{5}$$

where G is the total free energy of the system, G^B the value the free energy would have if all the atoms were in the environment they have in the bulk and A the surface area.

In the case of the soap film under consideration it is only the change in area which causes a change in free energy so dG can be equated to $d(G^s A)$:

$$dG = d(G^s A) = \gamma\,dA \tag{6}$$

therefore

$$\gamma = \partial(G^s A)/\partial A$$

therefore

$$\gamma = G^s + A(\partial G^s / \partial A)_{T,P} \tag{7}$$

Equation 7 gives the relation between surface energy G^s and surface tension γ. They are equal only when $(\partial G^s / \partial A)_{T,P}$ is zero.

The term $\partial G^s / \partial A$ is the rate of change of surface energy per unit area with area. Thus if G^s does not change as area increases γ and G^s are equal; if G^s changes with area they are different. If the local environment of the surface atoms remains the same, when area is increased, G^s is independent of area, and equals γ. This is what happens in a single-component liquid. In a solid or multi-component liquid the situation may be different. Increasing the area of a solid may involve plastic deformation of the surface, orientating the molecules and changing their local interactions. Then G^s will change with area. In a multi-component system changes in surface concentration of solute may occur.

In summary the definition of surface tension is given by Eqn 4, of surface energy by Eqn 5 and the relation between them by Eqn 7. (*Note:* Some authors develop these concepts in terms of the Helmholtz function rather than the Gibbs function, by supposing the soap film to be stretched at constant temperature and volume, rather than pressure.)

Practical usage

The thermodynamic treatment given above applies strictly to single-component anisotropic solids and liquids. To take account of multi-component systems and anisotropy further refinement is necessary.

Unfortunately (or perhaps fortunately) in many practical contexts, including adhesion, values for surface energies of the solids involved cannot be measured with sufficient precision to make such distinctions worth while. In fact, for this reason, most of the literature, this *Handbook* included, treats 'surface energy' and 'surface tension' as if they were interchangeable terms. The symbol γ is commonly used for both.

As dimensions of surface tension and surface energy are the same (Eqns 4 and 5), they may be expressed in the same units. The equivalent units of newtons per metre ($N\,m^{-1}$) or joules per square metre ($J\,m^{-2}$) can be used. Commonly millijoules per square metre ($mJ\,m^{-2}$) is used as it gives numbers of convenient magnitude and is identical to the dyne per centimetre or erg per square metre (dyne cm^{-1} or erg cm^{-2}) still met in the older literature.

Solids are sometimes classified as having 'high-energy surfaces' or 'low-energy surfaces' according to their surface energies. Thus low energy surfaces have γ up to about $100\,mJ\,m^{-2}$ and include organic surfaces especially polymers. High-energy surfaces have γ-values of hundreds or thousands of millijoules per square metre and comprise hard inorganic surfaces such as metals, oxides and ceramics.

These ideas are further developed in other articles, in particular **Surface characterization, Wetting and spreading, Wetting and work of adhesion, Contact angle and interfacial tension, Pretreatment of polymers** and **Pretreatment of metals**.

Select references

A W Adamson, *Physical Chemistry of Surfaces*, 4th edn, Interscience, New York, 1982.
B W Cherry, *Polymer Surfaces*, Cambridge University Press, 1981.

Surface energy components

D E PACKHAM

The magnitude of the **Surface energy** of a material depends upon the strength of the bonds which have to be broken to form the surface concerned. An extension of this idea is the suggestion of Fowkes that surface energy γ can be expressed as the sum of components associated with individual types of bond involved.[1] In general this can be expressed as

$$\gamma = \gamma^d + \gamma^p + \gamma^h + \cdots \qquad [1]$$

where the superscripts refer to **Dispersion forces** (d), **Polar forces** (p) and **Hydrogen bonding** (h). The number of terms in the equation and their form depend in any particular case on the structure of the substance concerned; it should be remembered that dispersion forces are universal, and so are always involved. Thus for mercury the surface energy could be expressed as

$$\gamma_{Hg} = \gamma_{Hg}^d + \gamma_{Hg}^m \qquad [2]$$

where m refers to metallic bonding.

An important point in the use of surface energy components is the realization that for non-polar liquids, such as alkanes, only dispersion forces act between molecules so

$$\gamma = \gamma^d \qquad [3]$$

Dispersion force interface

Alkanes cannot exhibit metallic bonding, so it can be argued that dispersion forces comprise the only type of bond that can act across a mercury–alkane interface. Fowkes has called such an interface a 'dispersion force interface', and argued that the interfacial tension was given by

$$\gamma_{12} = \gamma_1 + \gamma_2 - 2(\gamma_1^d \gamma_2^d)^{1/2} \qquad [4]$$

This equation should be compared with Eqn 2 in **Contact angles and interfacial tension**, where the subscript notation is explained.

One justification for the geometric mean term in Eqn 4 comes from the Good–Girifalco equation

$$\gamma_{12} = \gamma_1 + \gamma_2 - 2\phi(\gamma_1\gamma_2)^{1/2} \qquad [5]$$

If dispersion forces are the dominant bond type in both phase 1 and 2, and the molecules are similar in properties, ϕ will be close to unity. (This is discussed in the **Good–Girifalco interaction parameter**.) In these circumstances Eqn 1 can be regarded as a special case of Eqn 5.

Work of adhesion

From the definition of work of adhesion W_A and work of cohesion W_C given in **Contact angles and interfacial tension**, it can be seen that the equivalent of Eqn 1 can be written in terms of components of work of adhesion:

$$W_A = W_A^d + W_A^p + W_A^h + \cdots \qquad [6]$$

Further, the geometric mean assumption in Eqn 5 is equivalent to writing

$$W_A^d = (W_{C_1}^d \cdot W_{C_2}^d)^{1/2} \qquad [7]$$

This is further developed in discussing **Acid–base interactions**.

Evaluation of γ^d for liquids

For apolar liquids, the dispersion component is essentially the total surface tension (cf. Eqn 3). For polar liquids the dispersion component of surface tension can be obtained using Eqn 4 after measuring both surface tensions and interfacial tension, provided that γ^d of the other liquid is known.

Evaluation of γ^d for solids

Young's equation (see **Contact angles and interfacial tension** and the nomenclature described) relates **Contact angle** to surface energy terms. By measuring contact angles on a solid surface of liquids of known γ_L^d, γ_S^d can be calculated. The solid–liquid interfacial tension γ_{SL} is eliminated between Eqn 5 and the Young equation, giving

$$\gamma_S^d = \gamma_{LF}^2(1 + \cos\theta)^2/4\gamma_L^d \qquad [8]$$

In deriving this equation it has been assumed that the spreading pressure is zero: this and further relevant details are discussed under **Surface characterization by contact angles – polymers**.

Extension to polar components

Equation 4 applies to an interface across which only dispersion forces act: if other forces act as well there must be another negative term on the right-hand side giving the further lowering of interfacial energy (cf. **Contact angles and interfacial tension**, Eqn 2, and **Surface characterization by contact angles – polymers**, Eqn 4). It has commonly been assumed that, in the absence of primary bonding, the non-dispersion forces across an interface could be treated as polar, i.e.

$$\gamma = \gamma^{d} + \gamma^{p} \qquad [9]$$

and that their contribution to the lowering of interfacial tension was also based on a geometric mean relationship[2]

$$\gamma_{12} = \gamma_{1} + \gamma_{2} - 2(\gamma_{1}^{d}\gamma_{2}^{d})^{1/2} - 2(\gamma_{1}^{p}\gamma_{2}^{p})^{1/2} \qquad [10]$$

The theory of the **Good–Girifalco interaction parameter** can be developed to give some support for this relationship.[3] This has not been accepted by Fowkes, who has argued that the principal non-dispersion interactions are **Acid–base interactions** for which a geometric mean relationship does not apply. This approach is further considered in the article, **Acid–base interactions**. Despite these considerations much use has been made of Eqn 10.

Surface energies of solids

Once the dispersion component of surface energy of a liquid is known, the polar component can be obtained from the surface tension using Eqn 10. The approach based on Eqns 9 and 10 can then be used to estimate surface energies of solids, particularly polymers, very much in the same way as γ_{S}^{d} values are obtained from Eqn 8. Here γ_{12} is eliminated between Eqn 10 and the Young equation, giving

$$1 + \cos\theta = 2\frac{(\gamma_{S}^{d}\gamma_{L}^{d})^{1/2}}{\gamma_{LF}} + 2\frac{(\gamma_{S}^{p}\gamma_{L}^{p})^{1/2}}{\gamma_{LF}} \qquad [11]$$

This again assumes a spreading pressure of zero; see **Surface characterization by contact angles – polymers**. The contact angles of several liquids of known γ_{L}^{d} and γ_{L}^{p} are measured on the surface of the solid of interest giving a number of simultaneous equations of the form of Eqn 11. These contain only two unknowns, γ_{S}^{d} and γ_{S}^{p}, so they can be evaluated. Some values of surface energy components are given in Table 42, and are given in the literature.[2,3,5]

The concept of surface energy components provides one route to interfacial tensions and solid surface energies, others include the **Good–Girifalco interaction parameter** and **Acid–base interactions**.

Table 42. Values of dispersion (γ^d) and polar (γ^p)
components of surface energy

Substance	γ^d (mJ m^{-2})	γ^p (mJ m^{-2})
PTFE	18.6	0.5
Polyethylene	33.2	—
PVC	40	1.5
PMMA	35.9	4.3
Nylon 6,6	35.9	4.3

References

1. F W Fowkes et al., *Mater. Sci. Eng.*, **53**, 125 (1982).
2. D H Kaelble, *Physical Chemistry of Adhesion*, Wiley-Interscience, New York, 1971.
3. R J Good, *Surf. and Colloid Sci.*, **11** 1 (1979).
4. D Briggs, D G Rance, B C Briscoe in *Comprehensive Polymer Science*, Vol. 2, ed. G Allen, Pergamon Press, Oxford, 1989, p. 715.
5. J T Koberstein, *Encyclopedia of Polymer Science and Engineering*, Vol. 8, Wiley-Interscience, New York, 1987, p. 257.

Surface nature of polymers

D BRIGGS

It is very rare for the chemical nature of the surface of a polymer to be a mere extension of that of the bulk. Even when there are no deliberate additives, e.g. antioxidants, the surface can have a different molecular weight from the bulk due to segregation of low-molecular-weight 'tail' polymer and, in the case of segmented copolymers preferential segregation of the lower energy component may occur (see **Compatibility**).

The nature of a polymer surface has an important effect on many end uses: adhesive properties such as heat sealability (see also **Wetting and spreading, Printing ink adhesion**) and general processing and machine-handling properties (e.g. friction and 'blocking' or self-adhesion) are very sensitive to surface condition and optical properties are affected by haze, gloss and stains. Additives in polymeric materials include agents to prevent oxidation (thermal and photochemical), to neutralize acidity, to promote fire retardance and to aid processing. These are usually intended to be well distributed throughout the bulk, but in certain circumstances surface segregation may take place. Agents to lower surface friction, to increase surface conductivity and prevent 'blocking' are intended to migrate to the surface. In some cases they may not do so, in other cases they may migrate across interfaces through contact with another surface. Filler and pigments particles are often surface coated, and components of this coating may pass into the matrix and hence to the surface (see **Filled polymers**).

Surface segregation of emulsifier and stabilizer molecules used in some polymerization processes is a frequent occurrence (see **Release**). Common contaminants generally are lubricating oils and greases, mould release agents and various airborne materials, particularly in aerosol form.

Blooms and visually similar phenomena are sometimes seen on the surfaces of rubber mouldings. These may originate from the limited solubility of a particular component of the rubber compound, or from contamination, for instance by a **Release** agent from the mould. Some blooms are deliberately contrived – hydrocarbon wax gives some protection from ozone attack. Loadman and Tidd give a useful account of the classification and analysis of blooms on rubber.

'Engineering' polymer surfaces in order to change surface properties has a long history and is increasing in sophistication (see **Corona discharge treatment, Flame treatment, Plasma treatment**). Understanding structure–property relationships is vital and much research is carried out in this area using the range of **Surface analysis** techniques now available, particularly XPS and SIMS.

Select references

K L Mittal (ed.), *Surface contamination*, Vols 1 and 2, Plenum Press, New York, 1979.
M J R Loadman, B K Tidd in *Natural Rubber, Science and Technology*, ed. A D Roberts, Oxford University Press, Oxford, 1988.

T

Tack

D W AUBREY

When tack is significant, there is appreciable resistance to the separation of two materials brought momentarily into contact. The phenomenon may be subdivided into three distinctly different groups as given below:

1. *Cohesive tack* involves bulk flow of one or both materials during separation, and applies to materials like printing inks, paints, syrups, etc. Resistance to separation is governed by viscous flow according to the Stefan equation, as discussed by Wake.[1]
2. *Adhesive tack* involves a separation apparently at the original interface between the materials. This kind of tack applies particularly to **Pressure-sensitive adhesives**, although such adhesives may display cohesive tack under extreme conditions of rate or temperature.
3. *Autohesive tack* involves two elastomeric materials of essentially the same composition. In this case separation may be either adhesive (usually after short times of contact) or cohesive (usually after long times of contact). It is important in the use of contact adhesives and in plying together rubber surfaces in, for example, the manufacture of tyres.

Autohesive tack differs from the other types in that it involves mutual diffusion of polymer molecules across the interface; it is discussed elsewhere (see the articles on **Diffusion theory** and **Autohesion**). Adhesive tack between elastomeric materials and other, usually rigid, surfaces is considered in this article.

Mechanism of adhesive tack

The methods used for routine assessment of tack have been described elsewhere (see **Pressure-sensitive adhesives – adhesion properties**). These methods all involve the contact of two surfaces (adhesive and test substrate) under light pressure for a short time, followed by a separation step, the force (or energy) of separation being taken as a measure of tack. Thus, the tack phenomenon involved two processes – a bonding process in which the adhesive is allowed to deform or flow into contact with the surface for a time of the order $0.1–10$ s, and a debonding process in which the adhesive separates by peeling from the surface, involving a much more localized and shorter time-scale of deformation, typically $10^{-4}–10^{-6}$ s. Assuming that backing and adhesive thickness effects are eliminated, the tack value will depend on the following:

1.   The ease of deformation of the adhesive during the bonding stage, i.e. its shear compliance over the bonding period, which determines the degree of true interfacial contact achieved.
2.   The resistance to separation of the contacted areas of interface formed in (1), determined by interfacial attractive forces, a measure of which is the thermodynamic work of adhesion.
3.   The degree of deformation of the adhesive and hence the amount of energy dissipated as heat during the high-rate debonding stage, a measure of which would be tan δ at the appropriate rate. It depends strongly, of course, on (2).

Thus, attempts to rationalize the tack values obtained should involve:

for (1), knowledge of creep compliance at long times (which for high tack should be in the terminal region of the viscoelastic spectrum – see **Viscoelasticity**);
for (2), knowledge of the surface energetics of the materials involved (see **Surface energy**);
for (3), knowledge of tan δ (or equivalent parameters) at short times (which for high tack should be in the transition region of the viscoelastic spectrum – see **Viscoelasticity**).

The major studies of tack phenomena have concentrated on the probe method of measurement (see **Pressure-sensitive adhesives – adhesion properties**) and demonstrate clearly the relative importance of the above factors in determining tack.[2–4]

Mode of action of tackifier resins

Most of the study of tack has naturally involved attempts to understand the action of tackifier resins in increasing the tack of rubbers. For hydrocarbon rubbers (e.g. natural rubber, NR) and hydrocarbon tackifier resins (e.g. terpene resins such as poly-β-pinene), it is generally found that the effects of surface energy change ((2) above) are negligible compared

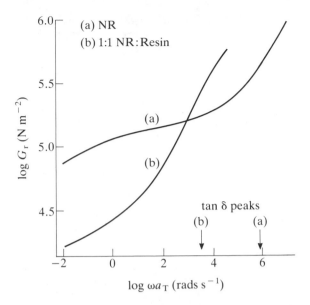

Fig. 129. Master curves of (log) storage modulus G'_r against (log) frequency ω at 296 K for (a) NR and (b) a 50/50 blend of NR with Piccolyte S115* tackifier resin.[5] (*Hercules Co. trade name)

with the influence of the tackifier resins on the viscoelastic behaviour at both high and low rates ((1) and (3) above). This may be illustrated in a master curve of dynamic shear modulus against (log) rate (frequency),[5] shown diagrammatically in Fig. 129. The addition of β-pinene resin to NR results in a reduction in modulus (increase in compliance) in the terminal zone (at rates corresponding to the bonding step) and a shift of the transition zone (where tan δ is high) to lower rates (corresponding to the debonding step). Thus the addition of a tackifier resin increases tack through its influence on both bonding and debonding processes.

References

1. W C Wake, Elastomeric adhesives, in *Treatise on Adhesion and Adhesives*, Vol 2, ed. R L Patrick, Marcel Dekker, New York, 1969, Ch. 4, p. 173.
2. C A Dahlquist in *Adhesion, Fundamentals and Practice*, Ministry of Technology, MacLaren, London, 1969, Ch. 5, p. 143.
3. R Bates, Studies in the nature of adhesive tack, *J. Appl. Polym. Sci.*, **20**, 2941 (1976).
4. P J C Counsell, R S Whitehouse, Tack and morphology of pressure-sensitive adhesives, in *Developments in Adhesives – 1*, ed. W C Wake, Applied Science Publishers, London, 1977, Ch. 4, p. 99.
5. D W Aubrey, The nature and action of tackifier resins, *Rubber Chem. Technol.*, **61** (3), 448 (1988).

Tensile tests

D E PACKHAM

The joint is also known as the 'butt joint' or 'polker chip' joint. Two solid cylinders are bonded end to end, and the joint is tested by applying a force along the common axis (Fig. 130).

At first sight it might seem a simple test to analyse with uniform tensile stress throughout the adhesive layer. In practice the stress distribution is not uniform: the disparity of modulus and Poisson's ratio between the cylinders and the adhesive means that shear stresses are introduced on loading. Thus the failure stress is not independent of the dimensions of the joint.

Adams, Coppendale and Peppiatt[1] have applied elastic **Finite element analysis** to the joint. Figure 131 gives some typical results. The bonded area comprises two different regions. First, in the central region, the shear stress is zero and the tensile stresses are uniform, the radial σ_r and circumferential σ_θ stresses being the same and related to the given by axial stress σ_z by

$$\sigma_r = \sigma_\theta = \left[v_a - \frac{E_a v_s}{E_s} \right]\left[\frac{\sigma_z}{1 - v_a} \right] \qquad [1]$$

where v is Poisson's ratio, E Young's modulus and the subscrips a and s refer to adhesive and substrate respectively.

Second, around the periphery of the joint both tensile and shear stresses act. As Fig. 131 shows, their magnitude depends on the aspect ratio of the joint, and also varies throughout the thickness of the adhesive layer.

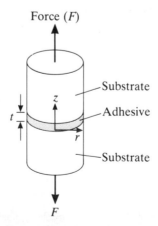

Force (*F*)

Substrate

Adhesive

Substrate

F

Fig. 130.   An axially loaded butt joint

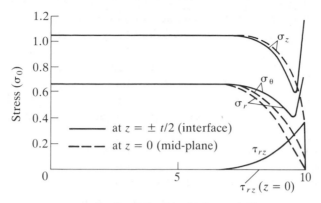

Fig. 131. An example of the stress distribution in an axially loaded butt joint after Adams *et al.*[1] (σ_0 = applied average axial stress; $E_a = 2.5$ GPa; $E_S = 69$ GPa; aspect ratio = 20). (Reprinted from ref. 3 with permission)

On the adhesive–substrate interface there is always a stress concentration at the edge of the substrate.

For a rubber adhesive Gent[2] found that he could use the WLF equation to obtain a master curve of breaking stress (see **Viscoelasticity – time–temperature superposition**). He used a **Fracture mechanics** analysis to link the breaking stress to critical strain energy density.

A more detailed discussion of this test may be found in ref. 3.

References

1. R D Adams, J Coppendale, N A Peppiatt, *J. Strain Anal.*, **13**, 1 (1978).
2. A N Gent, *J. Polym. Sci.*, **A2**, 283 (1974).
3. A J Kinloch, *Adhesion and Adhesives: Science and Technology*, Chapman and Hall, London, 1987.

Testing of adhesives

D A TOD

This article deals with the characterization of adhesives as polymers; the measurement of adhesion is dealt with in the article **Tests of adhesion**.

Structural adhesives such as epoxy resins can be treated as any rigid polymer and samples can be machined from cast sheets to produce test pieces. These can then be used to measure typical tensile properties such as failure stress and strain. Using accurate extensometry it is possible to characterize completely the uniaxial properties of an adhesive. The **Creep** of adhesive joints is especially important for structural adhesives maintained at high temperature. It is possible to determine the creep resistance of such materials by applying suitable loads at temperature to samples of the adhesive and to record the deformation with time. From such data it will soon be evident whether the adhesive is suitable for use or will cause a joint to deform with time.

The effect of temperature and rate upon the properties of an adhesive can be determined using the technique of dynamic mechanical analysis. In such a test small bars of the adhesive would be subjected to a small strain deformation at one end and the resultant torque at the other end of the sample measured. Using suitable stimuli and analytical techniques it is possible to resolve three properties of the material. These are the shear modulus, the loss modulus and the loss or phase angle $\tan \delta$ (see **Viscoelasticity**). When measured over a range of temperatures and rates such parameters give a unique fingerprint to the usefulness of any particular polymeric system.[1] The effects of rate and temperature upon these properties can be combined to give one master curve using the WLF equation (see **Viscoelasticity**). Probably the most important thermal transition in any adhesive system is the **Glass transition temperature**. Using dynamic mechanical analysis it is possible quickly to establish this transition and its sensitivity to various parameters such as moisture uptake. The degree of cure of an adhesive can also be established with this technique and this is shown for a typical epoxy resin in Fig. 132. This figure shows the major transition in the loss angle which is accompanied by a rapid loss in the shear modulus at the **Glass transition temperature**. As the time of cure is increased this transition in the adhesive increases by about 20 °C.

The cure of a thermosetting adhesive is usually measured by techniques such as cone and plate viscometry or by a vibrating needle system. From these measurements important characteristics such as the gelation point and the point of final cure can be determined. The time from mixing of an adhesive to the gelation point gives the useful life or 'pot life' of the material.

The technique of differential scanning calorimetry (see **Thermal analysis**) can also be used to provide valuable data on an adhesive. In this technique the rate of energy input to a sample is measured as a function of temperature. The degree of cure of a resin system and its **Glass transition temperature** can be established.

The degradation temperature is very important for adhesives used at high temperature. The onset of degradation is normally established by a technique such as thermogravimetric analysis (TGA). In this technique

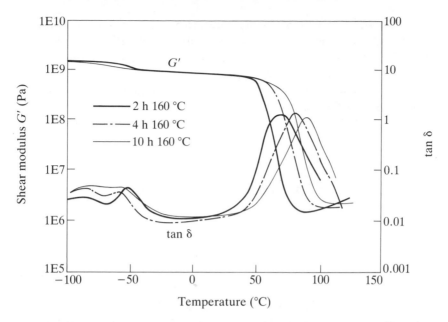

Fig. 132. Variation of shear modulus and loss angle with temperature for an epoxy resin cured for three different times at the same temperature

the weight of a sample is recorded as the temperature is increased. Eventually the sample will begin to decompose and this will be reflected in a weight loss.

The coefficient of expansion of a polymer is very much greater than that of a metal. Such a mismatch of expansion will introduce stresses within an adhesive joint especially if the joint is cured at an elevated temperature. The measurement of the expansion coefficient can be carried out by a variety of techniques and one of the simplest is thermomechanical analysis (TMA) where the dimensions of a small sample are measured over a range of temperatures. The change in the sample dimension gives the variation in expansion coefficient with temperature.

The degree of cross-linking in an adhesive will give an indication of its state of cure. Several techniques can be used to establish this and one of the commonest is the 'sol-gel' technique where the weight of an extracted portion of adhesive is compared to the unextracted material.

In treating adhesives as polymeric materials it is possible to apply **Fracture mechanics** tests to determine material properties independent of geometry. A typical sample would be the compact tension test piece where a sheet of material typically 6 mm thick and 100 mm square has a sharp crack inserted in one edge. The sample is then loaded perpendicular to the plane of the crack and the load at which crack propagation initiates recorded. These data can then be used to generate the fracture energy of the material which can then be used in the analysis of more complex shapes and also joint design.

An area of testing which is becoming more important is the **Impact resistance** of adhesives. The classic Charpy or Izod test methods are normally used for impact tests together with the falling or drop weight test. Generally such instruments are becoming more highly instrumented so that subtle changes in the adhesive's material properties with high rates can be followed.

There are many other tests which can be carried out on adhesives using standard test methods for polymeric materials[2–4] and these should be reviewed for the most appropriate procedure (see **Standards on adhesion and adhesives**, Appendix 1).

References

1. S V Wolfe, D A Tod, *J. Macromol. Sci. Chem.*, **A26** (1), 249 (1989).
2. R P Brown, *Handbook of Plastics Test Methods*, George Godwin/Plastics and Rubber Institute, London, 1981.
3. L E Nielsen, *Mechanical Properties of Polymers and Composites*, Marcel Dekker, New York, 1974.
4. A H Landrock, *Adhesives Technology Handbook*, Noyes Publications, Park Ridge, N.J., 1985.

Tests of adhesion

D A TOD

This article relates to methods of measurement of the bond strength between an adhesive and a substrate; the characterization of adhesives alone is considered in the article **Testing of adhesives**.

There are numerous different tests of adhesion which are used for five main purposes,

1. To check the quality of an adhesive to see if it falls within well-defined limits;
2. To determine the effectiveness of a surface pretreatment;
3. To gather data for the prediction of joint performance;
4. To select an adhesive from a group for a specific application;
5. To evaluate the effect of ageing.

It will often be necessary to undertake very complex testing to yield detailed results before an adhesive joint is used in service. Once this has been done, it may be adequate for quality control purposes to use a relatively simple test, if its results are sensitive to critical factors affecting performance of the joint. When a test procedure is chosen, it is necessary to consider the nature of the adhesive; a test suitable for a rigid structural adhesive is unlikely to be useful for an elastomeric system such as an adhesive tape.

There are three commonly used test configurations: tensile, shear and peel. Probably the commonest test piece used is the lap shear test piece. This comes in two principal forms, the single and the double lap joints. The advantage of this test piece is that it can be easily manufactured and quickly tested. In its simplest form it consists of two strips of metal approximately $100 \times 25 \times 1.6$ mm which are bonded together on their major surfaces with an overlap of 13 mm. After cure the joint is pulled apart in the axis of the bond and the failure load of the joint is then calculated. Although the practical manufacture and test of this joint is simple, the stress distribution developed is complex. This leads to a situation where the measured value of joint strength is of restricted value in the design of adhesively bonded components. However, in spite of its deficiencies this test does provide a rapid assessment of the shear strength of an adhesive.

The tensile test piece can take several forms; one simple configuration consists of two right circular cylinders whose ends are then bonded together. Such a joint is loaded to failure at right angles to the plane of the adhesive and the failure stress is determined from the loaded area and the failure load.

Peel tests are generally used for elastomeric or rubbery adhesives. A typical version of this test would be the 'T-peel test' where two strips of a rubber would be bonded together, face to face, and the force required to pull the strips apart would be recorded. The name of the test derives from the shape of the test piece during testing when the top of the 'T' are the two loaded arms and the vertical of the 'T' is the remaining bonded length.

Certain tests are specifically designed to assess the effects of environmental exposure upon adhesive joints. Such a test would be the **Wedge test** where an adhesive joint is formed out of two strips of metal bonded together, face to face. A wedge is then driven into the end of the joint to force the bonded surfaces apart and the joint is then immersed in water. The energy stored within the two strips provides a driving force for the adhesive to fail. The rate of failure of the joint will indicate the effectiveness of an adhesive and a surface pretreatment.

A completely different set of tests are referred to as **Fracture mechanics** test methods. In these tests attempts are made to measure true material properties of the adhesive joint independent of the geometry of test. Such test methodologies require careful preparation of samples so that all experimental variables are controlled. A typical parameter measured for such joints would be the adhesive fracture energy, and normally this would be determined as a function of some parameter such as the rate or temperature of testing.

The viscoelastic (see **Viscoelasticity**) nature of the adhesives means that all joints are affected by the rate and temperature of the test. A key parameter in the testing of such joints is the **Glass transition temperature** of the adhesive. When the joints are tested below this temperature the adhesive will be a low-strain rigid material; above this temperature it will adopt a more rubber-like nature.

A further important parameter is the adhesive thickness within a joint. In certain adhesive systems such as rubber-toughened epoxies there is an optimum thickness of adhesive within which energy-dissipation processes can take place. Above and below this critical size the adhesive fracture energy will be lower.

Specific tests of adhesion are described in more detail under the following articles: **Blister test, Climbing drum peel test, Fracture mechanics, Napkin ring test, Peel tests, Rubber to metal bonding – testing, Shear tests, Tensile tests, Wedge test** and in refs 1–3, see also **Standards on adhesion and adhesives** and Appendices 1 and 2.

All of these tests give a 'number', perhaps in the form of a force per unit area or an energy per unit area, which may loosely be termed the 'adhesion'. It is important to realize that this number reflects the properties of the joint–substrate, adhesive, interface and the interactions between them. Thus changes in the dimensions of a joint commonly change the value of 'adhesion' measured: different tests will not necessarily rank a series of adhesives or surface treatments in the same order. It is only **Fracture mechanics** tests which make any claim to measure a fundamental property, and, even here, great care is needed in interpreting the results (see **Adhesion – fundamental and practical**).

References

1. G P Anderson, S J Bennett, K L DeVries, *Analysis and Testing of Adhesive Bonds*, Academic Press, New York, 1977.
2. R D Adams, W C Wake, *Structural Adhesive Joints in Engineering*, Elsevier Applied Science Publishers, London, 1984.
3. A J Kinloch (ed.), *Developments in Adhesives – 2*, Applied Science Publishers, London, 1981, Chs 6, 7, 10.

Theories of adhesion

K W ALLEN

Introduction

At different times various theories have been advanced to account for the observed phenomena of adhesion. They start from the simplest ideas of the interlocking of a glue with the rough surface of the adherend, and have increased in complexity and sophistication through the last 60 or more years. A problem which has bedevilled this whole topic has been the idea that there should be one explanation which would encompass the whole range of examples. Few people were prepared to accept that their 'pet theory' which was, perhaps, quite adequate to explain the gluing of timber, was not useful or applicable to the whole range of varied applications, including the cementing together of the components of a compound lens for a microscope. Enthusiastic exponents have repeatedly argued for the extension of particular explanations beyond their range of validity.

It is only comparatively recently that it has come to be accepted that any particular instance of adhesion needs to be analysed and explained in terms of contributions from a variety of mechanisms. From the whole range which are available several will contribute, each one to a different extent.

Principal theories

The principal theories for the majority of cases can be divided into four groups: (1) mechanical theories, (2) adsorption theories, (3) diffusion theories, (4) electrostatic theories. Each of these four will be reviewed briefly and qualitatively before their relative significance and applicability are considered. More details are to be found in the specialized articles indicated.

For the special case of **Pressure-sensitive adhesion**, which is rather different from all the others, a fifth theory and explanation is necessary and this will be discussed after these others.

1. Mechanical theories The mechanical theories are the oldest and depend upon the intuitive ideas of the 'practical man'. They involve the interlocking of the solidified adhesive with the roughness and irregularities of the surface of the adherend, and have sometimes been described as a 'hook and eye' approach. Undoubtedly they are significant on a macroscopic scale for fibrous materials (e.g. paper, leather and wood) but not beyond them. In recent years a second range of significance has been recognized where the scale of roughness and interlocking is several orders of magnitude smaller, down to something approaching atomic dimensions. Many instances of useful commercial applications (e.g. in aerospace) depend for their strength and durability upon this microscopic interlocking to give an interphase which has the nature of a composite.

2. Adsorption of theories The adsorption theories encompass all the explanations which invoke chemical bonding and similar forces between the adhesive and the adhered. The most universal of these forces are the London **Dispersion forces** which are involved to some extent in every adhesive bond. The only requirement is that the two materials being joined are in sufficiently close and intimate contact. This is achieved by ensuring that at some stage during its application the adhesive is in a mobile liquid state. In some instances these ubiquitous forces are augmented by the development of primary chemical bonds, which may be essentially ionic or covalent. The acid–base interactions, which have attracted a good deal of attention quite recently, are a particular set of primary bonds and provide a valuable way of considering these interactions.

3. Diffusion theories For the specific case of the adhesion of two polymers, an interaction through the diffusion of polymer chain ends or segments across the interface, usually in both directions, becomes important. The concept is straightforward and easily grasped, although the theoretical development is both difficult and complex. However, it has been largely mastered and agreement has been achieved between theoretical predictions and practical results. This theory is not useful in considering adhesion between smooth and rigid materials where the molecules are essentially fixed and not mobile.

4. Electrostatic theories The existence of an electrical double layer at the interface between a metal and a polymer adhering to it has been satisfactorily demonstrated. Undoubtedly the electrostatic forces developed from this can contribute to the total adhesive bond strength in these cases, although the proportion is as yet obscure. There are some phenomena which cannot be explained without recourse to this explanation, even if some of the theories which have been invoked lack rigour and certainty.

Examples

A proper understanding of most practical examples of adhesion are now recognized to involve elements from several, if not all, these theories. This may best be understood by considering some examples.

First take the case of two sheets of aluminium alloy bonded with an epoxy structural adhesive in an aeroplane. The metal will have been carefully prepared by chemical etching and **Anodizing** to give a surface with a controlled oxide layer with micropores at the outside. The adhesive in its liquid state will penetrate into these pores, so there is a degree of mechanical interlocking, but it is on a very very small scale. The mechanism of this penetration may, perhaps, be considered as a diffusion process. Then there will develop forces across the interface between the adhesive and the metal. These will certainly and invariably include London **Dispersion forces** (van der Waals' forces) and may also involve some forces of primary chemical bonding. These last are probably the most significant in providing the necessary strength and durability which is demanded, but the other mechanisms are also involved in various proportions.

As a second example consider the case of wallpaper stuck to a plaster surface with a water-based cellulose adhesive. The adhesive certainly has to penetrate both surfaces by processes which are certainly a type of diffusion. As the water is lost by evaporation and diffusion through the two materials, there will be mechanical interlocking between the solidifying adhesive and the texture of paper and plaster. Finally, there will certainly be some chemical bonds developed within the paper and probably the plaster as well.

Pressure-sensitive adhesion The familiar pressure-sensitive tapes depend for their adhesion upon the continuing presence of a layer of a very viscous liquid between the tape and the surface to which it is adhering (see also **Rubber adhesion**). When the tape is first put on, this liquid is spread into a very thin layer by the pressure which is used, hence the name 'pressure sensitive'. The liquid is adsorbed on to the substrate, but in order to remove the tape this liquid has to flow, against the resistance created by its viscosity, into the gap created by the separation. So long as the adhesive remains a viscous liquid, considerable force is necessary to remove the tape. Once that liquid changes to a brittle solid, as it commonly does by various mechanisms under the action of sunlight, then the tape will peel away under its own slight weight. In addition to this viscous resistance to peeling, there is also the electrostatic force which has already been discussed, although the extent of its significance is not clear.

References for more detailed accounts are given in the more specialized articles referred to at the beginning of this article.

Thermal analysis

D E PACKHAM

Techniques of thermal analysis play a useful role in the characterization of polymers in general and of adhesives in particular.[1] In their most common mode, some property of the adhesive is recorded as temperature changes according to a chosen programme, but isothermal operation is usually possible, the property change with time being recorded. In thermogravimetric analysis (TGA) weight changes are observed; dynamic mechanical analysis (DMA) studies the storage and loss moduli and loss tangent (see **Viscoelasticity**). The term 'thermal analysis' refers _a fortiori_ to differential scanning calorimetry (DSC) which measures heat flow, and is the main subject of this article.

In a typical experiment a weighed quantity of the sample (usually around 10–15 mg) and an inert reference material are heated in the calorimeter. Any differences in thermal behaviour between the sample and reference is recorded, being obtained from the different quantities of electrical energy required to maintain both sample and reference material at the same temperature according to the selected rate of temperature change. A graph of heat flow (e.g. millijoules per second ($mJ\ s^{-1}$)) against temperature is produced.

A DSC trace might show some of the features shown in Fig. 133. The two peaks, A and B, represent enthalpy changes: the areas under them, subject to calibration of the instrument, give the enthalpy change associated with an endothermic (A) or exothermic (B) change.

Enthalpy changes
Typical endothermic changes include melting, boiling and sublimation.

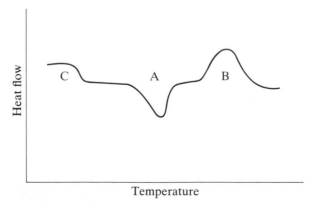

Fig. 133. A typical DSC trace showing an endothermic peak A, an exothermic peak B and a change in specific heat capacity C

Crystalline polymers are often encountered in the context of adhesion. An endothermic peak will occur at their crystalline melting-point, and it may be possible to calculate their percentage crystallinity from the peak area.[2] Other endothermic peaks in adhesives may be the result of melting or volatilization of a component of the adhesive itself or of an impurity.

Exothermic peaks are sometimes associated with crystallization. In some polymers crystallization is inevitable and will always be observed on cooling from above the crystalline melting-point. Polyethylene is an example of such a polymer. In others, some silicones and some polyesters, crystallization may be inhibited by rapid cooling so a sample may crystallize, giving an exotherm when heated in the DSC. Some polymers show more than one crystal form, each with its own crystallization and melting characteristics.

Another obvious source of exothermic peaks is chemical reaction in the sample. Curing in **Reaction setting adhesives** can be studied by DSC. Similarly the extent of cure of partially cured materials can be assessed. Degradation in its various forms is also likely to give rise to exothermic peaks, usually extended over a wide temperature range.

Changes in specific heat capacity

In the absence of enthalpy changes, an ideal DSC trace is a straight line the slope of which depends on the difference of specific heat capacity between sample and reference material. After calibration with a substance of reliably known specific heat (sapphire is sometimes used), the specific heat of the test material can be calculated.

In polymer materials a step may be observed in the DSC trace, see C in Fig. 133. This represents a change in specific heat capacity at a particular temperature and usually indicates the **Glass transition temperature** (T_g).

Vitrification is a complex process very sensitive to rates of heat and cooling. Lower cooling rates give lower values of T_g. The feature associated with the glass transition may not be the simple step of Fig. 133 but may take any of the forms shown in Fig.134.

The value of T_g is affected by polymer molecular weight (see **Glass transition temperature**) so if the polymer has a broad distribution of molecular weight the transition recorded by the DSC will tend to be broad rather than sharp (Fig. 134(b), cf. Fig. 134(a)). The glass transition temperature would usually be taken as a temperature at the point of inflection.

Differences between heating and cooling lead to an 'overshoot' (Fig. 134(c)) when the cooling rate is the slower and to an 'undershoot' when it is the faster (Fig. 134(d)). With some thermal histories, both features may be observed in the same curve (Fig. 134(e)).

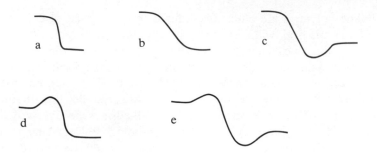

Fig. 134.   Forms that the DSC trace of a glass transition might take (axes
as in Fig. 133). See text for discussion of causes

Interpretation of DSC results

The results from an unknown or poorly understood adhesive may be
difficult to interpret. Curvature of the base line perhaps combined with
some of the features shown in Fig. 134(c)–(e)) may make it difficult to
distinguish enthalpy changes from specific heat changes. If the thermal
history is either unknown or uncontrolled the difficulty is compounded.

It is often desirable to repeat runs of the same sample to distinguish
irreversible changes such as volatilization and chemical reaction from
potentially reversible ones like crystalline melting and vitrification.
Similarly, results can be recorded on controlled cooling as well as on the
more usual heating runs.

An understanding of the principles of crystal nucleation and growth[3,4]
and of vitrification phenomena[5] will aid the interpretation and suggest
changes in heating and cooling rates which may clarify the situation
further.

It may well be impossible to interpret fully a complex DSC trace without
resort to results from other experiments; DMA will reveal glass transitions
and weighing the sample after the experiment (or better still TGA) will
establish whether volatile components are being lost.

Subject to these considerations, differential scanning calorimetry
provides a powerful aid to the understanding of the behaviour of adhesive
systems.

References

1. M J Richardson in *Comprehensive Polymer Science*, Vol. 1, ed. G Allen,
   Pergamon, Oxford, 1989, p. 867.

2. J P Runt, *Encyclopedia of Polymer Science and Engineering*, Vol. 4, Wiley-Interscience, New York, 1986, p. 488.
3. D C Bassett, *Principles of Polymer Morphology*, Cambridge University Press, Cambridge, 1981.
4. D E Packham, Adhesives and sealants, *Engineered Materials Handbook*, Vol. 3, ed. H F Brinson, ASM, Ohio 44073, 1990, p. 406.
5. G W Scherer, *Relaxation in Glass and Composites*, Wiley-Interscience, New York, 1986, Ch. 1.

Tie-layers

R J ASHLEY

A tie-layer or coextrudable adhesive has the function of improving interfacial forces between dissimilar polymers in the **Coextrusion** process. The layer is used in cases when resins coextruded are so incompatible that effectively no bond develops between them, or where the bond between two polymers is so limited as to be impracticable (see **Compatibility**).

Most of these tie-layers are based on modified polyolefins or their copolymers. Ionomers such as Surlyn may also be used as a tie-layer. The functionality of the adhesive layer derives from such comonomers as acrylic acid, methacrylic acid, maleic anhydride or vinyl acetate. Styrene–isoprene–styrene thermoplastic block copolymers blended with polyolefins can be used as tie-layers. The level of addition of these grafted units is generally between 0.5 and 15%. In the case of the polyethylene–acrylic acid materials the low acid concentrations are used in extrusion applications.

Methods for the incorporation of the graft unit may include irradiation of the base polymer and mixing with the comonomer, or reaction as solids or solutions with free radical agents and blending techniques (see **Addition polymerization, Radiation cured adhesives**).

Apart from their use as tie-layers in coextrusion the modified polymers can find other adhesive applications such as improving adhesion of extrusion coatings, thermal lamination interplies, and as dispersions or powder coatings.

Select reference

R J Ashley, *Adhesion 12*, ed. K W Allen, Elsevier Applied Science Publishers, London, 1988, Ch. 16.

Toughened acrylic adhesives

P CULLEN

Introduction

In recent years, uses of structural engineering adhesives for the production-line assembly of a wide range of components, such as cabinets, loudspeakers, motor parts, have shown significant growth. Much of this has been since the development and successful introduction of low-viscosity room-temperature curing and toughened adhesives, such as rubber-toughened acrylics.

Setting mechanism

Toughened acrylic adhesives are essentially monofunctional methacrylate monomers containing a dissolved rubber polymer added as a toughener, adhesion promoters, cross-linking agents, free radical stabilizers, cure accelerators and a free radical generator.[1] Cure is usually initiated by a reducing agent for the free radical generator dissolved in an organic solvent (surface activator for use with a single-part adhesive) or in a second adhesive component (two-part adhesive).

When using single-part adhesives with a solvent-based surface activator, in most applications it is only necessary to apply the activator to one of the surfaces to be bonded and the adhesive to the other. Once the bond is closed, cure is initiated. For close-fitting steel parts, handling strength is developed in a few minutes and full strength over a period of hours. While this approach can give very fast fixturing with good properties, a major limitation is inability to cure completely through gaps much above 1 mm, due to poor diffusion of the activator through the adhesive. This disadvantage is overcome by the use of two-part adhesives, where both parts are essentially high-viscosity adhesives, similar in appearance and are mixed either by static or dynamic mixing. The resultant mixed product is applied as a single continuous bead to one of the substrates to be bonded. Alternatively, a 'non-mix' approach, unique to acrylic adhesives, of applying a continuous bead of one component on top of a continuous bead of the other, can be used. Sufficient mixing to initiate cure takes place automatically on closing the bond. In general, this method is satisfactory if bond gaps are no larger than about 1–2 mm. Above this, mixing is recommended.

Methacrylate monomers chosen for tough acrylic adhesives are usually those that polymerize to give high-modulus, glassy-type polymers characterized by glass transition temperatures well in excess of room temperature (c. 100 °C). Adhesives based on these rigid, brittle polymers alone would carry high loads and give high tensile shear strength, but offer virtually no resistance to impact and peel forces. To overcome this the practice of adding a rubber to toughen an otherwise brittle polymer that has been used extensively for many years for the large-scale production of toughened engineering plastics, such as high impact-polystyrene (HIPS) and acrylonitrile–butadiene–styrene copolymers (ABS), has now been adopted for the manufacture of tough adhesives. Several examples of this are given in the patent literature.[2,3] Improved fracture toughness of these adhesives is attributed to the formation of a discrete secondary phase of small rubber particles (circumference 0.1–1.0 μm) dispersed throughout the brittle, continuous methacrylate polymer matrix. Toughness and crack resistance is derived from the ability of these flexible particles to absorb and dissipate energy associated with propagating cracks initiated during impact and cleavage loading of the bond.[4]

A similar structure would also be seen for a rubber-toughened epoxy adhesive. Table 43 compares the performance of toughened acrylic adhesives to standard **Anaerobic adhesives**.

Applications

Toughened acrylic adhesives can be formulated to have low, controllable viscosity, which makes possible rapid, precise dispensing (see **Dispersion of adhesives**) to parts to be bonded. The room-temperature curing or setting of these adhesives makes them particularly useful for the assembly of components in automatic or semi-automatic production lines. Examples of these are the bonding of loudspeakers (ferrite/metal and coil/cone), d.c. motors, bicycles, filing cabinets and general-purpose applications where some degree of toughness and impact resistance is required.

Table 43. Performance* of toughened acrylic adhesives

	Toughened acrylic	Standard anaerobic
Viscosity	3–100 000 mPa	1–3000 mPas
Speed (fixture time)	60–300 s	10–60 s
Strength		
Tensile shear (ASTM D1002-64) GB steel	15–30 N mm^{-2}	20 N mm^{-2}
Toughness		
T-peel (ASTM D1876-69T) GB aluminium	3–8 N mm^{-2}	0–2 N mm^{-2}
Impact strength (ASTM 0950-54)	30–50 kJ m^{-2}	10–20 kJ m^{-2}

*Data are taken from the technical literature and are believed to be typical.[5]

Table 44. Comparative benefits of industrial adhesives (toughened acrylics, anaerobics, epoxies)

	Toughened acrylics	Anaerobics	Epoxies	
			Two-part mixed	One-part heat cure
Bond strength				
Metals/ferrites	+ +	+ +	+	+ +
Metals/plastic	+ +	—	0	—
Toughness	+ +	0	—	+
Impact resistance	+ +	—	—	0
Temperature/humidity	+ +	+	+	+ +
Speed of bonding	+	+ +	0	—
Ease of cure				
Automatic application	+ +	+ +	—	0

Note: + + great benefit, + significant benefit, 0 no benefit, — significant disadvantage.

Advantages/disadvantages of toughened acrylic adhesives

Table 44 compares toughened acrylic adhesives to two other important classes of adhesive used widely in industrial assembly operations, **Anaerobic adhesives** and **Epoxide adhesives**. The speed of bonding of anaerobics makes them ideal for metallic substrates or where toughness and impact resistance are not critical. Epoxies provide good bond strength to many substrates, but are slower curing and require sophisticated mixing and/or dispensing equipment. Toughened acrylic adhesives provide better toughness and impact resistance than anaerobics and better speed and ease of use than epoxies.

Other articles of relevance are **Anaerobic adhesive, Epoxide adhesives** and **Toughened adhesives**. In **Structural adhesives**, properties of epoxide and **Polyurethane adhesives** are compared with those of toughened acrylics.

References

1. F R Martin, Acrylic adhesives, in *Developments in Adhesives – 1*, ed. W C Wake, Applied Science Publishers, London, 1977.
2. Lord Corp., US Patent 3 832 274, published 27 Aug. 1974.
3. E I Du Pont de Nemours and Company, US Patent 3 994 764, published 17 June, 1975.
4. C B Bucknall, Mechanism of rubber toughening, in *Toughened Plastics*, Applied Science Publishers, London, 1977, Ch. 7.
5. F R Martin, Acrylic adhesives, in *Structural Adhesives* – Developments in Resins and Primers, ed. A J Kinloch, Applied Science Publishers, London, 1986, Ch. 2.

Toughened adhesives

P CULLEN

Introduction

The advancement of toughened adhesive technology has transformed both acrylic and epoxy technology into high-performance adhesive packages providing important benefits in performance and durability. These benefits are derived from the evolution of improved adhesion control and toughening mechanisms. Toughened adhesives generally encompass both **Toughened acrylic** and **Epoxide adhesives**.

Toughening mechanism

Toughened adhesives, such as toughened acrylics and toughened epoxies are essentially monofunctional and difunctional monomers or resins containing a dissolved rubber polymer added as a toughener, together with adhesion promoters, stabilizers, cure accelerators and cure initiators.

In general, the methacrylate monomers or epoxy resins chosen for adhesives are usually those that polymerize to give high-modulus, glassy-type polymers, which offer virtually no resistance to impact and peel forces. To overcome this, the practice of adding a rubber to toughen an otherwise brittle polymer matrix has been used.

Rubbery materials, such as acrylonitrile–butadiene–styrene copolymers (ABS) and variations thereof have now been adopted for the manufacture of both **Toughened acrylic adhesives** and **Epoxide**.[1,2] The improved fracture toughness of these adhesives is attributed to the formation of a discrete secondary phase of small rubber particles dispersed throughout the brittle continuous methacrylate or epoxy polymer matrix. These flexible particles can absorb and dissipate energy associated with propagating cracks initiated during impact and cleavage loading of the bond.[3] A micrograph showing the structure of a toughened epoxide is given in the article **Scanning electron microscopy**.

The **Toughened acrylic adhesives** or **Toughened epoxide adhesives** are then capable of both high tensile strength and resistance to impact and peel forces.

Advantages/disadvantages of toughened adhesives

The evolution of toughened adhesives, both acrylics and epoxies means that adhesives which exhibit adhesion, hot strength and durability can now also maintain high impact resistance and elasticity, for example, at sub-freezing temperatures.

However, in contrast to **Anaerobic adhesives** and **Cyanoacrylate adhesives**, toughened adhesives, in general, tend to be slow curing and high in viscosity, although low-viscosity toughened acrylics are now emerging.

Other articles of relevance are **Toughened acrylic adhesives, Epoxide adhesives, Structural adhesive and acrylic adhesives**.

References

1. Lord Corp., US Patent 3 832 274, published 27 Aug. 1974.
2. E I Du Pont de Nemours and Company, US Patent 3 994 764, published 17 June, 1975.
3. C B Bucknall, Mechanism of rubber toughening, in *Toughened Plastics*, Applied Science Publishers, London, 1977, Ch. 7.

U

Underwater adhesives

M R BOWDITCH

Requirements of the adhesive

Materials suitable for use as underwater adhesives include carefully formulated cold-curing thermosets such as **Epoxide** and **Polyurethane adhesives**. Such multi-component systems are not significantly affected by the presence of water in the uncured condition and convert from the liquid to the solid state via **Addition polymerization** in which two or more coreactants form one polymeric product. The process does not involve the loss of solvents and does not require the presence of oxygen.

Perhaps the earlier materials found to have a useful capacity for adhesive bonding underwater depended upon the use of a stoichiometric excess of water-scavenging polyamide hardener in an epoxide-based adhesive. This approach can lead to the production of effective joints in the short term, but formulations of this type which are hydrophilic in the uncured state are also likely to absorb significant amounts of water in the cured condition. It is a widely accepted view that the extent of joint weakening in susceptible joints, quite apart from the consequences of plasticization, is a function of the water-uptake characteristics of the adhesive (see **Glass transition temperature**). The consequence is therefore likely to be that such joints will show poor **Durability** in the presence of water.

It is more satisfactory to choose an essentially hydrophobic adhesive in which water is only sparingly soluble and which has a low diffusion coefficient, and this was the approach chosen by the Admiralty Research Establishment when researching in this area.[1,2]

Wetting requirements

This requirement represents perhaps the greatest challenge in the underwater application of adhesives, particularly where high **Surface**

energy substrates are concerned. Such materials are typified by metals, metal oxides and ceramics and are representative of most useful structural materials with the notable exception of glass and other fibre-reinforced organic composites.

The nature of the problem, as well as the solution to it, is represented in Fig. 135. The figure shows the circumstances in which adhesive joints are normally made with the substrate and adhesive surrounded by air. In the making of adhesive joints the presence of air is seldom even considered, for the perfectly good reason that it does not represent a problem. However, the fact is that all surfaces are contaminated by the permanent gases, but they are only weakly adsorbed (see **Adsorption theory**) and are readily displaced by adhesive which then spreads freely and spontaneously

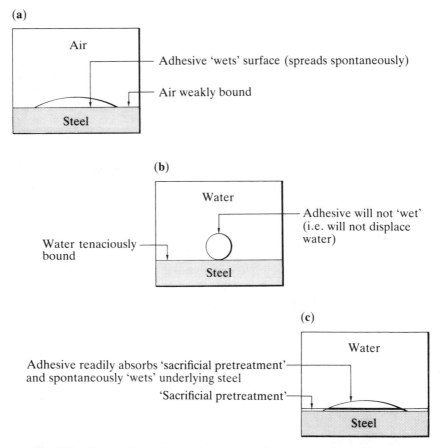

Fig. 135. Schematic representations of an adhesive in contact with a steel substrate (a) in air, (b) in water, (c) in water in the presence of a 'sacrificial pretreatment'

over the substrate surface. In this way intimate contact is achieved between liquid adhesive and solid substrate and adhesion results. Various **Theories of adhesion** are discussed elsewhere, but all require that adhesive and adherend are in intimate contact. For example, attractive intermolecular forces (van der Waals, see **Dispersion forces, Polar forces**) can operate only over short ranges (~ 1 nm) and chemical interaction between adhesive and adherend also requires the two reactants to be in close contact.

However, Fig. 135 also shows that when the fluid air is replaced by the fluid water a more serious problem is presented. Water, because of its relatively high surface tension (when compared with liquid organic-based adhesive for example), is much more strongly adsorbed by highly energetic surfaces and is a much more difficult contaminant to displace. The use of mechanical action during the application of the adhesive to displace water from these surfaces is extremely inefficient and weak, and variable joints are often the result.

One approach to the solution of this problem, adopted by the Admiralty Research Establishment at Holton Heath, is to apply a so-called sacrificial pretreatment to the substrate surface prior to the application of adhesive. The method involves the displacement of water and the deposition of a hydrophobic but adhesive-compatible (see **Compatibility**) film over the surface to be adhesively bonded.[3] The consequence is that those favourable circumstances, taken for granted when bonding in the atmosphere, are re-created under water and adhesive spreads spontaneously over the surface and the essential close contact between adhesive and substrate is achieved. The sacrificial pretreatment is so called because it is absorbed and/or displaced by the adhesive which then gains access to the substrate surface itself. Suitable pretreatments may be in the form of liquids which are made up of a mixture of suitable surfactants and other components with appropriate solubility parameters (see **Compatibility**). This may be applied to freshly cleaned, preferably grit-blasted surfaces by spraying through a suitably designed hood when it performs the dual function of providing a wettable surface for the adhesive and also of protecting the metal surface against corrosion for a limited period until the adhesive may be applied. Alternatively, they may be solids which can be applied simultaneously with the cleaning process if appropriate characteristics are possessed by the matrix materials used in the bristles of abrasive brushes for example. When this technique is used the thickness of the deposited layer is a function of the ratio of matrix to abrasive and also of application temperature.

Figure 136 shows, in the form of histograms, the strengths achieved with steel/steel tensile butt joints made under water both with and without the use of a sacrificial pretreatment. The lower of the two histograms

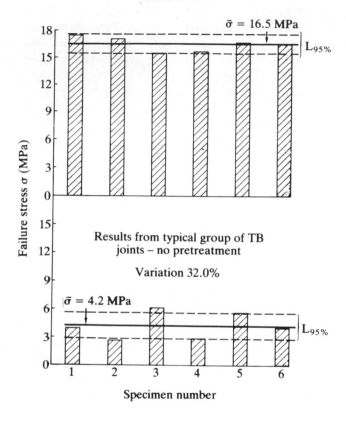

Fig. 136. Influence of a 'sacrificial pretreatment' for steel on the strength of tensile butt joints made under water with an epoxy-amine adhesive

shows that, where adhesive is applied under water immediately after grit-blasting, the achieved joint strength is low and extremely variable. These results clearly demonstrate the unpredictability of using purely mechanical means to displace water from energetic substrates (in this case mild steel) and the consequent incomplete wetting by the adhesive of the substrate. With joints made in this way failures appear to be exclusively interfacial between adhesive and adherend. In the upper histogram, Fig. 136 also shows that, where the displacement of water is energetically favourable after the application of a liquid sacrificial pretreatment, joint strengths are some three to four times higher with a greatly reduced

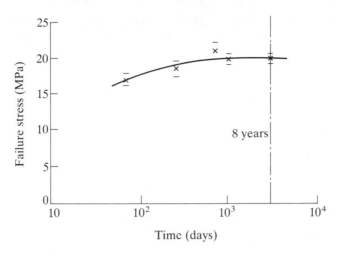

Fig. 137. Influence of exposure to sea water on the strength of tensile butt joints made between steel with an epoxy-amine adhesive

variation at a level of around 5%. The failure of joints made in this way is usually cohesive within the adhesive layer.

Where less energetic surfaces are to be bonded, such as those presented by polymeric materials used as matrices for fibre-reinforced composites or as surface coatings, the problems are much reduced as it will no longer be energetically favourable for water to occupy the surface. Provided the substrate is wettable by the adhesive there will be no great problem in producing a reliable adhesive joint. By way of illustration, untreated polyethylene will be no easier to bond under water than it is in the atmosphere, but glass-reinforced polyester will be readily jointed in this way. Even when saturated with water, useful adhesive joints may still be made with this latter material.

Durability considerations

The question of the **Durability** of adhesive joints in the underwater environment is clearly one of great importance because water is widely recognized as a major threat to their integrity. However, joints to polymeric substrates will generally experience degradation only as the result of a weakening of the adhesive following plasticization by dissolved water. Where energetic substrates are bonded, degradation may also occur as a result of displacement of adhesive by water.

The degree of disbonding by water may also be minimized, provided water gains access to the water-sensitive interface via diffusion through the adhesive, by selecting an adhesive with a low water solubility and a low diffusion coefficient as recommended earlier. This approach, coupled with careful joint design, may provide adequate durability, but where greater resistance to hydrolytic attack is required use of **Coupling agents** would be advised. Figure 137 shows graphically the effect of prolonged exposure to sea water on steel/steel tensile butt joints using the WRA4501 adhesive system, which is based on an amine-cured epoxide resin (see **Tensile tests, Epoxide adhesives**).

References

1. M R Bowditch, J D Clarke, K J Stannard, The strength and durability of adhesive joints made underwater, in *Adhesion 11*, ed. K W Allen, Elsevier Applied Science Publishers, London, 1987.
2. M R Bowditch, R A Oliver, An underwater adhesive repair method for offshore structures, in *Polymers in Offshore Engineering*, Conference proceedings, Plastics and Rubber Institute, London, 1988.
3. M R Bowditch, Formation of metal/resin bonds, UK Patent Specification 2 083 3778, 24 Mar. 1982.

V

Viscoelasticity

D W AUBREY

Viscoelasticity is concerned primarily with polymer deformation, not polymer or adhesive fracture processes. Its importance in adhesion processes lies in the fact that the deformation which accompanies adhesive fracture can make an important contribution to the overall joint strengths.

The simplest type of viscoelastic behaviour as shown by single-phase amorphous polymers is described in this article. Polymers which crystallize or form multiple phases such as blends, block copolymers, particulate or fibre-filled polymers, show more complex behaviour, since each amorphous region will show its own viscoelastic response to deformation.

As its name implies, viscoelasticity involves a blend of liquid-like viscous properties with solid-like elastic properties. In order to make any progress with characterizing viscoelastic behaviour, these two aspects of behaviour must be considered separately.

Types of response in polymers

Amorphous, uncross-linked polymers can exhibit various combinations of two kinds of elastic response and two kinds of viscous response to deformation. These responses are represented in Fig. 138 by mechanical analogues consisting of elastic springs and dashpots filled with viscous liquids.

Glassy response, dominated by extension and compression of secondary intermolecular bonds, is shown at the lowest temperatures. It is elastic in the same way as all rigid solids (e.g. snooker balls) are elastic and is represented in Fig. 138 as an 'ordinary elastic' spring of high Young's modulus E (stress/strain).

Leathery ('transition') response This is shown at somewhat higher temperatures, where segments of polymer chains, but not whole polymer

SIMPLE MODELS

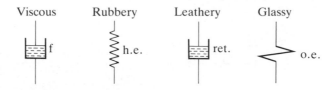

COMPOSITE MODELS
2-parameter:

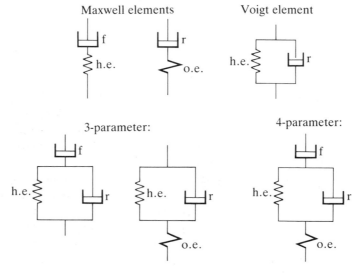

Fig. 138. Mechanical analogues for viscoelastic behaviour of polymers
(f = flow; h.e. = highly elastic; ret. = retardation; o.e. = ordinary
elastic)

chains, gain sufficient thermal energy to move under a deforming force.
This is a viscous process, the viscosity η (shear stress/shear strain rate)
depending on the frictional resistance to segmental motion. It is
represented by a retardation dashpot (Fig. 138). The temperature at which
the midpoint of this region falls is often taken as the **Glass transition
temperature**, T_g, and is a characteristic temperature for a polymer, enabling
different polymers to be compared in the same viscoelastic state.

High elastic ('rubbery') response This arises at still higher temperatures,
where the frictional resistance to segmental motion becomes negligible.
It involves recoverable deformation in which the polymer chains are

deformed from their equilibrium configurations against a network of molecular entanglements. It is represented by a high elastic spring (Fig. 138) which is of much lower modulus than the ordinary elastic spring.

Viscous flow (*'terminal'*) *response* This is seen at the highest temperatures. Under these conditions the entanglements have sufficient thermal energy to become undone, and the flow of whole molecules can occur. This is the same kind of flow which arises in all liquids and is represented by a flow dashpot (Fig. 138) of much higher viscosity than the retardation dashpot.

The modulus of an elastic spring is substantially independent of the temperature or rate at which it is measured and the work done in deforming it is all stored and recoverable. The stiffness of a dashpot, on the other hand, is markedly dependent on temperature or rate of deformation and the work of deformation is all dissipated as heat. Thus, depending on the relative contributions of springs and dashpots to its behaviour, a polymer will show various degrees of temperature- and rate-dependence in its properties, and various proportions of input energy dissipation as heat (hysteresis). Depending on the rate and temperature, it is common practice to approximate polymer behaviour by adopting one of the various composite mechanical analogues shown in Fig. 138. Of these, the four-parameter model is probably the most versatile, since this can be simplified to any of the other models by allowing the dashpot viscosities to range between infinitesimal and infinite.

Viscoelastic parameters

Viscoelastic characteristics of polymers may be measured by either static or dynamic mechanical tests. The most common static methods are by measurement of creep, the time-dependent deformation of a polymer sample under constant load, or stress relaxation, the time-dependent load required to maintain a polymer sample at a constant extent of deformation. The results of such tests are expressed as the time-dependent parameters creep compliance $J(t)$ (instantaneous strain/stress) and stress relaxation modulus $G(t)$ (instantaneous stress/strain) respectively. The more important of these from the point of view of adhesive joints is creep compliance (see also **Pressure-sensitive adhesives – adhesion properties**). Typical curves of creep and creep recovery for an uncross-linked rubber (approximated by a three-parameter model) and a cross-linked rubber (approximated by a Voigt element) are shown in Fig. 139.

According to the Boltzmann superposition principle, the final creep deformation caused by a series of step loading and unloading increments

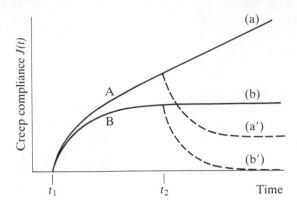

Fig. 139. Creep (curves (a) and (b)) and creep recovery (curves (a') and
(b')) for (A) an uncross-linked rubber and (B) a cross-linked
rubber (t_1 – load applied; t_2 – load removed)

such as those of Fig. 139 is predictable by the summation of the individual
creep responses from each increment.

Viscoelastic response of polymers can also be measured by dynamic
tests, in which the response of the polymer is measured under an imposed
sinusoidal oscillatory deformation of variable frequency. Because of the
viscoelastic nature of the polymer, there is a lag in phase angle δ between
stress and strain. The extreme values of δ would be zero, for a purely
elastic material, and $\pi/2$ for a purely viscous one. Typical instruments
for studying dynamic mechanical properties are the torsion pendulum,
the Weissenberg rheogoniometer, the dynamic mechanical thermal
analyser (DMTA) and the vibrating reed. Such methods give values for
viscoelastic parameters such as a 'complex' Young's modulus E^* or shear
modulus G^* and differentiate between an elastic contribution E' or G'
('storage' modulus) and a viscous contribution E'' or G'' ('loss modulus').
These are related by, for example, $E^* = E' + iE''$ where $i = (-1)^{1/2}$.
The 'absolute' modulus $|E^*|$ or $|G^*|$ is sometimes used. It is given by
$|E^*| = \sqrt{[E'^2 + E''^2]}$. The ratio E''/E' or G''/G', equals $\tan\delta$, the
dissipation factor (or loss factor), and gives a measure of energy dissipated
by hysteresis and is, of course, at a maximum when viscous (dashpot)
processes are involved. There is, of course, a close analogy between the
formal treatment of dynamic viscoelasticity and of alternating electric
current.

By application of the principle of time–temperature equivalence (see
Viscoelastic properties – time–temperature superposition) the results of
dynamic tests may be expressed as a master curve, either in the form of
a viscoelastic function (e.g. $\log G'$) against temperature T at constant

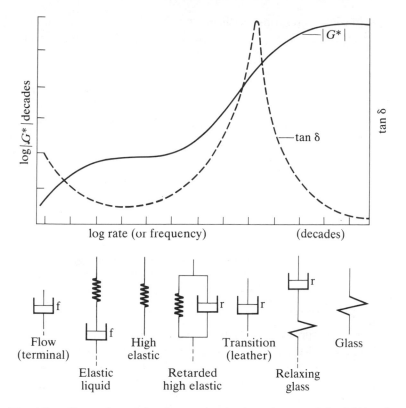

Fig. 140. Illustration of the changes in absolute shear modulus $|G^*|$ and
dissipation factor tan δ with frequency at constant temperature
for a simple amorphous polymer. Models show the changes in
viscoelastic state

frequency ω of measurement or in the form of viscoelastic function (e.g.
log G') against frequency (log ω) at constant temperature T. The latter
presentation is more useful for adhesive materials, since it is more common
to vary rates of joint fracture at constant temperature than vice versa.

We are now in a position to illustrate the various regions of viscoelastic
behaviour with an idealized curve of, for example, the magnitude of the
absolute dynamic shear modulus $|G^*|$ against frequency (log ω) at a
constant standard temperature (Fig. 140). On the same abscissa, the
energy dissipation (tan δ) and applicable mechanical models are also
shown.

Figure 140 illustrates the ideal viscoelastic behaviour for a simple
amorphous polymer. It shows how the regions of rate (and temperature)
independence coincide with elastic (spring) behaviour (where $|G^*| \approx G'$),
and regions of high rate (and temperature) dependence coincide with

viscous (dashpot) response (where $|G^*| \approx G''$). Adhesives should be designed to operate in the dashpot regions where the dissipation factor tan δ is highest.

Select references

J D Ferry, *Viscoelastic Properties of Polymers*, 3rd edn, John Wiley, New York, 1980.
L E Nielsen, *Mechanical Properties of Polymers and Composites*, Vol. 1, Marcel Dekker, New York, 1974.
I M Ward, *Mechanical Properties of Solid Polymers*, 2nd edn, John Wiley, New York, 1983.

Viscoelasticity – time–temperature superposition

D W AUBREY

Time–temperature equivalence

Because of the nature of viscoelastic relaxation processes (discussed briefly under **Viscoelasticity**), the effect of increasing the temperature is equivalent to allowing more time in a given test. The equivalence is expressed in general in relation to a chosen standard temperature T_0 by a shift factor a_T defined as the ratio

$$a_T = \frac{\text{Time } t \text{ for a given viscoelastic response at temperature } T}{\text{Time } t_0 \text{ for the same viscoelastic response at temperature } T_0}$$

Thus, a given creep compliance value $J(t)$ (see **Viscoelasticity**) for a cross-linked rubber may be obtained after a short time t at a higher temperature T or a longer time t_0 at a lower temperature T_0 (Fig. 141).

A typical shift is such that the effect of increase of 1 decade of t is approximately equivalent to an increase of $8-12°$ in T. It is therefore more convenient to use a logarithmic scale for the time (or rate) axis.

A knowledge of the shift factor–temperature relationship therefore allows a viscoelastic function to be plotted either against deformation time (or inversely against rate) at a standard temperature T_0 or against temperature at a standard time (or rate) of deformation. However, a modulus–(log) frequency curve (see Fig. 140 under **Viscoelasticity**) is not the exact inverse of the modulus–temperature curve because there are non-viscoelastic effects which also cause the modulus to change with temperature. The most important of these is the dependence of modulus on temperature in the high elastic region, and to correct for this the 'reduced' version of the measured viscoelastic function is used. For

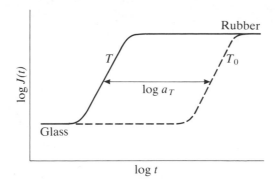

Fig. 141. Creep compliance at two different temperatures illustrating shift factor a_T

example, the reduced storage shear modulus would be given by

$$G'_r = G' \frac{T_0}{T} \frac{\rho}{\rho_0}$$

where T, T_0 are absolute test and standard temperatures, and ρ, ρ_0 the corresponding densities. The density component of this correction is very small and is often neglected.

Experimental time—temperature superposition

The experimental application of the time—temperature superposition principle will be described by reference to the formation of a shear modulus—(log) frequency master curve.

First, measurements are made of dynamic storage shear modulus G' over a range of frequencies at each of several temperatures, using an oscillatory method such as DMTA. Values are 'reduced' to G'_r (see above), then data points are plotted as $\log_{10} G'_r$ against $\log_{10}$ frequency ω on sheets of transparent paper, one sheet for each temperature. The abscissae for the various test temperatures T are then shifted horizontally along a common axis until data points for adjacent temperatures superimpose. The amount of horizontal shifting of each abscissa relative to that for the standard reference temperature T_0 gives the value of shift factor as $\log_{10} a_T$ for that particular temperature. The superimposed data points then form the master curve of $\log_{10} G'_r$ against $\log_{10} \omega a_T$ at the standard reference temperature T_0 (Fig. 142).

A great advantage of such master curves is that the dependence of viscoelastic function against frequency is shown over a much wider range

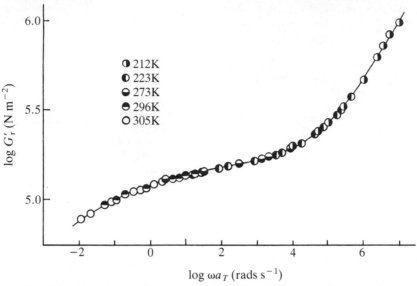

Fig. 142. Experimentally superposed results of (reduced) dynamic shear modulus obtained at various temperatures and frequencies

of rates (typically 10^{10}-fold) than are directly accessible with any viscoelastic measuring method (typically 10^3-fold). All viscoelastic functions including moduli, compliances, viscosities, loss functions, etc. are amenable to this treatment. In addition, fracture and adhesion processes in which viscoelastic deformations are prominently involved are commonly subjected to time–temperature superposition for clearer presentation and better understanding.

Use of the WLF equation in time–temperature superposition

If the values of shift factor $\log_{10} a_T$, obtained as above, are plotted against test temperature T a smooth curve is obtained. Williams, Landel and Ferry (1955) showed that the same shift factor–temperature relationship was obtained from the experimental shifting of results from a large number of amorphous polymers. The empirical relationship thus obtained is known as the WLF equation, and can be used in one of the two forms:

$$\log_{10} a_T = \frac{-C_1^s (T - T_s)}{C_2^s + T - T_s}$$

where $C_1^s = 8.86$, $C_2^s = 101.6$ K,

$$\log_{10} a_T = \frac{-C_1^g (T - T_g)}{C_2^g + T - T_g}$$

where $C_1^g = 17.44$, $C_2^g = 51.6$ K.

The values of C_1 and C_2 shown are referred to as universal, since they have been shown experimentally to apply to a large number of polymers. If natural logarithms are used, values for C_1^s or C_1^g are multiplied by 2.303.

Here T_s is a specific reference temperature, at which all amorphous polymers are in the same viscoelastic state. It was arbitrarily chosen by Williams, Landel and Ferry as $T_s = 243$ K for polyisobutylene and, in general, is given approximately by $T_g + 50$ K. The equation is now probably more used in the form with the glass transition temperature T_g as the viscoelastic reference temperature. It holds well over the range T_g to $T_s + 100$ K ($T_s \pm 50$ K).

The WLF equation can clearly be used as an alternative to the experimental shifting procedure described above, providing that a value for T_s (or T_g) is known. There is some risk attached to the direct use of the WLF equation, however, since there is variation of the values of C_1 and C_2 for different polymers, and T_s (or T_g) values may not be accurate. A better procedure would be to determine C_1 and C_2 values by experimental shifting, then to use the WLF equation, with these values, for subsequent shifts. Using this procedure, the WLF equation can be used with confidence for results in viscoelastic regions where experimental shifting fails (e.g. where an absence of slope makes experimental shifting impossible).

It will be clear that the direct use of the WLF equation will produce a master curve with either T_s or T_g as its reference temperature T_0. It may be, however, that the master curve is required at some other temperature (e.g. room temperature). In such a case it is a relatively simple matter to calculate the constants (C_1^0 and C_2^0) for any chosen reference temperature T_0.

The general form of the WLF equation may be predicted from considerations of free volume, and physical significance can be attached to the constants C_1 and C_2.

Select references

M L Williams, R F Landel, J D Ferry, *J. Amer. Chem. Soc.*, **77**, 3701 (1955).
J D Ferry, *Viscoelastic Properties of Polymers*, 3rd edn, John Wiley, New York, 1980.

W

Weak boundary layers

D M BREWIS

If a region of low cohesive strength exists at the interface between a substrate and a hardened adhesive, failure will occur at a low stress level. The region is termed a 'weak boundary layer'.[1] Likewise, such boundary layers can occur at other interfaces, including substrate–paint, substrate–printing ink and fibre–matrix in a composite.

The origins of weak boundary layers are varied and include mould release agent on a plastic, for example a silicone on nylon 66, an additive, for example a plasticizer, which has migrated to a plastic surface, a weak oxide on a metal, for example zinc, a protective oil on a metal and corrosion products at a metal surface.

When failure occurs at a low stress level, it is informative to examine the two sides of the failed structure using a surface-sensitive technique such as **X-ray photoelectron spectroscopy** or **Secondary ion mass spectrometry**. From the surface chemistries of the two surfaces it is possible to determine the **Locus of failure**, for example if a silicone is detected on both surfaces, then failure may be ascribed to an excessive quantity of this material. An example of a weak boundary layer is given in Fig. 143, where fluorinated material is transferred from polytetrafluoroethene (PTFE) to an epoxide at a low failure load.

Weak boundary layers are definitely the cause of many adhesion problems. However, there has been controversy on how widespread such boundary layers are, especially in the case of polyolefins.[2] For example, there is disagreement on whether some pretreatments for polyolefins are effective because they increase the functional groups in polyolefins, or because they eliminate weak boundary layers.

On most surfaces there will exist a layer of low-strength organic material. In some cases, for example, a protective oil on a metal, the layer will be deliberately present and will be relatively thick. High-energy solids such

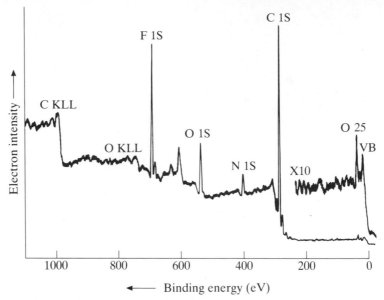

Fig. 143. XPS survey scan of the surface of an epoxide adhesive cured in contact with PTFE. The PTFE was detached at a very low failure load, but transfer of fluorinated material occurred. (From D Briggs in *Industrial Adhesion Problems*, eds D M Brewis, D Briggs, Orbital Press, Oxford, 1985, p. 15)

as metal will adsorb a thin layer of organic compounds from the atmosphere. Surfaces of polymers are likely to be covered by organic compounds of relatively low molecular weight (and cohesive strength) (see **Compatibility, Release**). Such compounds will include mould-release agents, anti-static agents, antioxidants and plasticizers.

Sometimes pretreatments are used to remove or at least reduce this contamination, but often no special effort is made to remove these compounds prior to bonding, printing or painting these surfaces and yet good adhesion is usually achieved. This means that the organic layers can usually be absorbed by the mobile phase, for example the adhesive. Thus the organic compounds only constitute a weak boundary layer if they cannot be displaced. The factors affecting displacement are discussed in the article **Displacement mechanisms**.

See also **Pretreatments of metals, Pretreatments of polymers, Surface energy**.

References

1. J J Bikerman, *Science of Adhesive Joints*, Academic Press, New York, 1961.
2. D M Brewis, D Briggs, *Polymer*, **22**, 7 (1981).

Weathering of adhesive joints

S TREDWELL

The practical importance and theoretical complexity of environmental deterioration of adhesive bonds has resulted in extensive research on many aspects of the problem.

Several different test methods have been used to investigate the **Durability** of bonded joints. By far the largest single problem facing adhesive scientists and technologists today is that of the long-term durability of adhesive bonds exposed to natural environments, i.e. weathering. **Structural adhesives** are used extensively in aerospace constructure (see **Aerospace applications**) principally with aluminium alloys, and aircraft are required to function throughout a service life of several decades, in many diverse natural environments ranging from low-temperature arctic conditions to tropical, high-humidity climates.

Much work has been undertaken on the **Accelerated ageing** of adhesive joints. Although the results of such work are valuable, there is no substitute for testing of joints after exposure to a range of actual climatic conditions. Among the most severe climates are tropical dry and tropical wet climates.

A typical example of such weathering trials[1,2] involved bonded aluminium double overlap joints, stressed and unstressed, peel and honeycomb specimens (see **Peel tests, Honeycomb structures**). These were exposed at the Royal Aircraft Establishment (RAE), Farnborough (temperate), and at the Joint Tropical Research Unit (JTRU) sites at Innisfail (hot–wet) and Cloncurry (hot–dry), Australia. Periods of up to 6 years were employed, exposing a variety of adhesive systems to the different climates.

The durability of the bonded joints was greatly influenced by the nature of the adhesive; the best performers in all climates were epoxy–novolak and nitrile–phenolic formulations. A tropical, hot–wet climate was the most damaging to bonded structures and the combination of high humidity and applied stress was particularly deleterious. During exposure to natural environments, the failure mode of aluminium joints was found to change gradually from wholly cohesive, within the adhesive, to include increasing amounts of interfacial failure (see **Locus of failure**).

Aluminium bonded joints have also been exposed to jungle weathering conditions near the equator in Surinam, South America, for 12 years.[3] Distinct differences were found between different jungle locations, with open jungle exposures being more severe than more protected, secluded jungle sites. The highest strength retentions were obtained with a room-temperature-curing epoxy as compared with a rubber-modified heat-cured epoxy.

A summary of the progress in evaluating adhesive bond permanence over a 25-year period has been published by Minford.[4] Further information can be found in other articles on **Durability, Epoxide adhesives, Phenolic adhesives** and **Surface pretreatments**.

References

1. M D G Hockney, *Technical Reports* (a) *70081* (1979); (b) *72100* (1972); (c) *73013* (1973), RAE, Farnborough.
2. J L Cotter in *Developments in Adhesives – 1*, ed. W C Wake, Applied Science Publishers, London, 1977.
3. J D Minford, *Int. J. Adhesion and Adhesives*, Jan., 25 (1982).
4. J D Minford in *Treatise on Adhesion and Adhesives*, Vol. 5, ed. R C Patrick, Marcel Dekker, New York, 1981, Ch. 3.

Wedge test

B M PARKER

The wedge test was developed as a very simple method of evaluating metal surface pretreatment, particularly with reference to predicting the **Durability** in service of bonded aerospace structures by **Accelerated ageing**.

Two thin adherend beams are bonded together with a short unbonded length at one end into which a wedge is driven. The general specimen type is shown in Fig. 144; dimensions given are as for ASTM D 3762-79, but

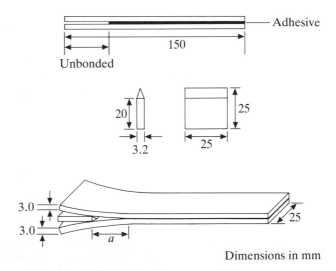

Dimensions in mm

Fig. 144. Wedge test specimen

metricated. Other dimensions have been used and overall specimen length can be reduced if it is known that the crack length is going to be short, or to save material. Ideally, adherend thickness *h* should be such that plastic deformation does not take place if fracture energy is to be calculated.

The initial crack length is measured between the shoulder of the wedge and the crack tip. The specimen is then exposed to the chosen environment, usually high **Humidity** and elevated temperature and the growth of the crack followed with time. Durability is inverse to crack growth.

The wedge test is a simplified version of the double cantilever beam test (see **Fracture mechanics test specimens**), where the displacement of the bonded beams is held constant, rather than being varied by increasing the load on the specimen, and is determined by the thickness of the wedge. Fracture takes place in mode I, crack opening, without a shear component. The critical strain energy release rate, also called fracture energy (see **Fracture mechanics**) can be calculated from the crack length and material parameters.

The formula developed by Mostovoy and Ripling for the fracture energy of an adhesive using the double cantilever beam specimen[1] and modified for constant rather than increasing displacement is

$$G_{1C} = \frac{y^2 E h^3 [3(a + 0.6h)^2 + h^2]}{16[(a + 0.6h)^3 + a^2 h]^2}$$

where G_{1C} is the fracture energy (kJ m^{-2}), *E* the adherend modulus (GPa), *h* the adherend thickness (mm), *a* the crack length (mm) and *y* the displacement at load point = (wedge thickness − adhesive thickness) (mm). The first term, in round brackets, is due to the bending of the beam, the second to the shear deflection, and 0.6*h* is an experimentally determined rotation factor.

While originally developed for testing the durability of **Pretreatment of aluminium**, it has since been applied to pretreatment of other metals, to **Fibre reinforced composite**,[2] to polymers,[3] to durability in other environments and in the **Selection of adhesives**[3] and **Primers**. An example of its use is given in **Pretreatments of titanium**.

Few comparisons with other **Tests of adhesion** have been made; ranking of adhesives[3] was the same as by lap **Shear tests** and of pretreatments compared with stressed durability tests,[4] but the wet T-peel test may be even more discriminating for metal surface pretreatments.[3,5]

Associated topics are **Anodizing, Etch primers, Fibre composite – joining, Coupling agents, Pretreatment of polymers, Standards, Stress corrosion, Tests of adhesion, Weathering of adhesive joints**.

References

1. S Mostovoy, E J Ripling, *US Government Report AD 874429*, 1970.
2. B M Parker in *Bonding and Repair of Composites*, Anon., Butterworths, London, 1989.
3. H Dodiuk, S Kenig, *Int. J. Adhesion and Adhesives*, **8** (3), 159 (1988).
4. R W Wolff, *Nat. SAMPE Symp.*, **28**, 1155 (1983).
5. C W Matz, *Nat. SAMPE Symp.*, **30**, 1088 (1985)

Wetting and spreading

M E R SHANAHAN

Depending on its affinity for the solid, a liquid placed on a substrate will show a tendency either to remain as a drop with a finite area of contact or to spread out and cover the solid surface available. In this article it is assumed that equilibrium has been attained; considerations of **Wetting kinetics** are given in a separate entry.

Consider a liquid drop at rest on an ideal solid surface (Fig. 145) and in particular the triple line of contact where solid S, liquid L and surrounding fluid F (either vapour of the liquid or a second liquid immiscible with the first) meet. To each interface solid–liquid, solid–fluid, liquid–fluid can be attributed a free interfacial energy or interfacial tension (respectively $\gamma_{SL}, \gamma_{SF}, \gamma_{LF}$). (The equivalence of these definitions is discussed in **Surface energy**.) Using the mechanical (tension) definition, a force (per unit length of triple line) acts along each interface. In order to assure equilibrium, a horizontal force balance must exist, and defining as θ_0 the **Contact angle** measured between the solid–liquid interface and the tangent

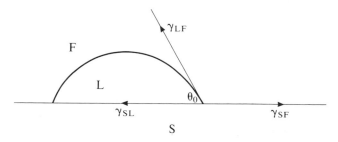

Fig. 145. Wetting and spreading. Sessile drop of liquid L exhibiting a contact angle θ_0 on solid S with surrounding fluid F which may be vapour of the liquid L or a second liquid

to the liquid–fluid interface at the triple line, Young's equation[1] must be respected:

$$\gamma_{SF} = \gamma_{SL} + \gamma_{LF} \cos \theta_0 \qquad [1]$$

This very simple demonstration of Young's fundamental relation is sometimes questioned but other, more rigorous derivations based on minimizing the overall free energy of the system solid–liquid–fluid lead to precisely the same result (e.g. ref. 2). Consideration of the vertical force component $\gamma \sin \theta_0$ leading to a 'wetting ridge' – local deformation of the solid of dimensions of the order of γ/E (E = Young's modulus of solid) is only necessary for very soft solids (e.g. soft polymers, gels) and does not modify the macroscopic validity of Eqn 1.[3]

The above applies to cases where $|\gamma_{SF} - \gamma_{SL}|/\gamma_{LF} < 1$. Under these conditions, finite **Contact angles** are obtained and the liquid is said to wet the solid. In the case where $\gamma_{SF} \leqslant \gamma_{SL} - \gamma_{LF}$, θ_0 will be π. This corresponds to lack of affinity between the solid and the liquid and contact will then be assured uniquely by external force fields (e.g. gravity). There is no wetting. This situation, however, arises rarely, if ever, when the surrounding fluid is vapour, although it can occur for a two-liquid system.

At the other extreme, we consider the situation when $\gamma_{SF} \geqslant \gamma_{SL} + \gamma_{LF}$. The contact angle θ_0 is then necessarily zero and the liquid is said to spread on the solid (some authors refer to wetting and spreading respectively as partial and complete wetting). We define the quantity S, the spreading coefficient of liquid L on solid S in the presence of fluid F:

$$S = \gamma_{SF} - \gamma_{SL} - \gamma_{LF} \qquad [2]$$

which may be interpreted as the net force (per unit length of triple line) provoking radial spreading of the liquid on the solid for a (hypothetical) contact angle of zero (see **Wetting and work of adhesion**). Referring to Young's equation, we see that if S is negative, we obtain a finite contact angle, whereas if S is positive, spreading ensues. If S is zero, Antonow's rule is respected. This corresponds to final equilibrium.

Each γ term, and as a consequence S, is both temperature and composition dependent. In general surface tensions decrease with temperature and several empirical expressions are known both for liquids and polymers.[4] Liquid surface and interfacial tensions may also be modified by altering the proportions of mixtures. Judicious changes of composition can be used to alter S for a given solid–liquid–liquid system in order to favour the wetting of the substrate by one or other of the liquids. Diagrams showing wetting propensity as defined by S as a function of components of surface tension are known as wetting envelopes. These can be of great use for such practical processes as lithographic printing.[5]

Importance of the spreading coefficient in adhesion

When an adhesive joint is produced, clearly intimate molecular contact between the adhesive and the substrate is desirable. Considering that at some stage in the process the adhesive will be liquid, it is clearly a good principle to choose a combination of surfaces with S positive. However, a second reason for favouring a high value of S also exists, as shown by the following simple argument. Although actual energies of adhesion may for various reasons be many factors larger than those predicted thermodynamically, it is generally accepted that high values of the latter are concomitant with good strength properties. Dupré's thermodynamic work of adhesion W_A for a liquid L (the adhesive) on a solid S (the substrate) in the presence of fluid F is given by

$$W_A = \gamma_{SF} + \gamma_{LF} - \gamma_{SL} \qquad [3]$$

(W_A is sometimes differently defined: see Eqn 4 in **Wetting and work of adhesion**).

The work of cohesion W_C of the liquid is

$$W_C = 2\gamma_{LF} \qquad [4]$$

Using Eqn 2 we have

$$S = W_A - W_C \qquad [5]$$

Thus, the higher the value of S the greater is the thermodynamic work of adhesion compared to the cohesive energy of the adhesive. Although after solidification of the adhesive these thermodynamic quantities are likely to be modified, the overall features will remain similar. Since the weak part of an adhesive bond tends to be the interface, a high value of S will favour cohesive failure within the adhesive rather than interfacial failure and in turn will tend to lead to a more reliable junction.

N.B. The above descriptions assume thermodynamic equilibrium to be attainable. In practice, it is common to observe the phenomenon of wetting hysteresis. Metastable rather than stable equilibrium is reached, as shown experimentally by a range of contact angles instead of a unique value. The causes are multiple: heterogeneity of the solid surface, local **Adsorption, Roughness of surfaces**, etc.

Related articles are **Adhesion, Contact angles and interfacial tension, Surface characterization, Surface energy** and **Wetting kinetics**.

References

1. T Young, *Phil. Trans. Roy. Soc. Lond.*, **95**, 65 (1805).
2. R E Johnson, *J. Phys. Chem.*, **63**, 1655 (1959).

3. M E R Shanahan, P G de Gennes in *Adhesion 11*, ed. K W Allen, Elsevier Applied Science Publishers, London, 1987, Ch. 5.
4. S. Wu, *J. Macromol. Sci. – Rev. Macromol. Chem.*, **C10** (1), 1 (1974).
5. D H Kaelble, P J Dynes, D Pav in Polymer Science and Technology, Vol. 9B, *Adhesion Science and Technology*, ed. L H Lee, Plenum Press, New York, 1975, p. 735.

Wetting and work of adhesion

J F PADDAY

Adhesion, by its definition, depends on the ability of two unlike phases to hold themselves together across a common interface. Physical adhesion must first take place before any other bonding processes such as chemical reaction can occur, and such physical adhesion depends on the strength of intermolecular force interaction,[1] on the area of contact and on the distance of separation separating the atoms forming the top layer of each surface (see **Dispersion forces, polar forces**). When both phases are undeformable, such as with two solids that are not atomically smooth, poor adhesion results because an insufficient area of each surface is in atomic contact with the other. When one phase is deformable, such as with a liquid of low viscosity, physical adhesion takes place at all parts of the surface. Physical adhesion with a liquid in contact with a solid leads to spreading and wetting; processes that now depend on the competition of adhesion forces with cohesion forces within the liquid.

Wetting is defined[2] as the process whereby a liquid when brought into contact with a solid displaces some of the fluid phase so that a stable triple-phase line of contact is formed, as seen in Fig. 146. Spreading, also

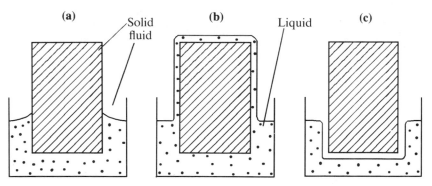

Fig. 146. A description of wetting (a), spreading (b) and spontaneous dewetting (c)

described in Fig. 146, is the process whereby the liquid, once in contact with the solid, displaces the third phase completely from all parts of the surface and at the same time creating a new liquid–air (third phase). Finally, one describes the process of dewetting (or spontaneous dewetting) as that wherein the third phase, usually a liquid, itself spreads and displaces completely the first liquid from the solid surface.

The free energies of adhesion, wetting and spreading are expressed in terms of the free energies of each interface, γ, such that

adhesion $\qquad\qquad W_{a_{123}} = \gamma_{13} + \gamma_{23} - \gamma_{12}$ $\qquad\qquad$ [1]

wetting $\qquad\qquad W_{e_{123}} = \gamma_{13} - \gamma_{12}$ $\qquad\qquad$ [2]

spreading $\qquad\qquad S_{c_{123}} = \gamma_{13} - \gamma_{12} - \gamma_{23}$ $\qquad\qquad$ [3]

The spreading energy is often termed the 'spreading coefficient',[3] see also **Wetting and spreading**. All these equations apply to ideally smooth surfaces: if the surface is rough they require modification by the Wenzel roughness factor (see **Contact angle**).

It should be noted that Eqn 1 defines work of adhesion in terms of γ_{13} the surface energy of the solid 1 in contact with fluid 3, which according to the circumstances may be a second liquid, may be a gas or may be air saturated with the vapour of liquid 2. Work of adhesion is often defined in a subtly different way:

$$W_a = \gamma_1 + \gamma_2 - \gamma_{12} \qquad\qquad [4]$$

where γ_1 is the surface energy of solid 1 and γ_2 that of liquid 2, both *in vacuo*, in equilibrium with their own vapour. In Eqn 4, W_a gives a measure of the energy required to cut cleanly along the 2–3 interface, removing the two phases into separate enclosures.

The treatment here is developed in terms of Eqns 1–3, introducing Young's equation[3] which relates the surface energies to θ, the **Contact angle**:[4]

$$\gamma_{13} - \gamma_{12} = \gamma_{23} \cos \theta \qquad\qquad [5]$$

so that

$$S_{c_{123}} = \gamma_{23}(\cos \theta - 1) \qquad\qquad [6]$$

$$W_{a_{123}} = \gamma_{23}(\cos \theta + 1) \qquad\qquad [7]$$

and

$$W_{e_{123}} = \gamma_{23} \cos \theta \qquad\qquad [8]$$

Finally we may divide these equations by γ_{23} to present them in dimensionless form and plot the reduced spreading coefficient $S_{c_{123}}/\gamma_{23}$ as a function of the reduced work of adhesion, $W_{a_{123}}/\gamma_{23}$ as seen in Fig. 147.

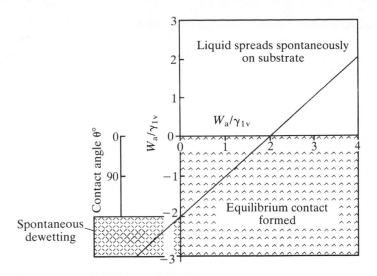

Fig. 147. Reduced spreading coefficient S_C/γ_{23} as a function of the reduced work of adhesion W_a/γ_{23}

It is important to note that the value of the spreading coefficient is not in any way limited in magnitude or sign. In Fig. 147 the three processes of Fig. 146 are now clearly defined by

$$0 < S_{c_{123}}/\gamma_{23} < \infty$$

Liquid 2 spreads spontaneously. No contact angle formed

$$-2 < S_{c_{123}}/\gamma_{23} < 0$$

Liquid 2 wets and a stable contact angle is formed

$$-\infty < S_{c_{123}}/\gamma_{23} < -2$$

Liquid 2 spontaneously recedes. No contact angle formed

The contact angle is thus only of significance when the reduced spreading coefficient is negative and lies between 0 and 2. The problem of measuring the spreading coefficient when there is no contact angle is overcome by measuring the disjoining pressure first described by Derjaguin.[5] The disjoining pressure π_{sl} is the pressure required to thin a spread layer by an element of thickness dt.

$$\int_{t_0}^{\infty} \pi_{12} \, dt = S_{c_{123}} \qquad [9]$$

where t_0 is the distance of closest approach of the surface atoms of the spreading liquid to those in the solid surface. The disjoining pressure is readily measured in a variety of ways and increases as the thickness of the spread layer is decreased by some external force such as gravitational drainage. Its value may also be calculated from Hamaker constants[1] for some systems where the physical interaction is derived principally from dispersion forces.

Thus it is deduced that effective adhesion depends critically on the liquid adhesive displacing the third phase and spreading into all the space between the two solid surfaces to be attached to one another. The angle of contact, the liquid surface tension and the disjoining pressure are the appropriate measurable properties that enable quantitative values of the free energy of spreading to be assessed.

References

1. J N Israelachvili, *Intermolecular and Surfaces Forces*, Academic Press, London, 1985.
2. J F Padday, *Wetting, Spreading and Adhesion*, Academic Press, London, 1978, p. 459.
3. A W Adamson, *Physical Chemistry of Surfaces*, 3rd edn, John Wiley, New York, 1976.
4. W A Zisman in Advances in Chemistry Series 43, *Contact Angle and Wettability*, American Chemical Society, Washington, 1964, pp. 1–51.
5. A D Zimon, *Adhesion of Dust and Powder*, trans. from Russian by Morton Corn, Plenum Press, New York, 1969.

Wetting kinetics

M E R SHANAHAN

When a liquid is put into contact with a solid surface, be it an adhesive, a paint, an ink or merely a drop of water, equilibrium will not generally be ensured immediately. **Wetting and spreading** will ensue spontaneously until either an equilibrium (stable or metastable) **Contact angle** is obtained or until the liquid has spread leaving a thin layer covering the solid surface available. Commonly observed situations usually invoke one or both of two main categories of the wetting process. If the solid is impervious the essential phenomenon will consist of the liquid drop spreading outwards, lowering its contact angle and increasing its radius of contact until equilibrium defined by Young's equation (see **Wetting and spreading**) is obtained. However, in the case of a porous solid, penetration of the liquid within the bulk may also occur – a phenomenon similar to capillary rise. In both cases the detailed kinetics are exceedingly complex,[1–3] but the following arguments may be applied in order to understand the basic underlying physical principles.

Spreading of a drop on a solid surface

Capillary regime[3] Consider the idealized case of a drop of liquid in the form of a spherical cap of small contact angle $\theta(t)$ spreading radially with time t to its equilibrium configuration represented by angle $\theta_0 [\theta_0 \leqslant \theta(t)]$. The solid is taken to be smooth, flat, horizontal isotropic and homogeneous. If the drop is small the force causing spreading will be essentially that due to the capillary imbalance at the triple line. Referring to the solid–liquid, solid–vapour and liquid–vapour interfacial tensions respectively as γ_{SL}, γ_{SV} and γ, the force (per unit length) of the triple line at the drop periphery and directed radially outwards, F, will be

$$F = \gamma_{SV} - \gamma_{SL} - \gamma \cos \theta(t) \sim \frac{\gamma}{2} [\theta^2(t) - \theta_0^2] \qquad [1]$$

where the simplified (second) expression is obtained from Young's equation and the approximate form of $\cos \theta$ for θ small.

If the drop spreads at radial speed $U(t)$, the work done (per unit time) is FU. This work will be dissipated by viscous effects due to the flow field within the liquid. Calculation of this flow field represents a formidable task, but a useful approximation was suggested by de Gennes[3] in which the 'lubrication approximation' is adopted. The liquid is regarded as a nearly flat wedge with a parabolic (Poiseuille) velocity profile parallel to the solid surface. (Boundary conditions of zero stress at the liquid–vapour interface and no slip at the solid–liquid interface, together with overall displacement rate U, allow the profile to be defined.) Viscous dissipation/unit volume takes the form ηv_z^2, where η is viscosity and v_z the local velocity gradient. Integration of this quantity using the velocity profile leads to a dissipation term $k\eta U^2 r/h$, where k is a numerical constant and r and h are drop contact radius and height respectively.

Bearing in mind that $U = dr/dt$ and taking for simplicity the case corresponding to $\theta_0 = 0$, we equate FU to the dissipation to obtain

$$\frac{dr}{dt} \sim \frac{2\gamma h^3}{k\eta r^3} \qquad [2]$$

Ignoring henceforth numerical constants and using the expression for constant drop volume ($V \sim \pi r^2 h/4$), Eqn 2 is simply solved to obtain the scaling law for drop spreading:

$$r \sim V^{3/10} \left(\frac{\gamma t}{\eta} \right)^{1/10} \qquad [3]$$

We thus see that the spreading of small drops where capillary forces dominate is proportional to $t^{1/10}$ and the overall rate is essentially governed by the ratio γ/η.

Gravitational regime[4] If a drop of liquid is sufficiently large, gravitational forces predominate over capillary effects. Defining a capillary length $K^{-1} = (\gamma/\rho g)^{1/2}$, where ρ is the liquid density (or more precisely the difference in densities between the liquid and the surrounding fluid) and g is the gravitational acceleration, the transition between the regimes occurs typically at $K^{-1} \sim 1.5$ mm.

For larger drops, we may reasonably ignore the effects of Eqn 1 and consider that the 'motive power' of spreading comes mainly from gravitational potential energy related to the height of the centre of gravity of the drop above the solid surface.

Again using dimensional arguments, we may consider the spreading force to be of the form $\rho g V$ and acting at a fraction f of the height h of the drop. The work done per second can then be expressed as

$$\dot{W} = -F\frac{d(fh)}{dt} \sim -\rho g V\frac{dh}{dt} \qquad [4]$$

Equating this to the dissipation term described above (multiplied by $2\pi r$ to take into account the entire periphery of the drop) and using the constant volume condition, we obtain

$$\frac{dr}{dt} \sim \frac{\rho g V^3}{\eta r^7} \qquad [5]$$

giving after integration

$$r \sim V^{3/8}\left(\frac{\rho g t}{\eta}\right)^{1/8} \qquad [6]$$

Thus in the gravitational regime, contact radius increases as $t^{1/8}$ and thus more rapidly than in the capillary regime. The overall rate is governed by the ratio $\rho g/\eta$. Note that in both cases the relative values of driving force to viscosity control the process, as may be expected intuitively.

Capillary rise

The penetration of a liquid within a porous medium can be compared to a first approximation to the phenomenon of capillary rise. We shall derive what is commonly known as Washburn's equation.[5] Consider a capillary tube touching a liquid surface with its axis vertical and internal radius r_0.

The overall force F causing capillary rise is given by the capillary effect acting round the circumference of the tube reduced by the effective weight of the column of liquid of height h in the tube.

$$F = 2\pi r_0 \gamma \cos \theta - \pi r_0^2 \rho g h \qquad [7]$$

(*N.B.* No distinction is made between dynamic and static contact angles in this elementary treatment.)

Force F is equilibrated by viscous drag within the tube. A Poiseuille velocity profile is assumed to exist due to the average pressure difference p across the height h and equal to $F/\pi r_0^2$. Standard fluid mechanical arguments lead to the volume increases per unit time in the capillary, dV/dt:

$$\frac{dV}{dt} = \frac{\pi p r_0^4}{8h\eta} = \pi r_0^2 \frac{dh}{dt} \qquad [8]$$

Substitution of Eqn 7 into 8 leads to

$$\frac{dh}{dt} = \frac{r_0^2}{8\eta h} \left[\frac{2\gamma \cos \theta}{r_0} - \rho g h \right] \qquad [9]$$

Although this simple equation is open to criticism on several grounds (constancy of θ, simultaneous assumption of regular meniscus and laminar flow), it serves as a basis. In the case of porous media, the gravitational terms will generally be neglected.

Related articles are **Contact angles and interfacial tension, Surface energy** and **Wetting and spreading**.

References

1. C Huh, L E Scriven, *J. Colloid Interface Sci.*, **35**, 85 (1971).
2. E B Dussan V., *Ann. Rev. Fluid Mech.*, **11**, 371 (1979).
3. P G de Gennes, *Rev. Mod. Phys.*, **57** (3), 827 (1987).
4. J Lopez, C Miller, E Ruckenstein, *J. Colloid Interface Sci.*, **56**, 460 (1976).
5. E W Washburn, *Phys. Rev.*, **17**, 374 (1921).

Wood adhesives – basic principles

D F G RODWELL

A major use of adhesives is for the bonding of wood, especially in the joinery and construction industries. This article surveys some of the principles, and adhesives used. More specialized accounts are given under **Wood adhesives – joints for furniture, Wood adhesives – veneering, Wood adhesives – edge banding, Wood adhesives – plastic laminating panels**. The adhesives used in wood-based boards, such as plywood and particle board are discussed in **Wood composites – adhesives**.

Principles

The basic mechanisms of wood adhesive bonding are still very poorly understood. This is mainly due to the complex chemistry and structure of the substrate. Wood is largely composed of cellulose and lignin which differ greatly in their reaction to adhesives.

In a newly made joint the wood and adhesive will be attached at various sites and by various means. At some sites the strong forces of covalent bonding may apply, while at others the weaker forces of van der Waals and **Hydrogen bonding** will operate (see **Absorption theory, Dispersion forces, Polar forces**). At yet others there will be mechanical interlocking of cured resin and wood surface (see **Mechanical theory**). It is thought, too, that there can be molecular penetration of the wood cell walls with subsequent polymerization (molecular entanglement).

A well-made joint will contain more than enough attachment sites, and at room conditions the total strength of the bonds will often exceed the cohesive strengths of the wood or cured resin. If the joint were tested to destruction at this stage then it would probably fail either in the adhesive or wood, whichever was the weaker; in fact one method of quantifying bond quality is to express the area of wood failure as a percentage of the joint contact area.

In service a joint is subjected to moisture and temperature fluctuations which induce internal stresses and these will interact at the bond line with externally applied loads. Certain moisture–temperature combinations can also promote chemical degradation of the constituent materials (see **Durability**). The weaker intermolecular bonds will be the first to fail, but their loss may not substantially reduce the overall strength of the joint. Continued ageing will eventually lead to a breakdown of the mechanical attachments and the long-term durability (especially in wet conditions) is now generally considered to be determined by the survival of sufficient covalent (high-energy) bonds or regions of molecular entanglement within the cell walls.

Types of wood adhesive

Animal glues were traditionally used and still find some employment in craft woodworking, but the most commonly used wood adhesives fall into two groups: (1) the formaldehyde-based synthetic resins of urea and phenol-resorcinol (see **Phenolic adhesives**) and (2) the various emulsions of polyvinyl acetate (see **Emulsion and dispersion adhesives**).

Within these groups are adhesives of widely varying properties and, especially where safety is an issue, it is important to select the right type for the conditions of use. Some adhesives can withstand moist conditions, some have gap-filling properties and others can withstand long-term loading. Several British Standards give information on the classification and requirements of wood adhesives, but these are soon to be replaced by European (CEN) Standards. When selecting an adhesive for a structural application it is particularly important to follow the recommendations of the relevant national building code or regulation. In the UK the three most important are BS 5268: 1984: Part 2 (The Structural Use of Timber), BS 4169: 1970 (Glued Laminated Timber Structures) and BS 6446: 1984 (Specification for Manufacture of Glued Structural Components).

Current British practice recognizes four categories of the formaldehyde-based adhesives:

1. Type WBP – weather and boil proof.
2. Type BR – boil resistant.
3. Type MR – moisture (and moderately weather) resistant.
4. Type INT – interior.

Urea–formaldehyde adhesives usually fall into the lower categories, melamine–formaldehyde and melamine–urea copolymers into the mid categories and phenol–resorcinol formaldehyde adhesives achieve the highest ratings.

Polyvinyl acetate adhesives tend to be placed into one of four categories according to a German (DIN) standard which has also found wide acceptance outside Germany. The categories are termed B1, B2, B3 and B4 in increasing degrees of durability. Water resistance is improved by the addition of a hardener to the basic emulsion.

One feature of polyvinyl acetate adhesive that makes it unsuitable for long-term loading applications is a tendency to 'creep' with time, but for many joinery applications they have found wide acceptance.

Making the joint

Bonding wood with a suitable adhesive is not normally a difficult task providing that certain requirements are met. Principally these are that the wood should be relatively dry (below 20% moisture content), the surfaces to be joined are freshly prepared and sufficient bonding pressure is applied. The quantity of adhesive to be applied and the curing time should be covered by the manufacturer's literature. Occasionally problems are encountered which are associated with the characteristics of a particular wood species. For example some timbers (notably teak) have a rather oily surface which can lead to poor surface wetting. Other timbers (obeche, for instance) are very absorbent and this can lead to 'starvation' of adhesive at the bond line.

Virtually all timbers are slightly acid. In some timbers (e.g. oak and western red cedar) the heartwood can approach pH 3 and this may interfere with the curing of some adhesives. There are ways of overcoming most bonding problems and Building Research Establishment Digest 314 gives advice in this area.

Select references

A Pizzi, *Wood Adhesives – Chemistry and Technology*, 2 vols, Marcel Dekker, New York, 1983 and 1989.

Building Research Establishment Digest 314, Building Research Establishment, Garston, Watford.

A Pizzi, Wood adhesives, present and future, *J. Appl. Polym. Sci., Appl. Polym. Symp.*, No. 40, 1984.

Wood adhesives – edgebanding

R F TOUT

This article is concerned with the use of adhesives in 'edgebanding' – the covering of edges of materials such as particle board (see **Wood composites – adhesives**). A background discussion on the bonding of wood can be found together with references to other related articles under **Wood adhesive – basic principles**.

In the conversion of chipboard into panels which are suitable for cabinet furniture, decorative panelling, etc. the process of covering the unsightly chipboard edges with a decorative material is a vital one. The operation, known as edgebanding, is carried out in furniture and related industries on continuous edgebanding machines. Two types are available, and they are based on either **Ethylene vinyl acetate copolymers** or polyvinyl acetate (PVAC) adhesives (see **Emulsion and dispersion adhesives**). On both types of machine the panels are edged 'on the run' as they are fed through the machine, but otherwise, the two operate in quite different ways.

Hot melt edgebanding

On all but the smallest machines, which require pre-glued edgings, the hot melt adhesive is heated to its liquid state in a heater tank and fed to an applicator head which coats the edge of the panel as it is fed through the machine. The edging strip, automatically fed from a spool or magazine, is brought into contact with and pressed against the coated panel edge in a pressure zone consisting of a number of pressure cylinders. It is here that the edgebond is formed. The edged panel then passes into a machining cubicle where excess edging material is removed.

For hot melt edgebanding, as with all hot melt bonding operations, the state and therefore the temperature of the adhesive at the point of bonding is critical. To obtain sound, durable bonds the applied adhesive film must be liquid enough when the panel enters the pressure zone to efficiently wet out and adhere to the back of the edging strip; the adhesive must then chill and solidify rapidly so that the bond develops sufficient coherent strength to withstand the impact of the cutters in the machining cubicle.

With hot melt adhesives, the rate of cooling of the applied adhesive is initially high, and this is ideal for high-speed production as bond strength develops rapidly. It can introduce problems, however, for tests have shown that even using sound edgebanding procedures on compact, high-speed edgebanders, adhesive applied at 200 °C will cool by 70–80 °C before bonding takes place. At these temperature levels edgebanding adhesives should still be capable of wetting out the edging material, but with poor edgebanding conditions, adhesive chilling will be even greater and bonding temperature could be too low to achieve good adhesion to the edging strip. A weak edgebond, susceptible to premature failure, will then result.

Several factors affect the temperature of the applied adhesive at bonding. The most important is adhesive application weight. Thick films will have a higher heat capacity than thin ones and will retain their heat longer. Therefore, within practical limits, thick adhesive spreads should be used. Squeeze-out can become a problem, however, if application weights are too heavy. The applied adhesive loses heat primarily into the chipboard edge, but also into the surrounding air. The use of cold panels and the siting of edgebanders in cold, draughty positions should be avoided, as they will result in greater adhesive chilling prior to bonding. Adhesive application temperature will also affect bonding temperature, and it is essential to select a temperature within the adhesive manufacturer's recommended range (generally 200–220 °C). It is important that the upper limit is not exceeded to compensate for excessive heat losses, as this will lead to adhesive degradation. Increasing machine speed will reduce heat losses from the applied adhesive, but feed rate has a smaller effect on bonding temperature than the other parameters mentioned.

A satisfactory bond also requires close contact between panel and edging, and sufficient pressure must be applied in the pressure zone to achieve this and to ensure a reasonable pressure on the glueline during bond formation.

PVAC edgebanding

On PVAC edgebanders, adhesive is applied at ambient temperature to the edges of the panels as they pass through the machine. Polyvinyl acetate adhesives cure by loss of moisture, either by evaporation into the air or by absorption into the adherends. Clearly, the rate of loss of moisture by natural means at normal ambient temperatures is too slow for normal edgebanding operations, and some forced drying of the applied adhesive is necessary. This is achieved by passing the panels in front of an IR heating zone, which removes moisture from the applied adhesive film, leaving it in a warm, tacky state. At the same time the edging (which has been pre-coated with PVAC and dried) is fed past a hot-air blower where the heat reactivates the adhesive. Edging and panel are then brought together in a pressure zone, where the two warm, tacky adhesives fuse together under pressure to form an instantaneous bond. Excess edging is then removed as the edged panel passes through a machining cubicle.

The act of bonding, therefore, is similar to that for a contact adhesive, and the requirements concerning the state of the adhesive films are also similar. If the applied adhesive is too moist at the point of bonding, the instantaneous bond that develops will be weak and likely to be damaged or even failed by the impact of the cutters during trimming. An inherently weak bond will also be obtained if the two adhesive films are too dry during bonding, and unable to fuse together efficiently. The first situation is likely to occur when heavy adhesive application weights and fast feed rates are used; low application weights and slow feed rates and/or insufficient heat reactivation of the adhesive on the edging would lead to a situation where good fusion of the two adhesive films was impossible.

Hot melt edgebanding produces excellent edge bonds at high speed when thin, flexible edgings are used. Although hot melts are thermoplastic, the temperature resistance of these bonds is perfectly adequate for most UK furniture uses.

Polyvinyl edgebanders were developed for areas where hot melt edgebonds are likely to be suspect. These include the bonding of the more rigid edging strips to panels which will be exposed to unusually high transportation or service temperatures, and for applying substantial wood lippings. The PVAC bonds should remain secure up to 100 °C, but the edgings need to be pre-coated with adhesive before bonding, and feed rates are slower than with hot melt edgebanding.

Select references

R F Tout, Examination of the Brandt PVAC edgebander, *FIRA Bulletin 77*, 2–4, March 1982.

R F Tout, Edgebanding with PVAC and hot melts, *Furniture Manufacturer*, 250–82, May 1983.

R F Tout, Some difficulties with lipping panels on a hot melt edgebander, *FIRA Bulletin 109*, 6–7, April 1990.

Wood adhesives – joints for furniture

R F TOUT

Adhesives are widely used to bond wood in joinery. Some of the principles are discussed in the article **Wood adhesives – basic principles**, where reference to related articles will be found. This article concentrates on the use of adhesives in joints for furniture.

The jointing systems currently used to assemble modern wood furniture are fundamentally the same as those used by the early furniture makers some 300 years ago. Today's furniture manufacturers have the benefit of highly sophisticated adhesives, machinery and techniques, yet no new jointing systems have emerged which seriously compete with the dowel and mortise and tenon joints for making chairs. Dowels are also used extensively for assembling cabinet furniture from veneered and laminated chipboard panels.

Dowel joints

The dowel joint derives its strength from the adhesive bond achieved between the dowel and the walls of the holes in the jointed parts, and from the strength of the dowel itself. The quality of the dowel and adhesive are therefore major factors to be considered when producing dowel joints, but other aspects, probably less obvious but equally important, require attention.

Polyvinyl acetate (PVAC) adhesives (see **Emulsion and dispersion adhesives**) are used for producing dowel joints. They are one-part systems which cure relatively quickly at normal ambient temperatures. The British Standard relating to PVAC adhesives, BS 4071, offers a recognized method for assessing the suitability of a PVAC for producing wood/wood joints. Of particular importance is the 'resistance to sustained load' test, as this will indicate how well the adhesive can cope with moisture-induced movement in the wood pieces after jointing.

Dowels should be machined from high-strength, straight-grain wood which is free from defects. Beech is an excellent timber in these respects, and beech dowels are used for both chair and cainet assembly.

Multi-grooved dowels are used extensively in modern furniture production as they offer real advantages over smooth dowels. In production the adhesive is normally applied only to the hole, and tends to be swept down to the bottom of the hole ahead of the dowel as the latter is driven home. As the dowel approaches the bottom of the hole adhesive is forced back up along the grooves so that adhesive is available to make a bond where it is most effective, i.e. between the dowel and the walls of the hole.

If the hole is too deep for the dowel, the adhesive will not be forced back along the grooves in the dowel, but will remain at the bottom of the hole, where it will contribute little to the strength of the joint. The combined depths of the two holes in the joint must therefore be carefully chosen and machined to match the length of the dowel, or the joint will be weak.

Hole diameter is also a critical dimension. Multi-grooved dowels should be a small interference fit in the holes. This will give the joint some immediate stability and ensure relatively thin adhesive films – both factors will help to reduce cramping time. With too tight a fit there could be a risk of the wood splitting, while a loose fit will result in thick adhesive films (and weaker bonds) and longer cramping times. These problems should not occur with multi-grooved dowels if the hole diameter is 0.5 mm smaller than the nominal dowel diameter.

Another most important factor is the moisture content of the wood when it is machined and also when the joint is assembled. Wood swells or shrinks if it gains or loses moisture, and the effect is highly directional. The wood should therefore be machined and assembled at a moisture content close to its equilibrium moisture level in service. If changes in moisture levels do occur after assembly, stresses will be set up in the adhesive film as the jointed pieces move relative to each other. If large enough, these moisture-induced stresses can cause joint failure over a period of time.

Mortise and tenon joints

The strength of a mortise and tenon joint comes from the adhesive bond between the faces of the tenon and the corresponding sides of the mortise. It will depend primarily on the area of the bond (i.e. the area of the tenon), but accuracy of machining and the moisture content of the wood will also influence joint strength.

Increasing the area of bond is one way of improving the strength of a joint. However, if tenon size is increased to such an extent that the corresponding increase in mortise size weakens the mortised member sufficiently to make it vulnerable to splitting when the joint is stressed, then the strength of the joint is effectively reduced. Subject to this limitation, however, the joint should be designed with as large a tenon as possible.

Tenon thickness is less important than surface area when considering area of bond, but it is critical in relation to the width of the mortise. Because adhesives perform better when used in thin films, it is vital that the tenon is a good fit in the mortise, and large mismatches between tenon thickness and mortise width must be avoided. An interference fit between the two could lead to splitting of the wood at the mortise during assembly, while a large clearance will result in thick gluelines and weak joint. A clearance of 0.1 mm is recommended.

The clearance between the tenon and the ends of the mortise is less critical, as these areas contribute little to joint strength. The same applies to the bottom of the mortise, although a large gap here could be more efficiently used by increasing the length of the tenon.

Because of the geometry of the joint, good control of wood moisture content throughout production is even more important with mortise and tenon joints than with dowel joints. Mortise and tenon joints are also normally assembled using PVAC adhesives. Double application (to both tenon and mortise) is considered to be necessary if optimum bond strength is to be achieved.

The benefits obtained from modern adhesives, machines and production techniques should not persuade manufacturers that there is no need to worry unduly about joint preparation. Rather the reverse is true, because modern trends – light designs, less rugged joints, dry atmospheres in offices and homes – have tended to make gluelines in wood joints more highly stressed, and so have increased the need to make joints with near to optimum strength.

Select reference

R F Tout, Dowel joints for furniture, *Eur. Adhesives and Sealants*, 17–18, Dec. 1988.

Wood adhesives – plastics laminated panels

R F TOUT

Particle board and other wood-based composites for interior use (see **Wood composites – adhesives**) are often faced with plastics laminates. In this article the use of adhesives for producing these laminated panels is discussed. A background discussion of the use of adhesives for wood is given under **Wood adhesives – basic principles**, where reference to related articles will be found.

Plastics laminate has been used for many years to provide durable and decorative surfaces on panels for such widely varying uses as domestic kitchens, office partitioning, shop fittings, etc. The introduction of post-forming and continuous laminates and the enormous range of designs which have become available over the last 10 years has increased their appeal and usage enormously.

Surface laminating

There are two main methods in current use for bonding laminates to wood-based substrates, and the choice depends usually on the equipment that is available and the type and volume of production that is required.

The most durable laminating bonds are achieved on a solid platen press. Generally, a polyvinyl acetate (PVAC) adhesive (see **Emulsion and dispersion adhesives**) would be used, but urea formaldehyde (UF) adhesives give equally satisfactory bonds. Pressing time and temperature would depend on the required production rate and type of press available. Typical pressing times for PVAC adhesives would be in the range 20–30 min at normal ambient temperatures, 2–3 min at 80 °C and approximately 1 min at 120 °C. Standard grades of PVAC can be used up to 80 °C, but at pressing temperature above 80 °C, a cross-linking PVAC must be used – these adhesives have enhanced temperature and moisture resistance properties compared to the standard PVACs.

Pressing temperature can also affect the flatness of laminated panels which have an unbalanced construction, i.e. different decorative materials on their two faces. The surfaces of these panels will be subjected to unequal forces due to the thermal contraction of the two surfacing materials as the panels cool down from the hot press, and this will cause immediate and permanent bowing in the panel. With seriously unbalanced panels, hot pressing above 60 °C should be avoided, particularly with thin panels (15–20 mm). Higher pressing temperatures can be used with balanced or thick (30–40 mm) panels.

If a solid platen press is not available for laminating, PVAC or UF adhesives cannot be used, and the choice would then be a contact adhesive, based on polychloroprene. These adhesives have a solvent carrier and need to be applied to both the surfaces being bonded. After a suitable open time to allow the solvent to evaporate from the applied adhesive films, the two coated surfaces are brought together under momentary pressure. A bond forms immediately as the two adhesive films fuse together.

This type of laminating is normally used by relatively small trade laminators and shopfitters, where variable panel sizes and short production runs are the norm. Bond quality is affected by adhesive coverage, open time, contact pressure and environmental conditions during bonding, and these can be controlled reasonably well within a factory where adhesive spraying equipment and a nip roll press are available. When panels are laminated on site, however, such equipment is unlikely to be available, and environmental conditions are rarely ideal.

When contact adhesives are used for laminating within a factory in conjunction with semi-automatic methods and good control of conditions, sound durable bonds should be obtained. When both adhesive and bonding pressure are applied manually, bond quality will be at a reduced level, particularly with large panels, and will be heavily dependent on the expertise of the operator.

Post-forming

Many advances have been made in the field of plastics laminates over the last decade, and one of the most significant has been the development of post-forming machines and techniques. Post-forming is the forming of a laminate which has already been bonded to the surface of a substrate panel, around a pre-profiled edge of the panel. Special post-forming grades of laminate are required for this process.

Post-forming was introduced some 30 years ago, and was initially carried out on static machines, with the panels stationary. These types of machines, which are still in use today, are versatile with respect to profile and panel geometry but limited in production rates. The introduction of continuous post-forming machines in the late 1970s enabled panels to be post-formed 'on the run', at reasonable feed speeds. Post-formed panels, with their rounded, seamless edges, offer great advantages over square-edged laminated panels both aesthetically and in the areas of performance and hygiene.

Whichever process is used, static or continuous, a special post-forming grade of the laminate must first be bonded to the surface of the panel, with the laminate overhanging the panel's pre-shaped edge. Local heating of the overhanging laminate is then required to soften it temporarily so that it can be formed around the profiled edge. The temperature which the laminate must be heated to is critical, and must be carefully controlled – too little heating and the laminate will crack, too much and it will blister and char. On both types of machine, IR heaters are used. Forming of the laminate around the edge is accomplished with an angled section on static machines, and with metal bars and a series of metal and rubber-shaped rollers on the continuous machines.

Contact adhesives are used to produce post-forming bonds on static machines. Adhesive would have already been applied to both the laminate and the profiled panel edge during surface laminating, as the contact adhesive would also have been used for this operation.

Polyvinyl acetate adhesives are normally used on continuous post-forming machines. Although the panels would have been surface laminated with PVAC adhesive prior to post-forming, no adhesive would have been applied to the post-forming surfaces during laminating. Adhesive, therefore, has to be applied to the back of the overhanging laminate and the profiled panel edge as the panel passes through the machine, normally via spray guns. The heaters which soften the laminate also drive moisture from the applied adhesive films, so that when the laminate is formed around and pressed against the edge, the two gluelines fuse together to form an instant bond, in a similar way to a contact adhesive. Because the adhesive is used in a different way to that used for surface laminating in a solid press, a special post-forming grade of PVAC must be used.

Select reference

T M Maloney, Composition board, in *Encyclopedia of Polymer Science and Engineering*, Vol. 4, Wiley-Interscience, 1986, p. 47.

Wood adhesives – veneering

A J SPARKES

This article is concerned with adhesion of wood veneers: general background and reference to related articles can be found under **Wood adhesives – basic principles**.

Veneering refers to the process of bonding a thin layer of decorative wood, to a less interesting base material. This process is as old as furniture itself with records of the use of veneers on chests and caskets dating back to the time of the pharaohs. At that time the wood veneers were bonded to solid wood cores using adhesive obtained from natural sources, blood albumen, casein or bone or hide glues (see **Animal glues and technical gelatins**).

Modern furniture production is based largely on the use of wood-veneered panels manufactured by hot pressing the veneers onto plywood, wood chipboard or medium-density fibre-board (MDF) cores using urea formaldehyde (UF) or polyvinyl acetate (PVAC) adhesive (see **Emulsion and dispersion adhesives**) as the bonding agent.

Wood veneers

Decorative wood veneers are normally cut with thicknesses in the range 0.5–0.6 mm. They will have a high moisture content at the time of slicing, but are dried down to a lower level, typically 8–10%, for furniture applications.

A wide range of decorative effects are available using veneers cut from different timbers – mahogany, oak and teak predominating in Europe. Enhanced decorative effects can be achieved using exotic veneers, burr walnut, curl mahogany or bird's-eye maple, for instance, or by the use of inlays or marquetry pieces. Veneer lay-ons are made by jointing veneer strips or inlay pieces using adhesive-backed paper tapes or by heat-rolling hot melt coated nylon thread across the joint lines.

Substrate materials

Veneered panels were originally manufactured using cores of lower-grade solid wood. The relatively large movement of solid wood in response to changes in atmospheric conditions can result in problems of distortion or wood splitting, defects which can be seen in original furniture pieces exhibited in museum collections. Wood-based sheet materials such as plywood, wood chipboard and more recently MDF manufactured from indigenous species are good substrates for wood veneering as they are relatively stable compared with solid wood, available in a wide range of thicknesses and board sizes for economic cutting, and supplied at moisture contents in the range 6–12% appropriate to present-day interior conditions.

Wood chipboard and MDF are normally manufactured with urea–formaldehyde resin binders with additives to impart special properties, for example moisture resistance and fire retardancy. Their surfaces, after sanding to achieve a thickness tolerance of not more than ± 0.3 mm, are compatible with most woodworking adhesives.

Veneering adhesives

Wood veneers are readily bonded with mixed-application urea–formaldehyde adhesive using a liquid resin, liquid hardener system with the acidity of the hardener varied to give a wide range of pressing times according to the temperature of the press plates. The viscosity of the mixed adhesive is an important characteristic controlled by the addition of an extender, often based on wood flour and starch. Low viscosity adhesives have a tendency to flow through veneers which are often highly porous with a consequent detrimental effect on finishing treatments. High-viscosity adhesives are more difficult to spread using the conventional doctor roll adhesive applicators at the recommended coating weight of approximately 120 g m^{-2}.

Polyvinyl acetate adhesives are sometimes used for wood veneering. Standard grades are used at room temperature or for warm pressing up to 70 °C, but above this temperature, cross-linked grades are recommended. These can be single-pack systems which are internally cross-linked or, less frequently, two-pack systems requiring the addition of a catalyst. The advantages of PVAC over UF adhesives are the elimination of mixing and easier cleaning of the application rollers.

Veneering conditions

Three types of presses can be used for hot press bonding wood veneers. Single-daylight through-feed presses are used by the larger companies. Most medium-size companies use multi-daylight presses with the number of press openings related to the scale of operation, four openings being the most popular. As an alternative, single-opening shuttle presses fitted with two trays can be used. One tray is used as a base for laying up the veneered panels prior to pressing while the other tray is under pressure in the press opening. Multi-daylight presses are considered to be the most appropriate for general veneering involving variable throughput with panels of different sizes. The press platens may be heated electrically or by oil or steam.

Platen temperatures up to 130 °C are used in single-daylight presses fitted with automatic loading and fast-closing. Multi-daylight presses generally operate in the range 80–100 °C with pressing times typically in the range 2–4 min. These lower temperatures are required to reduce the risk of adhesive precure when the adhesive on the panels placed in the bottom opening of the press starts to cure before the remaining openings are loaded and pressure is applied.

Assuming good control of thickness tolerance of the core material and veneers, veneering pressures are normally set in the range 350–600 kN m^{-2}. Pressures at the low end of the range are used to limit adhesive percolation when bonding porous veneers. Higher pressures are used to overcome core thickness variations or deficiencies in the flatness of the press platens. After bonding, the veneered panels should be stacked for at least 24 h on a flat base to allow the adhesive to cure fully and to avoid panel distortion.

Clearly, wood veneering is a specialized process, but with proper attention to substrate and veneer preparation, the selection of an appropriate adhesive and the control of the pressing operation, few problems are likely.

Select references

A J Sparkes, A guide to wood veneering, *FIRA Bulletin 96*, Dec. 1986.
A J Sparkes, The use of UF adhesives for veneering, *FIRA Bulletin 32*, Dec. 1970, pp. 84–5, 100.

Wood composites – adhesives

B H PAXTON

Adhesion between the phases is inevitably a vital aspect of all **Composite materials**. This article is concerned with adhesives in wood-based composite boards which provide economic and effective ways of using a natural material. Other articles referring to these materials are **Wood adhesives – edgebanding** and **Wood adhesives – plastic laminated panel**, background information on bonding wood can be found in **Wood adhesives – basic principles**.

The range of wood-based boards, and the applications in which they are used, is large, and increasing. Examples are the use of plywood in structural situations, such as I-beams or box beams, on the one hand, to the construction of short-lived products, such as coffins from veneered particle board. Not all types of wood-based boards are suitable for all uses, and although only a limited number of adhesives are used in their manufacture, choice of adhesive can be the most important factor governing performance in service.

British Standards[1–4] define property and durability levels for different types of boards, and give recommendations for use. Further advice on the selection of the most appropriate board for a particular end use is given in the *Building Research Establishment Digest 323*.

Types of adhesive

Adhesives of the aminoplastic (see **Condensation polymerization**) and phenol formaldehyde (see **Phenolic adhesives**) types are most widely used. Although basically similar, an adhesive for plywood manufacture will require a different formulation to one for particle board, or medium density fibre-board (MDF) since methods of application and processing differ. Thus, in plywood, large sheets of veneer must be uniformly coated with adhesive, usually by a roller or curtain coater; in particle board, chips or wafers must be coated with very fine adhesive droplets, while small bundles of wet fibres must be sprayed with adhesive in the manufacture of MDF. Hence formulation and production of resins has become a mixture of art and science, with resin manufacturers able to produce resins tailored for use in a particular board-manufacturing plant, or with a particular species of timber.

Urea–formaldehyde These are widely used for particle board and MDF manufacture and, to a smaller extent, for plywood. They are produced by the reaction of formaldehyde with urea, in molar ratios of between 1.2 and 1 and 2.0 and 1. Low molar ratios are preferred to minimize emission of formaldehyde in service; although low molar ratio resins have had inferior strength and water resistance and are slower curing, recent formulations have overcome these drawbacks to some extent. Curing occurs at elevated temperatures, with ammonium chloride, a common catalyst.

Typical adhesive spreads for plywood are 100–170 g m^{-2}, with curing temperatures of 100–160 °C, and bonding pressure of 1–1.6 MPa. For particle board, the adhesive content is around 8 g of solid resin per 100 g of dry chips, and 12 g solid resin per 100 g dry fibres in the case of MDF. Curing rates are very fast, typically 8–10 s mm^{-1} of board thickness for particle board.

Urea–formaldehyde bonded boards are suitable only for interior use, due to their limited water resistance. They undergo hydrolytic degradation in the presence of moisture, particularly at temperatures above 40 °C.

Melamine–formaldehyde They are rarely used alone, due to their high price. However, combining with urea–formaldehyde, either by mixing the two resins or preferably by producing as a co-condensate, gives a very substantial improvement in the durability of urea adhesives (see **Condensation polymerization**). A melamine content of at least 40% is needed to satisfy a standard method of accelerated ageing for moisture-resistant grades of particle board and MDF.[3,4].

Phenol–formaldehyde This adhesive is most widely used in the manufacture of plywood. The resin is usually a 'resol' type, prepared by the reaction of phenol with excess formaldehyde in the presence of an alkali catalyst. Curing occurs at elevated temperatures in the presence of an alkali. Phenolic adhesives need higher spreads, are slower curing and are more expensive than urea–formaldehyde, but they give the most durable 'weather and boil-proof' (WBP) bonding.

Although 'resol' types are also used in the manufacture of wood chipboard, other forms of particle board, such as waferboard and oriented strand board (OSB) are usually made using 'novolac' types. These resins are produced by reacting excess phenol with formaldehyde in the presence of an acid catalyst. The resin is converted to a fine powder, which is usually sprayed on to the large wafers along with molten wax, which helps the dry resin powder adhere to the wafers until it is cured under elevated temperatures of up to 200 °C. With this method, very small quantities of adhesive – as low as 2.5 g solid resin per 100 g dry wafers – can be used, while still achieving satisfactory bonding.

Isocyanates These are finding limited use in the manufacture of particle board, and avoid problems of formaldehyde release experienced with aminoplastic adhesives. They have the advantage of forming strong chemical (covalent) bonds with the wood, and so relatively low quantities of adhesive give high levels of bonding with good resistance to moisture. The high initial cost of the resins is offset by the small quantities needed, together with rapid curing rates at elevated temperature. Self-releasing isocyanates are now available which overcome the problems of the binder adhering to metal platens and cauls characteristic of early formulations.

Mineral binders Cement and gypsum have limited, but increasing, use as binders. Portland cement is used in producing wood wools slabs and cement-bonded particle board. Only a limited number of species, such as spruce, can be used, since most species contain chemicals which inhibit the curing of the cement.

In the manufacture of cement-bonded particle board, wood chips are mixed with Portland cement in the ratio 60% cement, 20% wood and 20% water. Curing times are very long – about 7 h at 70–80 °C, followed by a further 2–3 weeks conditioning at ambient temperature. The boards produced have high-density (1300 kg m^{-3}, double normal particle board), but have exceptional dimensional stability, moisture and fire resistance compared with other wood-based boards.

Gypsum is used as a binder in gypsum-bonded particle boards, and in fibre-boards from waste paper.

Other binders Adhesives based on tannins, occurring naturally in some timbers are likely to be more widely used in future as price and availability of oil-based products becomes less favourable. Usually small amounts of urea, phenol or isocyanate are incorporated with the tannin, to improve use and performance.

Natural lignin, present in the wood, is used as the binder in most fibre building boards (hardboard) which are produced by subjecting wet wood fibres to heat and pressure.

References

1. BS 6566: Part 8: 1985, Plywood.
2. BS 5268: Part 2: 1988, Structural Use of Timber.
3. BS 1142: 1989, Fibre Building Board.
4. BS 5669: 1989, Particleboard.

Select reference

A Pizzi, *Wood Adhesives – Chemistry and Technology*, 2 vols, Marcel Dekker, New York, 1983 and 1989.

X

X-ray photoelectron spectroscopy (XPS)

D BRIGGS

X-ray photoelectron spectroscopy is a very widely used surface analysis technique frequently referred to as ESCA (electron spectroscopy for chemical analysis).

The sample, in UHV, is bombarded with soft X-rays (usually MgKα or AlKα with energies of 1253.6 and 1486.6 eV respectively). The photoelectrons emitted are energy analysed to give a spectrum of the type shown below. Three types of peaks are superimposed on a secondary electron background: core level, Auger and valence band peaks. Core level peaks are due to photoelectrons emitted from the atomic (core) levels of the atoms present, e.g. Cls designates an electron from the 1 s level of carbon. The binding energies (E_B) of these electrons are obtained from the Einstein relation

$$E_B = hv - E_k - \phi$$

Where hv is the X-ray energy, E_k is the measured kinetic energy and ϕ is the sample work function. The E_B values are highly characteristic and allow identification of all elements except H. The peak intensities are proportional to the number of atoms sampled so that atomic composition can be calculated, with detection limits of typically 0.2 at%. Small variations in E_B, known as chemical shifts, occur for a given elemental core level in different chemical states. These can be determined from spectra obtained with high-energy resolution and allow some structural information to be derived.

A series of broader peaks, e.g. CKLL, are caused by Auger electrons. The process giving rise to these is described in the article **Auger electron spectroscopy**. Electrons emitted, with very low binding energy, from the valence shell, or bonding orbitals, appear in the valence band. These weak peaks are of very limited analytical value.

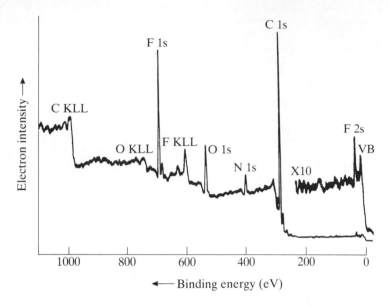

Fig. 148. XPS survey scan of the surface of an epoxy adhesive cured in contact with PTFE. The PTFE film became detached in a peel test at very low load, but some transfer to the epoxide has taken place. Characteristic core level peaks (e.g. F 1s), Auger peaks (F KLL) and valence band peaks (VB) are all present in this spectrum

The surface sensitivity of XPS is the result of the limited depth below the surface from which electrons can escape elastically (without energy loss). This depth depends on the kinetic energy of the electron, but typically the information comes from a region < 100 Å beneath the surface. For flat surfaces the sampling depth can be varied by changing the angle at which electrons leave the surface (the 'take-off' angle). For grazing exit angles the sampling depth can be reduced to ~ 10 Å.

The principal advantages of XPS are its low radiation damage rate and the lack of serious charging problems with insulators. Traditionally, spatial resolution has been very poor, but 'selected area' analysis of regions down to ~ 150 μm diameter is now routinely achievable. An imaging instrument with a spatial resolution of < 10 μm has recently become available. Wide use of XPS is made in the study of adhesion problems, e.g. in establishing failure planes (see **Locus of failure**), and it has contributed significantly to our understanding of polymer and metal **Pretreatment** processes and the associated mechanisms of adhesion enhancement; Fig. 148 gives an example.

Select references

D Briggs, M P Seah (eds), *Practical Surface Analysis*, 2nd ed, Vol. 1: *Auger and X-ray Photoelectron Spectroscopy*, John Wiley, Chichester, 1990.
J F Watts, *An Introduction to Surface Analysis by Electron Spectroscopy*, Oxford University Press, Oxford, 1990.

Standards concerned with adhesion and adhesives

(a) *British Standards relating to adhesives*

British Standard 5350 Standard methods of test for adhesives
List of parts

The methods already published or in preparation as Parts of BS 5350 are classified as follows:

Group A Adherends
Part A1   Adherend preparation
Part A2   Selection of adherends

Group B Adhesives
Part B1   Determination of density
Part B2   Determination of solids content
Part B4   Determination of pot life
Part B5   Determination of gelation time
Part B8   Determination of viscosity
Part B9   Determination of resistance to sagging (flow after application)

Group C Adhesively bonded joints: mechanical tests
Part C1   Determination of cleavage strength of adhesive bonds
Part C3   Determination of bond strength in direct tension

Part C4   Determination of impact resistance of adhesive bonds
Part C5   Determination of bond strength in longitudinal shear
Part C6   Determination of bond strength in direct tension in sandwich panels
Part C7   Determination of creep and resistance to sustained application of force
Part C9   Floating roller peel test
Part C10 90° peel test for a flexible-to-rigid assembly
Part C11 180° peel test for a flexible-to-rigid assembly
Part C12 180° 'T' peel test for a flexible-to-flexible assembly
Part C13 Climbing drum peel test
Part C14 90° peel test for rigid-to-rigid assembly
Part C15 Determination of bond strength in compressive shear

Group D Adhesively bonded joints: environmental tests

Part D4 Determination of staining potential

Group E Sampling and analysis of test data

Part E1 Guide to statistical analysis

Part E2 Guide to sampling

Group F Tests for flooring adhesives

Part F1 Performance tests for flooring adhesives

Group G Physical tests on anaerobic adhesives

Part G1 Determination of torque strength of anaerobic adhesives on threaded fasteners

Part G2 Determination of static shear strength of anaerobic adhesives

Part G3 Determination of ability of anaerobic adhesives to set on metal surfaces

Group H Physical tests on hot-melt adhesives

Part H1 Determination of heat stability of hot-melt adhesives in the application equipment

Part H2 Determination of low-temperature flexibility or cold crack temperature

Part H3 Determination of heat resistance of hot-melt adhesives

Part H4 Determination of maximum open time of hot melt adhesives (oven method)

British standards containing test methods of relevance to adhesives

BS 647 Methods for sampling and testing glues (bone, skin and fish glues)

BS 1203 Specification for synthetic resin adhesives (phenolic and aminoplastic) for plywood

BS 1204 Synthetic resin adhesives (phenolic and aminoplastic) for wood
Part 1 Specification for gap-filling adhesives
Part 2 Specification for close-contact adhesives

BS2782 Methods of testing plastics
Part 7 Rheological properties
Method 720A Determination of melt flow rate of thermoplastics
Part 8 Other properties
Method 835A Determination of gelation time of phenolic resins
Method 835B Determination of gelation time of polyester resins (manual method)

Method 835C Determination of gelation time of polyester and
      epoxide resins using a gel timer

Method 835D Determination of gelation time of thermosetting
      resins using a hot plate

BS 3046   Specification for adhesives for hanging flexible wallcoverings

BS 3424   Testing coated fabrics
      Part 7 Method 9. Method for determination of coating adhesion
      strength

BS 3544   Methods of test for polyvinyl acetate adhesives for wood

BS3712   Building and construction sealants
      Part 1 Methods of test for homogeneity, relative density,
          extrudability, penetration and slump
      Part 2 Methods of test for seepage, staining, shrinkage, shelf
          life and paintability
      Part 3 Methods of test for application life, skinning properties
          and tack-free time
      Part 4 Methods of test for adhesion in peel, tensile extension
          and recovery and loss of mass after heat ageing

BS 3887   Adhesive closing and sealing tapes
      Part 1 Specification for polypropylene, regenerated cellulose
      and unplasticized PVC tapes

BS 3924   Specification for pressure-sensitive adhesive tapes for electrical
      insulating purposes

BS 4169   Specification for glued-laminated timber structural members

BS 4346   Joints and fittings for use with unplasticized PVC pressure pipes
      Part 3 Specification for solvent cement

BS 4781   Specification for self-adhesive plastics labels for permanent use
      Part 1 General-purpose labels
      Part 2 Requirements for stringent conditions

BS 5131   Methods of test for footwear and footwear materials
      Part 1 Adhesives
      Section 1.1 Resistance of adhesive joints to heat and to peeling
      Subsection 1.1.1 Resistance to heat (creep test)
      Subsection 1.1.2 Resistance to peeling
      Subsection 1.1.3 Preparation of test assemblies for adhesion
          tests
      Section 1.4 Heat activation life of adhesives
      Section 1.6 Recommended environmental storage conditions
          for adhesive joints prior to heat resistance or peeling
          tests
      Section 1.7 The preparation of hot melt adhesive bonded
          assemblies for heat resistance (creep) and peel tests

Section 1.8 Rate of bond strength development in shear of hot
melt adhesives for lasting

Section 1.9 Measurement of green strength of adhesive joints

BS 5214　Specification for testing machines for rubbers and plastics
Part 1 Constant rate of traverse machines
Part 2 Constant rate of force application machines

BS 5270　Specification for polyvinyl acetate (PVAC) emulsion bonding
agents for internal use with gypsum building plasters

BS 5609　Specification for adhesive coated labels for marine use

BS 5980　Specification for adhesives for use with ceramic tiles and mosaics

BS 6138　Glossary of terms used in the adhesive industry

BS 6209　Specification for solvent cement for non-pressure thermoplastics pipe systems

DD 74　Performance requirements and test methods for non-structural
wood adhesives

DD 88　Method for the assessment of pot life of non-flowing resin
compositions for use in civil engineering

British Standard 3J 10　Specification for pressure-sensitive adhesive waterproof PVC tape

British Standard 3J 11　Specification for pressure-sensitive adhesive paper
masking tape

British Standard 2J 12　Specification for pressure-sensitive adhesive identification tape

(*b*) *American Society for Testing and Materials (ASTM): Standards
relating to adhesives*

Specifications for:

D 4689–90　　　　Adhesive, Casein-Type

D 1779–65 (1983)　Adhesive for Acoustical Materials

C 557–73 (1985)　Adhesives for Fastening Gypsum Wallboard to Wood
Framing

D 3498–91　　　　Adhesives for Field-Gluing Plywood to Lumber
Framing for Floor Systems

D 1580–60 (1984)　Adhesives, Liquid, for Automatic Machine Labeling
of Glass Bottles (Intent to Withdraw)

D 2559–84 (1990)　Adhesives for Structural Laminated Wood Products
for Use Under Exterior (Wet Use) Exposure
Conditions

D 3110–90	Adhesives Used in Nonstructural Glued Lumber Products
D 3930–90a	Adhesives for Wood-Based Materials for Construction of Manufactured Homes
D 2851–86	Liquid Optical Adhesive
D 4317–88	Polyvinyl Acetate-Based Emulsion Adhesives
D 3024–84	Protein-Base Adhesives for Structural Laminated Wood Products for Use Under Interior (Dry Use) Exposure Conditions
D 4690–90	Urea-Formaldehyde Resin Adhesives
D 1874–62 (1986)	Water- or Solvent-Soluble Liquid Adhesives for Automatic Machine Sealing of Top Flaps of Fiberboard Shipping Cases
D 4070–91	Adhesive-Lubricant for Installation of Preformed Elastomeric Bridge Compression Seals in Concrete Structures
C 916–85 (1990)	Adhesives for Duct Thermal Insulation
F 451–86	Acrylic Bone Cement
E 990–84	Core Splice Adhesive for Honeycomb Sandwich Shelter Panels
E 866–88	Corrosion-Inhibiting Adhesive Primer for Aluminum Alloys to Be Adhesively Bonded in Honeycomb Shelter Panels
D 1836–87	Hexanes, Commercial
F 500–77	Self-Curing Acrylic Resins Used in Neurosurgery
E 865–82	Structural Film Adhesives for Honeycomb Sandwich Panels

Test Methods for:

D 4300–88	Ability of Adhesive Films to Support or Resist the Growth of Fungi
D 3762–79 (1988)	Adhesive-Bonded Surface Durability of Aluminum (Wedge Test)
D 2919–90	Adhesive Joints Stressed by Tension Loading, Determining Durability of
D 1488–86	Amylaceous Matter in Adhesives
D 898–90	Applied Weight per Unit of Dried Adhesive Solids
D 899–89	Applied Weight per Unit of Liquid Adhesive
D 5040–90	Ash Content of Adhesives
D 4299–83	Bacterial Contamination, Effect of, on Permanence of Adhesive Preparations and Adhesive Films
D 1146–88	Blocking Point of Potentially Adhesive Layers

D 1713–65 (1986) Bonding Permanency of Water- or Solvent-Soluble Liquid Adhesive for Automatic Machine Sealing Top Flaps of Fiberboard Specimens (Intent to Withdraw)

D 1581–60 (1984) Bonding Permanency of Water- or Solvent-Soluble Liquid Adhesives for Labeling Glass Bottles (Intent to Withdraw)

D 1062–78 (1983) Cleavage Strength of Metal-to-Metal Adhesive Bonds

D 1781–76 (1986) Climbing Drum Peel Test for Adhesives

D 4680–87 Creep and Time Failure of Adhesives in Static Shear by Compression Loading (Wood-to-Wood)

D 2293–69 (1980) Creep Properties of Adhesives in Shear by Compression Loading (Metal-to-Metal)

D 2294–69 (1980) Creep Properties of Adhesives in Shear by Tension Loading (Metal-to-Metal)

D 3535–90 Deformation Under Static Loading for Structural Wood Laminating Adhesives Used Under Exterior (Wet Use) Exposure Conditions; Resistance to

D 1875–90 Density of Adhesives in Fluid Form

D 1994–91 Determination of Acid Numbers of Hot Melt Adhesives

D 4426–84 Determination of Percent Nonvolatile Content of Liquid Phenolic Resins Used for Wood Laminating

D 5113–90 Determining Adhesive Attack on Rigid Cellular Polystyrene Foam

D 2919–90 Determining Durability of Adhesive Joints Stressed in Shear by Tension Loading

D 4500–89 Determining Grit, Lumps, or Undissolved Matter in Water-Borne Adhesives

D 3931–90 Determining Strength of Gap-Filling Adhesive Bonds in Shear by Compression Loading

D 4497–89 Determining the Open Time of Hot Melt Adhesives (Manual Method)

D 3164–73 (1984) Determining the Strength of Adhesively Bonded Plastic Lap-Shear Sandwich Joints in Shear by Tension Loading

D 3163–73 (1984) Determining the Strength of Adhesively Bonded Rigid Plastic Lap-Shear Joints by Tension Loading

D 1304–69 (1983) Electrical Insulation, Adhesives Relative to Their Use as

D 4688–90 Evaluating Structural Adhesives for Fingerjointing Lumber

D 3166–73 (1979) Fatigue Properties of Adhesives in Shear by Tension Loading (Metal/Metal)

D 1579–86   Filler Content of Phenol, Resorcinol, and Melamine
                   Adhesives
D 3111–88   Flexibility Determination of Hot Melt Adhesives by
                   Mandrel Bend Test Method
D 4338–90   Flexibility Determination of Supported Adhesive
                   Films by Mandrel Bend Test Method
D 1184–69 (1986)   Flexural Strength of Adhesive Bonded Laminated
                   Assemblies
D 3167–76 (1986)   Floating Roller Peel Resistance of Adhesives
D 2183–69 (1982)   Flow Properties of Adhesives
D 5041–90   Fracture Strength in Cleavage of Adhesives in Bonded
                   Joint
D 4502–85 (1990)   Heat and Moisture Resistance of Wood-Adhesive
                   Joints
D 4498–85 (1989)   Heat-Fail Temperature in Shear of Hot Melt Adhesives
D 4499–89   Heat Stability of Hot-Melt Adhesives
D 1583–86   Hydrogen Ion Concentration of Dry Adhesive Films
D 950–82 (1987)   Impact Strength of Adhesive Bonds
D 1101–89   Integrity of Glue Joints in Structural Laminated Wood
                   Products for Exterior Use
D 4027–81 (1991)   Measuring Shear Properties of Structural Adhesives
                   by the Modified Rail Test
D 3983–81 (1991)   Measuring Strength and Shear Modulus of Nongrid
                   Adhesives by the Thick Adherend Tensile Lap
                   Specimen
D 2739–90   Measuring the Volume Resistivity of Conductive
                   Adhesives
D 1151–90   Moisture and Temperature, Effect of, on Adhesive
                   Bonds
D 1995–91   Multi-Modal Strength of Autoadhesives (Contact
                   Adhesives)
D 1489–87   Nonvolatile Content of Aqueous Adhesives
D 1490–82 (1987)   Nonvolatile Content of Urea-Formaldehyde Resin
                   Solutions
D 1582–86   Nonvolatile Content of Phenol, Resorcinol, and
                   Melamine Adhesives
D 4339–89   Odor of Adhesives, Determination of
D 903–49 (1983)   Peel of Stripping Strength of Adhesive Bonds
D 1876–72 (1983)   Peel Resistance of Adhesives (T-Peel Test)
D 2558–69 (1984)   Peel Strength of Shoe Sole-Attaching Adhesives,
                   Evaluating
D 1916–88   Penetration of Adhesives

D 2979–88	Pressure Sensitive Tack of Adhesives Using an Inverted Probe Machine
D 3121–89	Pressure-Sensitive Adhesives by Roll Ball
D 896–90	Resistance of Adhesive Bonds to Chemical Reagents
D 1183–70 (1987)	Resistance of Adhesives to Cyclic Laboratory Aging Conditions
D 816–82 (1988)	Rubber Cements
E 229–70 (1981)	Shear Strength and Shear Modulus of Structural Adhesives
D 4501–85	Shear Strength of Adhesive Bonds Between Rigid Substrates by the Block-Shear Method
D 4562–90	Shear Strength of Adhesives Using Pin-and-Collar Specimen
D 1337–91	Storage Life of Adhesives by Consistency and Bond Strength
D 906–82 (1987)	Strength Properties of Adhesives in Plywood Type Construction in Shear by Tension Loading
D 905–89	Strength Properties of Adhesive Bonds in Shear by Compression Loading
D 3807–79 (1984)	Strength Properties of Adhesives in Cleavage Peel by Tension Loading (Engineering Plastics-to-Engineering Plastics)
D 1002–72 (1983)	Strength Properties of Adhesives in Shear By Tension Loading (Metal-to-Metal)
D 2295–72 (1983)	Strength Properties of Adhesives in Shear by Tension Loading at Elevated Temperature (Metal-to-Metal)
D 2557–72 (1983)	Strength Properties of Adhesives in Shear by Tension Loading in the Temperature Range of -267.8 to $-55°C$ $(-450$ to $-67°F)$
D 2339–82 (1987)	Strength Properties of Adhesives in Two-Ply Wood Construction in Shear by Tension Loading
D 3165–73 (1979)	Strength Properties of Adhesives in Shear by Tension Loading of Laminated Assemblies
D 3528–76 (1981)	Strength Properties of Double Lap Shear Adhesive Joints by Tension Loading
D 2674–72 (1984)	Sulfochromate Etch Solution Used in Surface Preparation of Aluminum, Analysis of
D 1383–64 (1987)	Susceptibility to Attack by Laboratory Rats of Dry Adhesive Films
D 1382–64 (1987)	Susceptibility to Attack by Roaches of Dry Adhesive Films
D 897–78 (1983)	Tensile Properties of Adhesive Bonds

D 2095–72 (1983)   Tensile Strength of Adhesives by Means of Bar and
                   Rod Specimens
D 2556–91   Viscosity, Apparent, of Adhesive Having Shear Rate
                   Dependent Flow Properties
D 1084–88   Viscosity of Adhesives
D 1714–65 (1983)   Water Absorptiveness of Fiberboard Specimens for
                   Adhesives (Intent to Withdraw)
D 1584–60 (1984)   Water Absorptiveness of Paper Labels (Intent to
                   Withdraw)
D 1338–91   Working Life of Liquid or Paste Adhesives by
                   Consistency and Bond Strength
D 3236–88   Apparent Viscosity of Hot Melt Adhesives and
                   Coating Materials
D 1000–88   Pressure-Sensitive Adhesive Coated Tapes Used for
                   Electrical Insulation

Practices for:

D 3632–77 (1990)   Accelerated Aging of Adhesive Joints by Oxygen
                   Pressure Method
D 2918–71 (1987)   Adhesive Joints Stressed in Peel, Determining
                   Durability of
D 1828–70 (1987)   Atmospheric Exposure of Adhesive-Bonded Joints
                   and Structures
D 904–57 (1981)   Artificial (Carbon-Arc Type) and Natural Light,
                   Exposure of Adhesive Specimens to
D 1780–72 (1983)   Conducting Creep Tests of Metal-to-Metal Adhesives
D 3932–80   Control of the Application of Structural Fasteners
                   When Attached by Hot Melt Adhesives
D 3310–90   Corrosivity of Adhesive Materials
D 3482–90   Electrolytic Corrosion of Copper by Adhesives
D 3929–80 (1984)   Evaluating the Stress Cracking of Plastics by
                   Adhesives Using the Bent-Beam Method
D 1879–70 (1987)   Exposure of Adhesive Specimens to High-Energy
                   Radiation
D 3433–75 (1985)   Fracture Strength in Cleavage of Adhesives in Bonded
                   Joints
D 3434–90   Multiple-Cycle Accelerated Aging Test (Automated
                   Boil Test) for Exterior Wet Use Wood Adhesive
D 3933–80   Preparation of Aluminum Surfaces for Structural
                   Adhesives Bonding (Phosphoric Acid Anodizing)
D 2094–69 (1980)   Preparation of Bar and Rod Specimens for Adhesion
                   Tests

D 2651–90	Preparation of Metal Surfaces for Adhesive Bonding
D 2093–84	Preparation of Surfaces for Adhesive Bonding
D 3808–79 (1984)	Qualitative Determination of Adhesion of Adhesives to Substrates by Spot Adhesion Test Method
D 4783–90	Resistance of Adhesive Preparations in Container to Attack by Bacteria, Yeast, and Fungi
D 1144–89	Strength Development of Adhesive Bonds, Determining
D 3658–90	Torque Strength of Ultraviolet (UV) Light-Cured Glass/Metal Adhesive Joints
E 874–89	Adhesive Bonding of Aluminum Facings to Nonmetallic Honeycomb Core for Shelter Panels

Guide for:

D 4800–87	Classifying and Specifying Adhesives
D 4896–89	Use of Adhesive-Bonded Single Lap-Joint Specimen Test Results

Terminology Relating to:

D 907–91b	Adhesives

GENERAL METHODS OF TESTING

Specifications for:

E 95–68 (1987)	Cell-Type Oven with Controlled Rates of Ventilation
E 145–68 (1987)	Gravity-Convection and Forced-Ventilation Ovens
D 1193–77 (1983)	Reagent Water
E 171–87	Standard Atmospheres for Conditioning and Testing Materials

Test Methods for:

E 74–91	Calibration of Force-Measuring Instruments for Verifying Time-Load Indication of Testing Machines
D 618–61 (1990)	Conditioning Plastics and Electrical Insulating Materials for Testing
E 4–89	Load Verification of Testing Machines
E 70–91	pH of Aqueous Solutions with the Glass Electrode, Test for
E 337–84 (1990)	Relative Humidity by Wet- and Dry-Bulb Psychrometer, Determining
E 83–90	Verification and Classification of Extensometers
E 96–90	Water Vapor Transmission of Materials in Sheet Form, Test for

Practices for:

E 104–85 (1991)   Maintaining Constant Relative Humidity by Means
                       of Aqueous Solutions
E 105–58 (1989)   Probability Sampling of Materials
E 122–89   Sample Size to Estimate the Average Quality of a Lot
                       or Process, Choice of
E 29–90a   Significant Digits in Test Data to Determine
                       Conformance with Specifications

Terminology Relating to:

E 41–86   Conditioning

*(c) International Organization for Standardization (ISO): Standards
relating to adhesives*

ISO 4578:1990   Adhesives—Determination of peel resistance of high-
                       strength adhesive bonds—Floating roller method
ISO 4587:1979   Adhesives—Determination of tensile lap-shear
                       strength of high strength adhesive bonds
ISO 4588:1989   Adhesives—Preparation of metal surfaces for
                       adhesive bonding
ISO 6237:1987   Adhesives—Wood-to-wood adhesive bonds—
                       Determination of shear strength by tensile loading
ISO 6238:1987   Adhesives—Wood-to-wood adhesive bonds—
                       Determination of shear strength by compression
                       loading
ISO 6922:1987   Adhesives—Determination of tensile strength of butt
                       joints
ISO 7387:1983   Adhesives with solvents for assembly of u PVC pipe
                       elements—Characterization—Part 1: Basic test
                       methods
ISO 8510:1990   Adhesives—Peel test for a flexible-bonded-to-rigid
                       test specimen assembly—Part 1: 90 degree peel
ISO 8510-2:1990   Adhesives—Peel test for a flexible-bonded-to-rigid
                       test specimen assembly—Part 2: 180 degree peel
ISO 9142:1990   Adhesives—Guide to the selection of standard
                       laboratory ageing conditions for testing bonded joints

ISO 9653:1991	Adhesives—Test method for shear impact strength of adhesive bonds
ISO 10123:1990	Adhesives—Determination of shear strength of anaerobic adhesives using pin-and-collar specimens
ISO 10354:1992	Adhesives—Characterization of durability of structural-adhesive-bonded assemblies—Wedge rupture test

Standard test methods for adhesive joints

(Reproduced with permission from *Adhesion and Adhesives, Science and Technology*, by A J Kinloch, Chapman and Hall, 1987.)

Joint geometry	Standard	Comments
Axially loaded (tensile) butt joints		
	ASTM D 897-78	For substrates in 'block' form
	ASTM D 2094-69 and D 2095-72	Specifically for bar- and rod-shaped substrates
	BS 5350: Part C3: 1979	UK version of above
	ASTM D 429-73	Rubber-to-metal bonding
	ASTM D 816-82	Specifically for rubbery adhesives
	ASTM D 1344-78	Cross-lap specimen specifically for glass substrates
Lap joints loaded in tension		
	ASTM D 1002-72	Basic metal-to-metal single-lap joint test giving the 'single-lap-shear strength'
	BS 5350: Part C5: 1976	UK version of above test
	ASTM D 2295-72	Single-lap joint test for metal-to-metal joints at elevated test temperatures
		As above but at low temperatures
	ASTM D 2557-72	Single-lap joint for rigid plastic substrates
	ASTM D 3163-73	
	ASTM D 3164-73	Single-lap joint where a small rectangle of plastic substrate is sandwiched between metallic substrates
	ASTM D 3528-76	Basic metal-to-metal double-lap joint test giving the 'double-lap-shear strength'
	BS 5350: Part C5: 1976	UK version of test

Joint geometry	Standard	Comments
	ASTM D 3165-73	Metal-to-metal laminate test for large bonded areas but may also be used with plastic substrates
	ASTM D 906-82	Specifically for adhesives used in plywood laminate constructions
	ASTM D 2339-82	Specifically for wooden (two-ply) laminate substrates
	ASTM D 3983-81	Thick substrates used in a single-lap joint; shear modulus and strength of adhesive determined
Lap joints loaded in compression	ASTM D 905-49	Mainly intended for wooden substrates
	ASTM D 4027-81	For measuring shear modulus and strength of adhesive; uses 'rails' to maintain only shear load
Torsional shear	ASTM D 3658-78	Specifically for ultraviolet light-cured glass-to-metal joints
	ASTM E 229-70	Uses a 'napkin ring' test for determining shear modulus and shear strength of structural adhesives
Cleavage	ASTM D 1062-78	Metal-to-metal joints
	ASTM D 3807-79	For engineering plastics substrates
	ASTM D 3433-75	Flat and contoured cantilever-beam specimens for determining the adhesive fracture energy, G_{Ic}
Peel joints	ASTM D 3167-76 BS 5350: Part C9: 1978	Floating roller test

Joint geometry	Standard	Comments
	BS 5350: Part C10: 1976	90° peel test for flexible-to-rigid joints
	BS 5350: Part C14: 1979	90° peel test for rigid-to-rigid joints
	ASTM D 903-49, BS 5350: Part C11: 1979	180° peel test
	ASTM D 1876-72 BS 5350: Part C12: 1979	'T' peel test
	ASTM D 1781-76	Climbing drum peel test for skin-sandwich assemblies
	ASTM 429-73	For rubber-to-metal bonding
	ASTM D 2558-69	For shoe-soling materials
Impact resistance	ASTM 950-82	Uses a block shear joint
Disc shear strength in compression	ASTM D 2182-78	Determines adhesive shear strength on compression
Creep resistance	ASTM 1780-72, ASTM D 2294-69	Single-lap joint loaded in tension employed
	ASTM D 2293-69	Single-lap joint, having a long overlap and between 'rails', loaded in compression
	BS 5350: Part C7: 1976	Various test geometries permitted
Fatigue	ASTM D 3166-73	Single-lap joint loaded in tension employed
Environmental resistance	ASTM D 896-84	General method for assessing resistance of joints to chemicals; may use any ASTM standard test geometry
	ASTM D 904-57	General method for assessing resistance of joints to artificial and natural light; may use any ASTM standard test geometry
	ASTM 1151-84	General method for assessing resistance of joints to moisture and temperature; may use any ASTM standard test geometry

Joint geometry	Standard	Comments
	ASTM D 1183-70	General method for assessing resistance of joints to a cyclic laboratory ageing environment (moisture and temperature); may use various ASTM standard test geometries
	ASTM D 1828-70	General method for assessing resistance of joints to natural outdoor ageing; may use any ASTM standard test geometry
	ASTM D 1879-70	General method for assessing resistance of joints to high-energy irradiation; may use any ASTM standard test geometry
	ASTM D 2918-71	Method for assessing the effect of *stress* and moisture and temperature; uses a peel joint test
	ASTM D 2919-71	As above but uses single-lap joints loaded in tension
	ASTM D 3762-79	As above but uses a wedge (cleavage) test
	ASTM D 3632-77	Effect of a high-pressure oxygen environment on wood-to-wood and wood-to-metal joints; uses a lap joint test
Pressure-sensitive tack		
	ASTM D 2979-71	Inverted probe test
	ASTM 3121-73	Rolling-ball test

A selected bibliography on adhesion
(see article **Literature on adhesion**)

A. Books – general texts

1. W C Wake, *Adhesion and the Formulation of Adhesives*, Applied Science Publishers, 1982.
   Divided evenly between fundamentals of adhesion and description of particular adhesive types and applications.
2. D M Brewis, D Briggs, *Industrial Adhesion Problems*, Orbital Press, Oxford, 1985.
   Good treatment of theoretical fundamentals of adhesion and interfaces followed by application in different industrial contexts.
3. A J Kinloch, *Adhesion and Adhesives: Science and Technology*, Chapman and Hall, 1987.
   Thorough and well-written work of value as a textbook and handbook for the research worker.
4. H F Brinson (ed.), Adhesives and sealants, *Engineered Materials Handbook*, Vol 3, ASM, Ohio, 1990.
   A comprehensive reference work of nearly 900 000 words, for engineers. Many topics of industrial relevance placed firmly in a scientific context.

B. Books – more specialized works

5. G P Anderson, S J Bennett, K L De Vries, *Analysis and Testing of Adhesive Bonds*, Academic Press, 1977.
   Survey of tests relating them to the mechanics and chemistry on which they are based.
6. A J Kinloch, *Durability of Structural Adhesives*, Applied Science Publishers, 1983.
   Discussion of the fundamentals and their application to different materials.
7. R D Adams, W C Wake, *Structural Adhesive Joints in Engineering*, Elsevier Applied Science Publishers, 1984.
   Mechanics of adhesive bonds related to the materials science of the adhesives.
8. A J Kinloch, *Structural Adhesives: Developments in Resins and Primers*, Elsevier Applied Science Publishers, 1986.
   Eight chapters each discussing advances in different materials.
9. R F Wegman, *Surface Preparation Techniques for Adhesive Bonding*, Noyes Publications, 1989.

Treatment methods for a wide range of metals and polymers are discussed, and detailed procedures are given.

C. *Books – selection and manufacturing*

10. H. Monternot, *Guide du Collage*, CTIM, 60300 Senlis, France, 1978. Clear guide to selection and use of adhesives.
11. W A Lees, *Adhesives in Engineering Design*, Springer-Verlag, 1984. Useful sections on design, surface preparation and adhesive selection.
12. J Shields, *Adhesive Handbook*, 3rd edn, Butterworths, 1984. Vast source of information on joint design, selection of adhesives and surface preparations, testing and properties of adhesives, trade sources, lacks index.
13. A. H. Landrock, *Adhesives Technology Handbook*, Noyes Publications, 1985. Useful source of information on joint design, surface preparation, adhesive types, bonding processes, durability and testing.
14. I Skeist (ed.), *Handbook of Adhesives*, 3rd edn, Van Nostrand Reinhold, 1990. Over 40 chapters covering fundamentals, adhesive materials, substrates, bonding technology and selection.

D. *Review articles in series devoted to adhesion and in works of wider scope*

15. *Developments in Adhesives – 1*, ed. W C Wake, 1977; *Developments in Adhesives – 2*, ed. A J Kinloch, 1981, Applied Science Publishers.
16. *Treatise on Adhesion and Adhesives*, Vols 1–6, 1966–89, ed. R L Patrick, Vol. 7, 1991, ed. J D Minford, Marcel Dekker.
17. *Adhesion 1* (1977) to *Adhesion 15* (1991), ed. K W Allen, Proceedings of annual Adhesion Conference at the City University, London, Elsevier Applied Science Publishers.
18. *Encyclopedia of Polymer Science and Engineering*, Wiley-Interscience, 19 vols 1985–90. Many references can be found through the index, but see especially F J Meyers, Adhesion and bonding in Vol. 1, pp. 476–546.

E. *Abstracting periodicals*

Adhesives Abstracts
Chemical Abstracts

F. *Trade magazines*

Adhesives Age
Adhesion
Assemblages adhésifs
European Adhesives and Sealants

G. *Research journals devoted to adhesion*

International Journal of Adhesion and Adhesives
Journal of Adhesion

Journal of Adhesion Science and Technology
Journal of the Adhesion Society (Japan)

H. *Other research journals which have a significant adhesion content*

Applied Surface Science
Composites
European Polymer Journal
Industrial and Engineering Chemistry
Journal of Applied Physics
Journal of Composite Materials
Journal of Materials Science
Journal of Polymer Science
Journal of Applied Polymer Science
Journal of Colloid and Interface Science
Journal of the Electrochemical Society
Polymer
Polymer Engineering and Science
Rubber Chemistry and Technology
SAMPE Journal
Surface and Interface Analysis
Surface Science
Transactions of the Society of Rheology
Thin Solid Films

Index

Titles of articles are given in **bold letters**: the pages on which they start are shown in **bold numerals**.